Neuroimaging, Software, and Communication

Edison Bicudo

Neuroimaging, Software, and Communication

The Social Code of Source Code

palgrave
macmillan

Edison Bicudo
School of Global Studies
University of Sussex
Brighton, UK

Department of Sociology
University of Sao Paulo
Sao Paulo, Brazil

ISBN 978-981-13-7062-5   ISBN 978-981-13-7060-1   (eBook)
https://doi.org/10.1007/978-981-13-7060-1

This Palgrave Macmillan imprint is published by the registered company Springer Nature Singapore Pte Ltd.
The registered company address is: 152 Beach Road, #21-01/04 Gateway East, Singapore 189721,
Singapore

The fish trap exists because of the fish;
once you've gotten the fish,
you can forget the trap […]
Words exist because of meaning;
once you've gotten the meaning,
you can forget the words.
Master Chuang-Tzu

Foreword

Dr. Bicudo's book is a broad study of the social and political context of software in brain imaging. I think Bicudo is right to claim that his analysis is an instructive case, to help us think about role of software in science, and in society.

A friend and colleague once told me that it used to be said, that the world runs on oil, but it is now truer to say, that the world runs on software. People write software, and they write in social and political contexts with social and political motivations. Contexts and motivations guide the software that is written, and our access to it. Software has a great influence on what we can do, and even how we can think, so we must consider these contexts and motivations, if we want to choose the kind of science we want to build, and the society we build it in.

Programmers are not usually well equipped to reflect on these matters. As a programmer myself, I find that my peers can have a fierce, even violent disregard for discussions on subjects that are not strictly technical, even when it is clear that the problem at hand is social or political. Through many discussions, I have often remembered the famous joke about the drunk under the lamppost. A passerby sees the drunk man, and asks him what he is doing. He says "I am looking for my keys, that dropped out of my pocket". The passerby: "Did you drop them here?". "No, I dropped them in the alley behind me, but the light is better under the lamppost". Dr. Bicudo has done us a great service, by going to look for the metaphorical keys in the alley, which is the only place they can be found.

There are several aspects of Bicudo's analysis that are particularly important.

The first is his discussion of "semi-flexible software". Semi-flexible software is software that has relatively strong technical and social barriers to contributions from outside a single institution. These are flagship projects for particular labs. They serve to concentrate power, reputation, and influence in those few labs that could collect and sustain a team to write such software. Semi-flexible packages dominate the field. They have become so ordinary, that those of us in the field may not see the harm that they can do. Here, Bicudo has many advantages in his analysis. He looks at us with a fresh eye, from outside, and does not suffer from our deadening habits of thought. He considers the political, when many of us will not. He starts his analysis from his native Brazil, and listens to the researchers from smaller labs, with less funding. He describes their feeling that they must interact like satellites, or colonies, with the larger labs, that own the packages, in North America and Europe.

This model of semi-flexible software has deep roots in an earlier, simpler age of software development. At the time that these packages began, even scientists thought of serious software as something that belonged in a corporate world. Serious software earned money and cost money; it was best written by professionals. Scientists also wrote software, but it was not serious software. You can see this idea return many times in the revealing interviews throughout the book.

If scientific software is not serious software, there is no need for the heavy-duty tools of the software engineer, such as careful code design, code review, and continuous testing. The code is an afterthought to the research that bore it; it dumbly does, but it does not speak; it is seen, but not heard.

Bicudo shows this is a political landscape. This old software stands for an old system. The system is semi-flexible; it contains certainties, of the institutions at the top, and those that must follow. The software illustrates and enforces the system. If we want to change the system, we must change the software.

This leads Bicudo to his discussion of free and open-source software. The free software movement has its roots in "antiestablishment sentiment". Many developers in open source have an explicit desire to break the monopolies that are easily made by commercial software companies. They write code to give the rest of us greater agency.

Now we see this landscape, we have to ask, with Bicudo, is the science we have now, the science we want? Is the software we have now, the software we want? Do we have to accept our constraints as natural to the condition of our politics, or can we change?

I think we can change, and we will. We are lucky to have passed the age that could not accept that free software was serious software, or that scientists could not write good code. Now, in the age of open source, it would be foolish to bet against the work of committed volunteers, building communities around open code. Now, we have the tools to work efficiently, to collaborate, and to teach each other how to write better code. Free languages, such as Python, and R, have grown very quickly in science, and already dominate some fields, such as machine learning and data science. As their communities grow, so do their code libraries, with greater depth, range, and quality. Free languages bring open source culture and practice. In this culture, code must speak, because, in order to grow, it must be easy to understand, and so, easy to fix, easy to extend, and easy to teach.

Change will come, but it will take many hands. Some of the work will be technical. Scientists, some alone, some in teams, will write the new code, and replace the old. Some of the work will be social, and political, as we analyse and reflect on the system that we have, and the things that keep it running. I am glad to see Dr. Bicudo's book. It is a serious part of this second, essential form of work.

Birmingham, UK

Dr. Matthew Brett
School of Psychology
University of Birmingham

Dr. Matthew Brett is a lecturer in the School of Psychology, University of Birmingham, UK. With a background in psychology and medicine, he became involved with computer programming at an early age. He is the author of a *Toolbox* called MarsBar that is one of the most frequently cited in the neuroimaging literature. Since 2005 he has been working to expand the use of Python and open source coding methods in neuroscience. He is the co-founder of the NIPY community, dedicated to Python in neuroscience, and the author of *Nibabel*, a Python library for neuroimaging that is the foundation of many other Python applications in neuroscience.

Acknowledgements

This project was supervised by Dr. Sylvia Garcia, for whose academic support I am very grateful.

I thank Professor Alex Faulkner and Professor Brian Salter for their unfailing support and guidance.

I am very grateful to all my interviewees, both those who decided to disclose their names and those who preferred to remain anonymous. I would like to mention some people who proved particularly welcoming towards my project: Professor Luiz Murta Junior, Professor Carlos Garrido, Professor Alexandre Franco, Professor Li Li Min, Professor Fernando Paiva, Professor Kelly Braghetto, Dr. Matthew Brett, and Professor Rainer Goebel.

Finally, I would like to thank two people who provided me with an invaluable institutional support at key moments of the fieldwork: Debora Lima (D'Or Institute) and Tildie Stijns (Donders Institute for Brain, Cognition and Behaviour).

Apart from the fieldwork in the UK, Netherlands, and Portugal, this study was funded by the Coordination for the Improvement of Higher Education Personnel (Capes).

Contents

List of Charts

chapterFour (Owning Code: Institutional Aspects of Software Development) {

chapterFive (Using Code: The Social Diffusion of Programming Tasks) {

List of Tables

List of Maps

chapterFive (Using Code: The Social Diffusion of Programming Tasks) {

List of Boxes

chapterOne First Words

Observing the Brain

Spoken language, a basic resource with which people engage in social life, may come to mix up different aspects of human existence. This is what happens, for example, with the word "observation." An observation is made when somebody watches something happening. In this sense, a person would "observe" the rocking waves of the sea. An observation can also be made when somebody collects empirical evidence in support of an argument or theory. Hence the "observations" made by somebody studying how the waves move. In yet another usage, an observation is made when somebody comments on a certain event or thing in the world. It is in this sense that a person would "observe" that waves have never moved so strongly. Thus the same word points to three aspects of human experience: visuality, materiality, and communicability.

Moving between these three realms constitutes an important part of scientists' activities. By observing the world, they gather scientific observations in order to observe that some phenomena can be explained more accurately. For realizing these three kinds of observations, scientists have always relied on instruments. Many of the explicative deeds of

© The Author(s) 2019
E. Bicudo, *Neuroimaging, Software, and Communication*,
https://doi.org/10.1007/978-981-13-7060-1_1

science would be very difficult, if not impossible, without the help provided by several instruments with which scientists scrutinize the world (Arendt 1998; Shapin and Schaffer 1985).

Initially, those instruments were concrete objects such as microscopes and chronometers. Then scientists managed to make organisms and molecules become instruments, creating techniques such as synthetic chemistry and biotechnology. More recently, very particular research instruments were brought to light: algorithms, software, computer code, in a word the whole set of computational productions that can be described as digital technologies. In addition to the physical, electrical, chemical, and organic layers with which scientists cover the world, there is now a digital layer that emerges as an additional instrument helping scientists observe everyday phenomena.

In biomedicine, those processes have been taking place for very long, leading to the formation of a domain called medical imaging. This is an area where researchers and physicians produce and analyse images of different parts of the human body, in an attempt to understand its functions and disfunctions. In neuroimaging, a section of medical imaging, researchers focus on the brain. At an early phase of medical imaging and neuroimaging, it became clear that instruments had to be sought in order to expand the capacities of the observers' eyes. From a period when cutting instruments were used to directly observe the body's inner components, one passed to a period marked by the use of visual instruments such as microscopes (Kevles 1998). Centuries had to go by before physicians and scientists could begin to have recourse to computational resources. This happened only after the Second World War (Laal 2013), but mainly in the 1970s with the creation of computed tomography (Blank 2007; Bradley 2008). The study and comprehension of the body became increasingly dependent on the production of images that had the putative capacity of not only mirroring the body but also displaying its functioning and structures in the clearest manner.

In the late 1970s, a new revolution was initiated with the invention of magnetic resonance imaging (MRI) (Filler 2009; Prasad 2014). The emergence of this sophisticated system of image collection was made possible by findings from quantum mechanics and the knowledge of the properties of subatomic particles (Zimmer 2004; Prasad 2005).

Like computed tomography, MRI enabled the observation of the body's inner components in a non-invasive way, something that had long been sought by researchers and physicians (Puce et al. 1995; Savoy 2001; Rose 2016). Another predecessor of MRI was positron emission tomography (PET), a technology that shares two characteristics with MRI: both rely on the principles of quantum mechanisms and enable non-invasive observations. PET was "an incredibly complex, expensive [...] set of techniques and technologies" (Dumit 2004, p. 3), which MRI also is, but to a lesser degree. Moreover, MRI generates more detailed images, which is key in medical imaging and especially neuro-imaging because of the brain's complexity. These features allowed MRI to replace PET as the preferred technique for the collection of medical images (Bandettini, n.d.; Savoy 2001; Rose 2016). Consistently, MRI technology was enhanced and refined in the 1980s and 1990s (Filler 2009; Prasad 2014; Savoy 2001), leading Savoy (2001, p. 36) to speak of "a collection of exciting developments on the technological front."

Neuroimaging has been one of the areas on which MRI has provoked the strongest impact. One of the main upheavals was the discovery of functional magnetic resonance, which amounts to the use of MRI scanners to detect changes in the brain, in terms of either blood or oxygen concentration, when research subjects perform certain physical or cognitive tasks, like playing music or performing mathematical calculations. Hence the excitement provoked by MRI scanners, which are said to "capture snapshots of thought" (Zimmer 2004, p. 275), reveal "the mental operations carried out by the human brain" (van Horn et al. 2005, p. 58), or enable scientists "to access the contents of the human mind" (Rose 2016, p. 158). Thanks to the research possibilities opened up by magnetic resonance and especially functional magnetic resonance imaging, the technology was sought by many research centres across the world, a trend that had reached notable force by the end of the 1990s (Gold et al. 1998).

However, as scientists' visual demands became increasingly strong, the 2D, black-and-white images produced by MRI scanners rapidly proved limited. They are very different from the colourful and attractive brain maps that people have become accustomed to see in magazines, conferences, and the television. The following images illustrate the differences I am referring to:

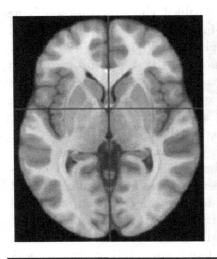 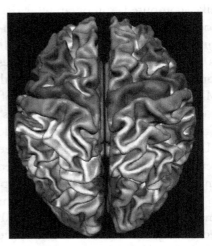

A	B
A MRI-like imagea (*Source* Screenshot of the SPM software (Wellcome Centre for Human Neuroimaging))	3D brain map (*Source* Screenshot of the BrainVoyager software (Brain Innovation))

Image A was produced by an MRI scanner whereas image B derives from the procedures described by Bruehl (2015, p. 2): "With the improvement of MRI acquisition speed, data transfer techniques, and computation capacities and algorithms, it is now possible to directly transfer the raw data from the MRI scanner to a PC/laptop computer as soon as they are acquired. This computer can then process the fMRI data [...]" (2). Not all computers have the capacity to realize such data processing, though; in order for this to be possible, specialized software must be installed. In the case of figure B, the application is called BrainVoyater. The difference between black-and-white, 2D scans, and colourful brain maps is not an issue of visual beauty only; it has also to do with analytical precision and research potentialities. For the data processing performed by a computer software can extract more from data, hence producing images that show more details of the brain, enabling scientists to explore poorly understood neurological aspects.

One can easily perceive that in order for cutting-edge brain studies to be carried out, software becomes a decisive tool. This book focuses on neuroimaging software, exploring the social and political dimensions of this computational resource. Whenever neuroimaging researchers wish to realize their "observations" (that is when they "observe" what happens with the brain's constituent parts; when they make empirical "observations," collecting data to scrutinize its functions; and when they "observe" that a new comprehension of the brain has been achieved), software packages have to play a central role.

Neuroimaging Through Software

When positron emission technology (PET) was the main technique used to generate brain images, very powerful computers were in need for data processing. With the advent of MRI technology, data became more flexible, which allowed the use of computers of lower performance. This is why Plaja, Vendrell and Pujol (1995, p. 57) claimed that "[...] MRI is technologically simpler than PET, as it depends more on advances in the field of 'software' than 'hardware' [...]." Therefore, the diffusion of MRI technology in brain studies led to a parallel diffusion of software packages. This is how Aguirre (2012, p. 765) described the situation of the late 1990s: "Like a streak of bacteria on an agar plate, independently grown, freely-distributed neuroimaging software bloomed in this environment." In the first years of the twenty-first century, it was already possible to claim that neuroimaging researchers had become highly dependent on software (van Horn et al. 2005).

The progress of computer applications has been consistently fostered by advances in MRI scanners. For example, MEDx, a commercial neuroimaging software package initially developed in 1993, in the United States, "was regularly updated to incorporate advances in neuroimaging methods" (Aguirre 2012). Another example is SPM, a package released in 1991, in the UK, to process PET data. Subsequently, it was modified to also process MRI data, which required constant reworking to accommodate progresses in MRI technology (Ashburner 2012).

One of the key shifts in MRI technology was the production, in 2005, of so-called ultra-high field scanners whose magnetic strength reaches very high values. Even though some researchers challenge the visual benefits of such scanners (Maubon et al. 1999; Rutt and Lee 1996), and some have doubts pertaining to their safety for human beings (Kolk et al. 2013; Moser 2010), most neuroimaging researchers consider that this kind of machine makes it possible to realize very detailed and valuable observations of the human brain. Software packages had to be modified so as to process the data generated by ultra-high field MRI scanners. This is what happened with the BrainVoyager package, produced in the Netherlands. In the early 2010s, its main developer wrote: "Several new tools for analyzing high-resolution fMRI data [...] are already available in BrainVoyager" (Goebel 2012, p. 753). According to one of my interviewees, a neuroimaging researcher based in the Netherlands, ultra-high field devices enable more detailed visualizations; "this means there is more information data, but this means also, the software analysis tools have to be treated to deal with that additional information."

The evolution of neuroimaging software was not fostered by only the progress in MRI technology, though. Parallel technical factors also had considerable impact on these computational resources. For example, when personal computers began to be largely used in scientific analysis, in the late 1990s, adjustments to software had to be implemented (Ashburner 2012). Recently, new changes to software packages have been fostered by resources like cloud technology, the internet, and high-performance computer processors. A researcher and Professor whom I interviewed in Portugal is working towards incorporating virtual reality approaches into neuroimaging software. In 2012, Goebel (2012, p. 753) predicted that "[...] smart phones and tablets will raise increasing interest in visualizing and analyzing neuroimaging data on mobile devices." In effect, the production of brain maps in mobile devices has been made possible by some software packages, including the one designed by Goebel himself.

Nowadays, neuroimaging researchers can simply not imagine future advances in their field without imagining simultaneous advances in data analysis software. As claimed by van Horn and colleagues (2005, p. 74):

"[...] a more thorough understanding of the brain and its cognitive processes will necessitate increased computational infrastructure, novel software technology to accelerate data analysis and to mine vastly larger amounts of data, and the sharing of primary research data." The need for enhanced neuroimaging software packages is to a large degree justified with reference to the growing size of datasets. According to a neuroimaging researcher and university Professor I interviewed in Brazil, "if you carry out a study where you have a hundred people with a certain disease, you collect brain images, and a hundred sound people (the control group), the high-resolution computational data are very heavy, so to say." As research samples of this size, and even larger, becomes frequent in neuroimaging research, the automatic procedures made possible by software packages can only become increasingly central. "Without these supporting technologies, the use of images as the representational form would be next to impossible" (Joyce 2006, p. 18).

As part of my research project, a literature review of neuroimaging papers was carried out, making it clear that software packages have become key. In some papers, such as the one published by Cutanda, Moratal and Arana (2015), the software package is even cited in the paper's title. Another piece of research literature in which software packages are frequently cited is the research project. It has become common practice to indicate, in research proposals, the set of software packages that will be used in data analysis. In an interview in the UK, with a renowned researcher and Professor, the ways in which people write neuroimaging research proposals were addressed: "We normally give a level of detail and say something like, you know, we will... we will use program x to normalize everything to a standard space and then we will use this program, this program, this program to calculate an image and this program to do the final statistical analysis."

The presence of software permeates neuroimaging research from the writing of research projects through data analysis to the preparation of final publications. Therefore, when talking about brain scans and brain maps, we are dealing with the "representational possibilities that are unimaginable without the help of a computer" (Prasad 2005, p. 302). In this book, these possibilities will be focused on, not from a technical perspective, but from a sociological and communicative point of view.

However, in order to begin to comprehend how this viewpoint can be applied to the universe of software, it is necessary to ask: how were neuroimaging studies conducted before the advent of specialized software packages?

Neuroimaging Through Handicraft

MRI technology began to be developed in the early 1970s and the first images were obtained in the final years of that decade. In their turn, the first neuroimaging software packages were released only in the early 1990s. What happened in the 1980s, when researchers were producing images without relying on computer applications? To answer this question, let us consider the following example from a paper by Mohamed and colleagues (2005, p. 181): "The images were reviewed by a blinded, experienced observer for the presence of cortical abnormalities, hippocampal atrophy or signal change in the temporal lobes." In this paper, even though it was published in 2005, an old-style image analysis, so to say, was conducted, as an analyst, by using visual skills and experience, searched for signals of abnormalities in brain images.

In this sort of traditional analysis, the hand was a frequent ally of the eyes. According to Bandettini (n.d., p. 1), early neuroimaging researchers devoted much effort to drawing regions of interest. In the mid-1970s, van Essen (2012, p. 758) did a postdoctoral study in the UK; in his account: "[…] I spent most of my postdoctoral year developing and using a manual 'pencil and tracing paper' method for generating cortical flat [brain] maps […]." When software packages appeared, this manual process was abandoned.

Once again, we are not simply dealing with an aesthetic difference. Computational approaches enable a more accurate observation, in addition to being generated in seconds of computer processing. In its turn, images manually produced were subjected to human imperfection, as this approach "[…] was tedious, demanded considerable expertise, and involved much trial-and-error — the eraser was as important as the pencil!" (van Essen 2012, p. 758). Therefore, the neuroimaging researcher, before the advent of neuroimaging software in the 1990s, was a craftsman, in the sense intended by Sennet (2008), because manual skills, expertise, and repetition were in need.

In the twenty-first century, it is very hard to find studies such as the one published by Haznedar and colleagues (2005, p. 735) where manual techniques were employed and two researchers "outlined the caudate, putamen, and thalamus on contiguous axial MRI slices." The vast majority of today's neuroimaging studies are conducted with the help of software, which brings about speediness and quantitative precision to the research process. In this way, a different stance is cultivated which differs from the patient persistence required in the 1970s and 1980s. As shall be analysed in this book, a new attitude eventually emerges in which the reliance (or over-reliance) on software, algorithms, and computer code threatens to become an uncritical adherence to standardized research protocols. This state of things is the outcome of the rapid technological progresses outlined in the sequence.

Neuroimaging Software's Trajectory

As already explained, neuroimaging software enables researchers to process images collected from large populations of research subjects. Frequently, those studies aim to generate a single image, which summarizes the average features of the study population. Based on statistical techniques such as normalization, researchers created and refined normalized spaces on which findings are depicted, generating so-called brain atlases. Such representational techniques began to be developed in the 1980s (van Essen 2012), being subsequently incorporated by software packages. Indeed, normalization techniques were the main or exclusive task performed by the earliest packages. This is why, in 1995, Lancaster and collaborators (1995, p. 210) referred to neuroimaging software as "spatial normalization software."

These normalization packages started appearing in the early 1990s. Some degree of uncertainty and improvisation marked this decade. Many researchers did not have knowledge of, or access to, pioneering packages, thus having recourse to applications created for non-scientific purposes. The neuroimaging paper published by Warach and colleagues (1995, p. 233) constitutes a good example: "Figures were prepared with the aid of Adobe Photoshop to create montages, adjust brightness, contrast, and size, average images acquired with different diffusion gradient directions, and erase background noise outside of the brain [...]."

In the end of the 1990s, some specialized packages had managed to gain scientific solidity and reputation. At that moment, Gold and colleagues (1998) published a study on neuroimaging software. By looking at the methodology section of neuroimaging papers, and performing searches on the internet, they identified eight packages being used by neuroimaging researchers. With great accuracy, they concluded: "The list of choices [in terms of software packages] will probably become larger as the field continues to grow and with new laboratories developing research efforts" (Gold et al. 1998, p. 83).

Fulfilling this prophecy, an actual proliferation of packages took place in the 2000s and 2010s (van Essen 2012). Some of these applications incorporate quite ambitious and creative techniques, as exemplified by the AFNI package, which produces animations depicting brain responses after a stimulation (Saad and Reynolds 2012). As shall be explored in this book, this multiplication of analytical tools has derived, to a large degree, from the variety of research needs voiced by neuroimaging researchers based in several countries. In the current historical phase, it is possible to point to the existence of "many popular neuroimaging packages" (Nieuwenhuys et al. 2015, p. 2570). They have been of pivotal utility for both clinicians desirous of diagnosing brain diseases and researchers willing to generate a deeper knowledge of the brain. At this point, the somewhat different purposes of these two groups of neuroimaging software users have to be disentangled.

Brain Images, Brain Data

The interest in images has been a growing tendency in biomedicine. In brain studies, this is manifested through the multiplication of images whose presence has become more and more frequent in scientific publications and conferences (Henson 2005). Previous studies showed that for medicine and radiology students, it is crucial to develop skills allowing accurate image analysis (Good 1994; Prasad 2005). The emphasis on visuality is not exclusive to biomedicine, though. According to Joyce (2006, p. 3), "a turn toward visualization" has characterized the contemporary, Western culture, with images permeating several aspects of human life.

The reference to visual aspects is present in the name of the scientific domain focused on in this book: neuro-*imaging*. However, an important qualification must be advanced here. This book focuses on the research side of neuroimaging, that is all the prospective, speculative, and theoretical activities aimed at generating new knowledge of the human brain. Moreover, it highlights the production of software packages needed for these research endeavours. Interestingly, when one focuses on these research tasks, and not so much on clinical tasks, then images are subsumed into a larger and more important category labelled as "data." From the viewpoint of software, images are nothing but a collection of data that, once processed, become visual outputs on the computer screen. In this sense, brain images are actually obtained via "the translation and compilation of numerical values via computer software" (Joyce 2006, p. 5).

Every visual representation seen on a computer screen is composed of numerical data, on the one hand, and a visual manifestation of these data, on the other (Waldby 2000; Prasad 2005). "These images should truly be called *image data* because they can conveniently slide between being data or images" (Prasad 2005, p. 292). In the 1990s, hence at the dawning of neuroimaging software, this twofold nature was somewhat obscure because as soon as computers completed the data processing operations necessary for the production of brain images, only these latter were kept by the machine while the underlying data were discarded by most software packages (Saad and Reynolds 2012). Currently, data are never disposed of, revealing that images and data are two sides of the same coin.

The gist of this story, from the viewpoint of neuroimaging software, is that the visual dimension is less decisive than the numerical dimension. In other words, even though one of the main purposes of neuroimaging research is the production of visual representations of the brain, quantitative data are more important than images when one delves into the inner technicalities of software packages. In this way, apart from a short discussion in chapterFive, this book shall not explore the visual dimensions that previous studies explored in detail

(Joyce 2006, 2011; Prasad 2005, 2014; Blume 1992; Kevles 1997; Pasveer 1989; Waldby 2000; Yoxen 1987; Dumit 2004) but the role of data and computer code.

Neuroimaging researchers know that the relevance of data is valid not only in the scope of my sociology study: it is an aspect that plays a major role in brain studies. Since the 1990s, it has become increasingly central, as the datasets dealt with by researchers have grown very large. In 1993, when Cox (2012, p. 743) became data analyst at the Medical College of Wisconsin, United States, he was given the responsibility of saving the biophysics department where people were "'drowning in data'" after the introduction of MRI scanners. As MRI technology was diffused globally, several research centres began to experience this urge to cope with an expanding amount of research data (Aguirre 2012). In the twenty-first century, such tendencies have been reinforced, as noted by van Horn and colleagues (2005, p. 57):

> Future investigations of in vivo brain function using fMRI can be expected to continue gaining in sophistication, as the questions being asked about the brain, and the methods themselves, become more elaborate. With possibly hundreds or thousands of data files, each recorded under conditions that vary over time, the amount of data may soon overwhelm the abilities of investigators to keep track of this information from even a single fMRI investigation.

Therefore, this book, in addressing neuroimaging research, deals with a question that is attracting many people's attention: the growing relevance of data processing and data analysis. Academic research, but also a growing range of human activities, become sources of data, opening up much leeway for the creation of packages and applications whose utility lies, precisely, on their capacity to process and organize data.

As is common in social life, a certain activity can only gain solidity and meaning when there is a group of social agents responsible for its realization. They grasp the complexities of the issues at stake, negotiate the directions that the activity should follow, and produce meaningful discourses to justify their choices before other social groups. In the case of software, we are dealing with social actors whose power and relevance have gradually become clear: computer programmers.

The Socialization of Computer Programming

In times when many actions and interactions are mediated by software packages (Berry 2011), which come to form a dense "codespace" (Mackenzie 2006, p. 17), the work of software developers (or programmers) necessarily acquire a special relevance. As Galloway (2012, p. 38) pointed out:

> In our present information society, computer programmers and their practice are at the heart of the infrastructure that supports and constrains nearly all the information activities in which we are involved and engaged [...] For this reason we should be interested in programmers, their practice, and the history of that practice.

By scrutinizing the social dimensions of computer programming, this book joins the interpretive efforts aimed to grasp "[...] the way in which certain social formations are actualized through crystallization in computer code" (Berry 2011, p. 37). In spite of their relevance, the tasks performed by programmers and other "computer specialists" have for decades been "opaque to outsiders" (Ensmenger 2010, p. 1), as most social agents would consider software development as a mysterious, inscrutable activity. According to Ensmenger, programmers have historically been, in the eyes of many people, a category of difficult social and disciplinary classification. "Even to this day, their occupational expertise remains difficult to clearly define or delineate" (Ensmenger 2010, p. 30).

As a delineation is needed in the scope of this book, a simple notion will be adopted: programmers are all the people who write computer code in order to implement some computational functionalities.[1] As shall be seen, this definition, albeit apparently simple, does not go without underlying difficulties, because different categories of computer code do exist, and many people would reasonably dispute that the

---

[1] By means of code, programmers make the computer realize structured and repetitive operations called "algorithms." These latter frequently receive some inputs, which are processed and converted into outputs. If algorithms are correctly implemented, the same input will always generate the same output.

production of some of these categories would entitle somebody to be called a programmer. However, this initial definition suffices to provide us with some references to begin our journey into the universe of computer programming. Thus let us assume that programmers are people who write highly formalized texts with which instructions can be sent to the computer so this latter behaves in a specified way. Frequently, the mastery of such programming skills enables code writers to eventually design new software, and this is why "programmers" may also be called "software developers."[2]

Most accurately, Ensmenger (2010, p. 3) noticed that a few programmers have turned into geek celebrities, which is frequently due to their commercial success, whereas blind eyes have been turned to "[…] the vast armies of largely anonymous engineers, analysts, and programmers who designed and constructed the complex systems that make possible our increasingly computerized society." If professional software developers have long laboured in the shadows, even more obscure have been the crowds of programming students and amateur programmers. Nowadays, there is a growing number of people with intermediate programming knowledge who are nonetheless capable of designing computer applications of considerable usefulness. "The volume of new code created by non-professional programmers today by far out-weights that of the professionals" (Skog 2003, p. 307). This is so because of both technical and social factors. As for technical factors, it will be seen that programming languages (which software developers use to "converse" with the computer) have slowly assimilated elements of everyday language, thus becoming less intricate for people with no computer science or engineering background. In terms of social factors, there is the aforementioned relevance gained by computers and computer applications in social life, attracting the interest of people desirous of joining this computational saga. As a result, computer code writing and software development have increasingly been carried out by non-professional

---

[2] The word "hacker" will also be used in this book. Even though this word has recently gained a negative connotation (somebody who realizes criminal actions on the internet), it is worth trying to contribute towards reviving the original usage of this term, making reference to people who possess outstanding programming skills.

programmers, a trend that manifests itself in the multiplication of adventurous programmers, or quasi-programmers, who in spite of having limited programming skills, endeavour to launch computer or mobile phone apps, frequently searching for commercial rewards.

A cognitive and technical division can be pointed to. On the one hand, there are the programmers whose work comply with the tenets of software engineering, established as a discipline with scientific standing. On the other, there are the code writers whose abilities are based on personal effort, technical improvisation, and try-and-error experiences. Professional programmers, and mainly those who are based in companies, would surely make reference to their knowledge authority to secure privileged positions, keeping low-skilled code writers at bay. This "tension between the messy tinkering of real-word computing and the clean abstractions of academic minded computer scientists" (Ensmenger 2010, p. 31) has been going since early moments in the history of programming.

In this book, attention will be mainly given to non-professional programmers. Even though neuroimaging software is designed by scientific researchers working in academic settings, those people frequently lack solid programming knowledge. To a large extent, their programming endeavours are triggered by everyday research necessities. Once they need to realize some data processing and appropriate software is not available for those purposes, they frequently set out to learn some programming techniques so as to design a software package (or, more frequently, a piece of software package) to reach their goals. It will be shown that these needs are frequently fulfilled by means of non-systematic learning tactics, a circumstance that eventually generates knowledge gaps and deficiencies. As a result, neuroimaging software development is, to a surprisingly large extent, a product of much unorthodox programming work and technical improvisation.

Generally, it is assumed that people design software for either having monetary gains or enjoying a creative task. In this vein, Weber (2004, p. 37) asked: "[…] do people write software to make money, or to create and experiment as true artists do?" Previous studies have scrutinized commercial software development (Ensmenger 2010; LaToza et al. 2006) as well as the personal rewards and creative joys of code writing

(Mackenzie 2003; Brooks 1995; Torvalds and Diamond 2001). This book highlights a third reason why people write computer code: to carry out research, observe the body more accurately, understand brain process, and generate new knowledge. Attention is drawn to academic programming (or scientific programming) as it is performed by people having background in areas such as biology, medicine, psychology, and even social sciences. Therefore, whenever I speak of neuroimaging researchers, I am referring to people with varied scientific background, carrying out research in academic institutions, and sharing the interest in the brain and the need for using neuroimaging software. We are dealing with manifestations of a process of socialization of programming whereby different professionals are invited to join a global code writing endeavour. Another specific trait of the analysis provided here lies in its theoretical foothold, as software development is looked at from a communicative perspective.

Software Development and Communication

When it comes to interpreting the social dimensions of software, algorithms, and computer code, two approaches have been frequently subscribed to. On the one hand, some analysts, especially those focusing on the early phases of the computing history, highlight the cultural and political impacts of programmers' work (Atal and Shankar 2015; Lakhani and Wolf 2005; Raymond 2001; Stallman 2002b; Torvalds and Diamond 2001; Weber 2004; Barbrook 2003; Ensmenger 2010). Human and social aspects are foregrounded as analysts consider that "[…] the computer revolution of the previous century didn't just happen. It had to be made to happen, and it had to be made to happen by individual people, not impersonal processes" (Ensmenger 2010, p. 25). A key notion is that software developers either disrupt the business environments they penetrate or challenge the commercial dominance that companies try to impose. Hence the emphasis on the "antiestablishment sentiment" (Torvalds and Diamond 2001, p. 161) disseminated by developers. Many analysts, as explained by Ghosh (2005, p. 26), wish to grasp "the motives of developers" and comprehend "why hackers do what they do" (Lakhani and Wolf 2005). Because of the political and ideological aspects highlighted in this approach, but also because some of these analysts draw on ideas from the social exchange theory (Blau 2006), great interest was devoted to open source software, its non-commercial logics, and the formation of communities around it. Hence the reference to "[…] the importance of social and cultural aspects of design communities" (Heliades and Edmonds 1999, p. 394).

On the other hand, some analyses, frequently influenced by the actor–network theory (Latour 1996, 2005), have promoted an interpretive inversion, putting the software and algorithms at the heart of the interpretation (Kitchin 2017; Kitchin and Dodge 2011; Kittler 1997; Lash 2007; Mackenzie 2006; O'Reilly 2013). If in the previous approach software was the product of revolutionary hackers, it is now presented as a social actor in its own right, endowed with logic and power. "Occupying the position of subject at the beginning of sentences, algorithms may be said to 'adjudicate,' 'make mistakes,' 'operate with biases,' or 'exercise their power and influence'" (Ziewitz 2016, p. 5). Eventually, algorithms are in this approach said to be outstanding figures in political disputes, a notion that led to the formulation of expressions such as "algorithmic regulation" (O'Reilly 2013), "algorithmic power" (Kitchin 2017, p. 16), and "power through the algorithm" (Lash 2007, p. 71). It would then be possible to point to the "ongoing triumph of software" (Kittler 1997, p. 151). The links between developers, as well as the tensions between programmers and companies, so important in the previous approach, are replaced with an interest in sociotechnical networks, "software engineering's human/social side" (Devanbu 2009, p. 69), and "the contingent, relational and contextual way in which algorithms and software are produced" (Kitchin 2017, p. 25).

In this book, software development is not regarded as a domain exclusively constituted by free choices taken by revolutionary code writers, like in the first approach. For it is admitted that programmers' actions are frequently shaped by the dynamics of their institutions and the academic standards of their scientific disciplines. They are certainly insightful people, but: "Insights do not arise arbitrarily but are the result of a rule-governed process" (Habermas 2008, p. 157). Nor is software development described here as the kingdom of a transcendental algorithmic force. Programmers are surely subjected to scientific standards, institutional mandates, and technical rules; yet they continue to be capable of tinkering with such constraints, negotiating reasons, and promoting creative solutions. At any rate, if algorithms have come to dominate programmers (their creators), this is at least partially based on voluntary acceptance of algorithmic dominance, and not completely on sheer coercion. Thus algorithmic regulation is not different from legal regulation where the "[…] communicative freedom of citizens can assume a form that is mediated in a variety of ways by legal institutions and procedures, but it cannot be completely replaced by coercive law" (Habermas 2008, p. 33).

Therefore, in this book the design of neuroimaging software is framed as a manifestation of communicative processes. Programmers need to exchange clarification and support, provide peers with justifications pertaining to their organizational choices, make their coding ideas clear in the code itself, and other communicative efforts, each of which can be seen "as a rational activity on the part of creatures to whom can be ascribed *intention* and *purpose*" (Dummet 1993).

Social actions have gained a particular nature over the last decades. On the one hand, the web of social actions or, to be more precise, "systems of actions" (Santos 2000) are increasingly dependent on a variety of "systems of objects" (Santos 2000). In the case of software development, these *systems of objects* are constituted by computer hardware but also by hidden infrastructures such as networks of optical fibers. On the other hand, social actions, which always manifest a certain rationality, are increasingly shaped by rationales formulated in distant places. "[…] in today's world, so-called rational actions assume this classification because of foreign rationalities […]" (Santos 2002, p. 81). Thus more than speaking of "communicative actions" (Habermas 1984, 1987, 1996), it is now possible to speak of "systems of communicative actions," a concept that I proposed elsewhere (Bicudo 2014) and will be further explored here.

In order to properly explore the communicative side of software development, it will be necessary, on many occasions, to scrutinize some of the coding approaches and techniques employed by developers.

In the communicative relations in which they engage, computer code frequently constitutes a pivotal mediation, being the carrier of expressive efforts and scientific needs. Thus Berezsky and colleagues (2008, p. 6) are right to claim that "[...] to understand the contemporary world, and the everyday practices that populate it, we need a corresponding focus on [...] computer code [...]."

In the same way that it is hard to classify the work carried out by developers, the interpretive and philosophical classification of their existence is not straightforward. In her classic interpretation, Arendt (1998) described the four spheres composing "the human condition": *labour, work, action,* and *thought*. Software development (considered as a creative task, not a job) is not *labour*, because it is not oriented towards the needs and cycles of the body. It is not *work*, because it is not primarily aimed at producing material things. It has some aspects of *action*, because the creation of a software package may influence people's behaviours. Finally, it has some aspects of *thought*, because, as claimed by Brooks (1995), programming is highly dependent on creativity and logical reasoning. In this way, programmers' work would occupy a grey zone between the realm of action and the realm of thought. This book is also aimed to shed some light on such hybrid form of existence.

Open Source Software and the Internet

In a study focusing on programming, it is obviously necessary to deal with issues that are crucial for programmers, including the issue of open source software. What characterizes a software package as open source is not only free access (absence of fees) but also the publication of computer code, enabling other developers to study the code, modify, and adjust it. One might think that open source software is a recent trend. However, the practice began in the 1950s and actually preceded the commercial exploration of software (Schwarz and Takhteyev 2010). The success of Unix, an open source operating system largely used in the 1970s and 1980s, conferred an ideological enthusiasm onto open source. "It was a time of rampant idealism. Revolution. Freedom from authority" (Torvalds and Diamond 2001, pp. 57–58).

Two routes of software development (the open source model and the commercial model) coexisted in harmony until the mid-1970s when tensions began to emerge (Weber 2004). These tensions ushered in a "crisis of public software in the 1980s" (Schwarz and Takhteyev 2010, p. 629), characterized by the proliferation of applications designed by large corporations, threatening the space previously conquered by open source. In spite of the economic power gained by those companies, the 1990s witnessed an upsurge of open source software whose most illustrious representative is the Linux project. Launched in 1991, the initiative attracted large numbers of developers who contributed code to this open source operating system and eventually formed the "Linux community" (Torvalds and Diamond 2001, p. 134). Ideological and political motives were not cultivated by Linus Torvalds himself, the community's leader, but, during that period, many programmers engaged in different currents of non-proprietary software development were greatly inspired by the desire to revolutionize or democratize the software domain. Among them, one outstanding figure was Richard Stallman, founder of the Free Software Foundation and strong advocate of software development as a promoter of an egalitarian society (Stallman 2002b, d, e). Irrespective of their political contents (or absence thereof), several non-proprietary projects were launched, which was also true for the neuroimaging area. "There are many examples of neuroimaging software projects that embraced the open-source ideal and bloomed during the 1990s" (Aguirre 2012, p. 766).

In the twenty-first century, even though some corporations have managed to secure a tight control over certain software niches, open source packages continue to flourish and proliferate (von Hippel and von Krogh 2003; Wagstrom 2009). Some domains have even come to be almost completely, if not completely, dominated by open source software. This is the case of neuroimaging. It will be seen that nowadays, considering the dozens of packages commonly cited in the neuroimaging literature, only a tiny fraction is produced by companies, has protected source code, and requires the payment of licence fees.

The success of the open source model is explained by a range of factors, which depend on the specific features of the areas where software is used. Variations notwithstanding, it is sure that non-proprietary software attracts large groups of users not only because of the absence of economic costs (Schwarz and Takhteyev 2010). For those having ideological motives, the main reason for adhering to non-proprietary software is the possibility to be freed from the standards and strategies of companies. According to Stallman's (2002a, p. 41) principle: "'Free software' is a matter of liberty, not price." In addition to ideology and price, what explains the robust diffusion and popularity of non-proprietary software is technical performance. As explained by Stallman (2002b, p. 166) himself, "[...] what people began to note, around 1990, was that our software was actually better. It was more powerful, and more reliable, than the proprietary alternatives." This is so because whereas commercial software is produced by a limited number of programmers hired in companies, there is frequently a larger group of people contributing code to non-commercial projects, many of whom are talented programmers working out of spontaneous excitement towards the initiative.

In terms of neuroimaging software, more specifically, a central characteristic has determined the dominance of and preference for open source packages: access to computer code. As academic researchers need to have full control over and knowledge of what happens with data during data processing phases, it is crucial that code be analysable. Moreover, without access to code it would be impossible to modify some of the software's functionalities or implement new features, which is frequently done in scientific programming.

Therefore, open source software, in neuroimaging and many other fields alike, has opened up new ways to organize data analysis. Furthermore, it has brought about new ideological, interpersonal, and legal schemes to software development. As for legal aspects, the most remarkable achievement was the creation, by Richard Stallman and its Free Software Foundation, of the General Public License (GPL) approach whereby software packages released as public goods, as well as modified versions of them, remain legally protected from commercial exploration (Atal and Shankar 2015; Stallman 2002b, c; von Hippel and von Krogh 2003). GPL is precisely the most commonly used licence in neuroimaging software.

However, in the scope of this book, the most meaningful characteristic of open source software is actually the particular communicative schemes defined by its development. As noted by von Hippel and von Krogh (2003, p. 212):

[…] open source software projects present a novel and successful alternative to conventional innovation models. This alternative presents interesting puzzles for and challenges to prevailing views regarding how innovations 'should' be developed, and how organizations 'should' form and operate.

In this book, an analysis is provided of not only how these views are changed when software is collectively produced but also of how programmers have created practical solutions so these views can be put into practice. On this point, another crucial issue for software developers is detected, one which shall also be addressed in different parts of the book: the use of the internet in collaborative programming projects.

According to Aguirre (2012, p. 765), the open source tendency gained traction in tandem with the expansion of the internet, triggering the "decentralization of software development": "Anyone with an internet connection could produce, modify, and share software, backed by a powerful philosophical movement to celebrate the activity." One of the main deeds of the internet, from the viewpoint of software development, has been the establishment of connections between programmers, as it has made it possible for them to exchange technical information efficiently and cheaply (Weber 2004). As noted by Raymond (2001, p. 51), the diffusion of the internet enables the expansion of developer communities, which were previously constituted as small, "geographically compact communities." Moreover, it will be seen that internet connections have been largely used in the everyday negotiations on which today's software development projects rely. The internet has then been one of the main material components of the systems of communicative actions that underpin software development.

Research Methods

At the end of this book, a Methodological Appendix is provided, detailing all the research methods and instruments mobilized in the conduct of this research project. Therefore I am providing here only a very brief description.

The study is based on five methods. First, neuroimaging papers published in 2015, 2005, 1995, and 1985 were analysed. Looking at the methodological sections of these publications, I identified the software packages used by neuroimaging researchers and the locations where they are designed. Second, by collecting information on the institutions in which the authors of these papers are based, it was possible to realize a geoprocessing analysis and produce maps depicting the geographical reach of software packages. Third, these analyses enabled me to identify six groups (or levels) of countries, defined by the number of software packages developed locally. Based on this classification, four countries were selected for the conduct of fieldwork: Brazil (a country belonging to the level with the lowest production of neuroimaging packages), Portugal (in the level just above that of Brazil), Netherlands (belonging the level of the most productive areas), and the UK (in the level just below that of Netherlands). In this way, different geographical situations could be accounted for in this study.

The two final methods consisted in internet surveys. The first survey (identified as Survey 1 in different parts of this book) relied on the participation of users of neuroimaging software packages, identified through the analysis of neuroimaging papers. I received 119 questionnaires from 25 countries, using descriptive statistics to display the results, and subjecting those quantitative data to some statistical tests. Finally, the second survey (to which I refer as Survey 2), had the participation of neuroimaging software developers. This time 23 questionnaires were received from 8 countries. Descriptive statistics and some statistical tests were also used here.

Once again, I invite the readers interested in knowing the details of these methods to consult the Methodological Appendix.

Structure of the Book

This book is organized in two parts composed of two chapters each. The analysis of a quite specific social aspect of computer programming is delivered in each of these four chapters.

In the 1stPart (comprising chapterTwo and chapterThree), I foreground processes that yield integration, fostering collaboration and interdependence between programmers. In chapterTwo, I focus on the practical solutions and instruments they have mobilized to engage in collaborative software development. In section1, the worth of collaboration for software development is explicated, and modalities of collaborations outlined. In section2, there is an analysis of modularization, a development technique that has turned into a preferred approach to software development, as well as its communicative implications. In this same section2 I introduce the concept of software flexibility, which allows the classification of software packages based on communicative criteria. I move on to deliver, in section3, an analysis of the different forms in which computer code is shared, transferred, inherited, and received. In section4 I discuss two approaches that were frequently used in previous analyses (which I name "software development as state of nature" and "software development as culture"), making the case for an alternative, communicative approach.

In chapterThree, the goal is to analyse the rationales behind software development, stressing how they create the possibilities for a series of social relations, collaborations, and competitions. In section6 it is claimed that publication of computer code, characteristic of open source software, amounts to an externalization of thought whereby much potential for clear communication is generated. This discussion continues in section7, in which it is claimed that computer programming is not only the realm of control, as there is much space for creative and uncertain solutions that constitute the roots of communication in software development. The discussion of section8 focuses on the expressive dimensions of open source software, the point being made that publication of computer code makes room for self-display through computer code. Finally, section9 deals with the ways in which the idea of "community" emerges in the discursive trajectories followed by software developers.

The 2ndPart is dedicated to the analysis of processes generating inequalities and dependencies between different programmers and research institutions. In chapterFour I highlight the geographical and institutional dimensions of software development. In section11 I show how computer code and software can be appropriated by academic institutions, revealing that concerns with ownership are not exclusive to companies. I carry on discussing this issue in section 12, in which the importance of computer code ownership for universities is highlighted. The financial costs and inequalities present in software development are focused on in section 13, which explains how such issues have impacts on the uneven geographical diffusion of packages. This chapter also deals with inequalities in terms of scientific prestige between institutions developing neuroimaging software. A social and political phenomenon is described in section 14: today's most popular neuroimaging packages represent what I name "semi-flexible packages" whose production does not follow actual collaborative patterns, a circumstance that eventually slows down progresses in software development. The institutional and cognitive hierarchies of software development are further explored in section 15, in which the fundamental difference between "code" and "script," as well as its communicative implications, is exposed. In section 16 the particular case of SPM, the world's most popular neuroimaging package, is focused on, including an analysis of the hierarchical and collaborative trends generated by the package.

This book sheds light on the situation of programmers. However, there is no over-specialization in academic settings. Not only do those people design certain software packages, but they also use their own computer productions, in addition to using many other packages designed by other people. In this way, the figures of software developer and software user are very often mixed up in neuroimaging. This is why chapterFive stresses software use. In section 18 I emphasize the current willingness, manifested by a growing number of programmers, to design intuitive, user-friendly packages. In section 19 I focus on the relations between developers and users, showing that, to a certain degree, users can at times become a sort of co-developers. In discussions about computational resources, a division between "the virtual" and "the real" is frequently assumed. In section 20 I propose a communicative refinement, arguing that one should rather speak of a theoretical division between the "virtual realm" and the "actional realm," in order to account for the empirical division between potentialities and concrete phenomena. In section 21 I delve into the collective and individual tactics mobilized by neuroimaging researchers to use and design software packages while coping with their frequently limited computing knowledge. A seeming paradox is the focus of section22: if on the one hand, some software packages are turning into gold standards of neuroimaging analysis, on the other hand a persistent lack of standardization is verified in this disciplinary domain. The concept of loop, so crucial in code writing, is the point of departure for section23, which continues to stress software users' cognitive limitations, in addition to discussing the grey zones and hidden bugs of software packages. A discussion that is most interesting from the viewpoint of the so-called "big data" issue is offered in section24, where I show that generation of research data is not necessarily a sign of scientific excellence, as technical and theoretical expertise pertaining to data analysis may be lacking. In this situation, academic researchers, who are supposed to possess a potential for creative actions, threaten to fall prey to imitative behaviours. Each of these four chapters closes with a brief analysis of an empirical example, offering an additional exploration of a specific issue.

Finally, some last words are advanced, revisiting key ideas introduced throughout the book and refining the understanding of communication. In this way, I attempt to contribute to the comprehension of contemporary societies in their entanglements with digital technologies, algorithms, and software. If communication is the main phenomenon that characterizes social life, as assumed in the theoretical approach adopted here, then it is crucial to understand in which ways social negotiations and debates come to be either reinforced or disturbed when computer code, its production, and its collective use turn into central communicative mediations.

References

Aguirre, Geofrey K. 2012. "FIASCO, VoxBo, and MEDx: Behind the code." *NeuroImage* 62:765–767.

Arendt, Hannah. 1998. *The human condition*. 2nd ed. Chicago: University of Chicago Press.

Ashburner, John. 2012. "SPM: A history." *NeuroImage* 62:791–800.

Atal, Vidya, and Kameshwari Shankar. 2015. "Developers' incentives and open-source software licensing: GPL vs BSD." *B.E. Journal of Economic Analysis and Policy* 15 (3):1381–1416.

Bandettini, Peter. n.d. A short history of Statistical Parametric Mapping in functional neuroimaging.

Barbrook, Richard. 2003. "Giving is receiving." *Digital Creativity* 14 (2):91–94.

Berezsky, Oleh, Grigoriy Melnyk, and Yuriy Batko. 2008. "Modern trends in biomedical image analysis system design." In *Biomedical engineering: Trends in electronics, communications and software*, edited by Anthony N. Laskovski, 461–480. Rijeka: InTech.

Berry, David M. 2011. *The philosophy of software: Code and mediation in the digital age*. New York: Palgrave Macmilllan.

Bicudo, Edison. 2014. *Pharmaceutical research, democracy and conspiracy: International clinical trials in local medical institutions*. London: Gower and Routledge.

Blank, Robert H. 2007. "Policy implications of the new neuroscience." *Cambridge Quarterly of Healthcare Ethics* 16:169–180.

Blau, Peter Michael. 2006. *Exchange and power in social life*. New Brunswick: Transaction.

Blume, Stuart S. 1992. *Insight and industry: On the dynamics of technological change in medicine*. Cambridge: MIT Press.

Bradley, William G. 2008. "History of medical imaging." *Proceedings of the American Philosophical Society* 152 (3):349–361.

Brooks, Frederick P. 1995. *The mythical man-month*. Reading: Addison-Wesley.

Bruehl, Annette B. 2015. "Making sense of real-time functional magnetic resonance imaging (rtfMRI) and rtfMRI neurofeedback." *International Journal of Neuropsychopharmacology* 18 (6):1–7.

Cox, Robert W. 2012. "AFNI: What a long strange trip it's been." *NeuroImage* 62:743–747.

Cutanda, Vicente, David Moratal, and Estanislao Arana. 2015. "Automatic brain morphometry and volumetry using SPM on cognitively impaired patients." *IEEE Latin America Transactions* 13 (4):1077–1082.

Devanbu, Prem. 2009. "Study the social side of software engineering." *IEEE Software* 26 (1):69.

Dumit, Joseph. 2004. *Picturing personhood: Brain scans and biomedical identity*. Princeton: Princeton University Press.

Dummet, Michael. 1993. *The seas of language*. Oxford: Oxford University Press.

Ensmenger, Nathan. 2010. *The computer boys take over: Computers, programmers, and the politics of technical expertise*. Cambridge and London: MIT Press.

Filler, Aaron G. 2009. "The history, development and impact of computed imaging in neurological diagnosis and neurosurgery: CT, MRI, and DTI." *Nature Precedings*:1–76. Available at http://dx.doi.org/10.1038/npre.2009.3267.5.

Galloway, Patricia. 2012. "Playpens for mind children: Continuities in the practice of programming." *Information & Culture* 47 (1):38–78.

Ghosh, Rishab Aiyer. 2005. "Understanding free software developers: Findings from the FLOSS study." In *Perspectives on free and open source software*, edited by Joseph Feller, Brian Fitzgerald, Scott A. Hissam, and Karim R. Lakhani, 23–46. Cambridge: MIT Press.

Goebel, Rainer. 2012. "BrainVoyager: Past, present, future." *NeuroImage* 62:748–756.

Gold, Sherri, Brad Christian, Stephan Arndt, Gene Zeien, Ted Cizadlo, Debra L. Johnson, Michael Flaum, and Nancy C. Andreasen. 1998. "Functional MRI statistical software packages: A Comparative analysis." *Human Brain Mapping* 6:73–84.

Good, Byron. 1994. *Medicine, rationality, and experience*. Cambridge: Cambridge University Press.

Habermas, Jürgen. 1984. *The theory of communicative action, vol. 1: Reason and the rationalization of society*. Boston: Beacon Press.

Habermas, Jürgen. 1987. *The theory of communicative action, vol. 2: Lifeworld and system*. Cambridge: Polity Press.

Habermas, Jürgen. 1996. *Between facts and norms: Contributions to a discourse theory of law and democracy*, Studies in contemporary German social thought. Cambridge, MA: MIT Press.

Habermas, Jürgen. 2008. *Between naturalism and religion*. Cambridge: Polity Press.

Haznedar, M. Mehmet, Francesca Roversi, Stefano Pallanti, Nicolo Baldini-Rossi, David B. Schnur, Elizabeth M. LiCalzi, Cheuk Tang, Patrick R. Hof, Eric Hollander, and Monte S. Buchsbaum. 2005. "Fronto-thalamo-striatal gray and white matter volumes and anisotropy of their connections in bipolar spectrum illnesses." *Biological Psychiatry* 57 (7):733–742.

Heliades, G. P., and E. A. Edmonds. 1999. "On facilitating knowledge transfer in software design." *Knowledge-Based Systems* 12:391–395.

Henson, Richard. 2005. "What can functional neuroimaging tell the experimental psychologist?" *The Quarterly Journal of Experimental Psychology* 58A (2):193–233.

Joyce, Kelly. 2006. "From numbers to pictures: The development of magnetic resonance imaging and the visual turn in medicine." *Science as Culture* 15 (1):1–22.

Joyce, Kelly. 2011. "On the assembly line: Neuroimaging production in clinical practice." In *Sociological reflections on the neurosciences*, edited by Martyn Pickersgill and Ira van Keulen, 75–98. Bingley: Emerald Group.

Kevles, Bettyann Holtzmann. 1998. *Naked to the bone: Medical imaging in the twentieth century*. New York: Basic Books.

Kevles, Daniel. 1997. *The physicists: The history of a scientific community in modern America*. Cambridge: Harvard University Press.

Kitchin, Rob. 2017. "Thinking critically about and researching algorithms." *Information, Communication & Society* 20 (1):14–29.

Kitchin, Rob, and Martin Dodge. 2011. *Code/space: Software and everyday life*. Cambridge: MIT Press.

Kittler, Friedrich A. 1997. "There is no software." In *Literature, media, information systems: Essays*, edited by Friedrich A. Kittler and John Johnston, 147–155. Amsterdam: OPA.

Kolk, Anja G. van der, Jeroen Hendrikse, Jaco J. M. Zwanenburg, Fredy Visser, and Peter R. Luitjen. 2013. "Clinical applications of 7T MRI in the brain." *European Journal of Radiology* 82:708–748.

Laal, Marjan. 2013. "Innovation process in medical imaging." *Procedia—Social and Behavioral Sciences* 81:60–64.

Lakhani, Karim R., and Robert G. Wolf. 2005. "Why hackers do what they do: Understanding motivation and effort in free/open source software." In *Perspectives on free and open source software*, edited by Joseph Feller, Brian Fitzgerald, Scott A. Hissam, and Karim R. Lakhani, 3–22. Cambridge: MIT Press.

Lancaster, Jack L., Thomas G. Glass, Bhujanga R. Lankipalli, Hunter Downs, Helen Mayberg, and Peter T. Fox. 1995. "A modality-independent approach to spatial normalization of tomographic images of the human brain." *Human Brain Mapping* 3 (3):209–223.

Lash, Scott. 2007. "Power after hegemony: Cultural studies in mutation?" *Theory, Culture & Society* 24 (3):55–78.

Latour, Bruno. 1996. *Aramis, or the love of technology*. Cambridge: Harvard University Press.

Latour, Bruno. 2005. *Reassembling the social: An introduction to Actor-Network Theory*. Oxford: Oxford University Press.

LaToza, Thomas D., Gina Venolia, and Robert DeLine. 2006. "Maintaining mental models: A study of developer work habits." Proceedings of the 28th International Conference on Software Engineering, New York, USA.

Mackenzie, Adrian. 2003. "The problem of computer code: Leviathan or common power?" Available at http://www.lancaster.ac.uk/staff/mackenza/papers/code-leviathan.pdf.

Mackenzie, Adrian. 2006. *Cutting code: Software and sociality*. New York: Peter Lang.

Maubon, Antoine J., Jean-Michel Ferru, Vincent Berger, Marie Colette Soulage, Marc DeGraef, Pierre Aubas, Patrice Coupeau, Erik Dumont, and Jean-Pierre Rouanet. 1999. "Effect of field strength on MR images: Comparison of the same subject at 0.5, 1.0, and 1.5T." *RadioGraphics* 19:1057–1067.

Mohamed, Armin, Stefan Eberl, Michael J. Fulham, Michael Kassiou, Aysha Zaman, David Henderson, Scott Beveridge, Chris Constable, and Sing Kai Lo. 2005. "Sequential I-123-iododexetimide scans in temporal lobe epilepsy: Comparison with neuroimaging scans (MR imaging and F-18-FDG PET imaging)." *European Journal of Nuclear Medicine and Molecular Imaging* 32 (2):180–185.

Moser, Ewald. 2010. "Ultra-high-field-magnetic resonance: why and when?" *World Journal of Radiology* 2 (1):37-40.

Nieuwenhuys, Rudolf, Cees A. J. Broere, and Leonardo Cerliani. 2015. "A new myeloarchitectonic map of the human neocortex based on data from the Vogt-Vogt school." *Brain Structure & Function* 220 (5):2551–2573.

O'Reilly, Tim. 2013. "Open data and algorithmic regulation." In *Beyond transparency: Open data and the future of civic innovation*, edited by Brett Goldstein and Lauren Dyson, 289–300. San Francisco: Code for America.

Pasveer, Bernike. 1989. "Knowledge of shadows: The introduction of X-ray images in medicine." *Sociology of Health and Illness* 11 (4):360–381.

Plaja, Carme Junqué i, Vera Vendrell, and Jesús Pujol. 1995. "La resonancia magnética funcional: una nueva técnica para el estudio de las bases cerebrales de los procesos cognitivos." *Psicothema* 7 (1):51–60.

Prasad, Amit. 2005. "Making images/making bodies: Visibilizing and disciplining through magnetic resonance imaging." *Science, Technology & Human Values* 30 (2):291–316.

Prasad, Amit. 2014. *Imperial technoscience: Transnational histories of MRI in the United States, Britain, and India*. Cambridge: MIT Press.

Puce, A., R. T. Constable, M. L. Luby, G. McCarthy, A. C. Nobre, D. D. Spencer, J. C. Gore, and T. Allison. 1995. "Functional magnetic resonance imaging of sensory and motor cortex: Comparison with electrophysiological localization." *Journal of Neurosurgery* 83 (2):262–270.

Raymond, Eric S. 2001. *The cathedral & the bazaar: Musings on Linux and open source by an accidental revolutionary*. Sebastopol: O'Reilly.

Rose, Nikolas. 2016. "Reading the human brain: How the mind became legible." *Body & Society* 22 (2):140–177.

Rutt, Brian K., and Donald H. Lee. 1996. "The impact of field strength on image quality in MRI." *Journal of Magnetic Resonance Imaging* 1:57–62.

Saad, Ziad S., and Richard C. Reynolds. 2012. "SUMA." *NeuroImage* 62:768–773.

Santos, Milton. 2000. *La nature de l'espace: technique et temps, raison et émotion*. Paris: L'Harmattan.

Santos, Milton. 2002. *A natureza do espaço: técnica e tempo, razão e emoção, Milton Santos collection 1*. Sao Paulo: Edusp.

Savoy, Robert L. 2001. "History and future directions of human brain mapping and functional neuroimaging." *Acta Psychologica* 107:9–42.

Schwarz, Michael, and Yuri Takhteyev. 2010. "Half a century of public software institutions: Open source as a solution to hold-up problem." *Journal of Public Economic Theory* 12 (4):609–639.

Sennett, Richard. 2008. *The craftsman*. New Haven: Yale University Press.

Shapin, Steven, and Simon Schaffer. 1985. *Leviatahn and the air-pump: Hobbes, Boyle, and the experimental life*. Princeton, NJ: Princeton University Press.

Skog, Knut. 2003. "From binary strings to visual programming." In *History of Nordic computing*, edited by Janis Bubenko Jr., John Impagliazzo, and Arne Solvberg, 297–310. Boston: Springer.

Stallman, Richard M. 2002a. "Free software definition." In *Free software, free society: Selected essays of Richard M. Stallman*, edited by Joshua Gay, 41–44. Boston: GNU Press.

Stallman, Richard M. 2002b. "Free software: Freedom and cooperation." In *Free software, free society: Selected essays of Richard M. Stallman*, edited by Joshua Gay, 155–186. Boston: GNU Press.

Stallman, Richard M. 2002c. "Releasing free software if you work at a university." In *Free software, free society: Selected essays of Richard M. Stallman*, edited by Joshua Gay, 61–62. Boston: GNU Press.

Stallman, Richard M. 2002d. "Why software should be free." In *Free software, free society: Selected essays of Richard M. Stallman*, edited by Joshua Gay, 119–132. Boston: GNU Press.

Stallman, Richard M. 2002e. "Why software should not have owners." In *Free software, free society: Selected essays of Richard M. Stallman*, edited by Joshua Gay, 45–50. Boston: GNU Press.

Torvalds, Linus, and David Diamond. 2001. *Just for fun: The story of an accidental revolutionary*. New York: HarperCollins.

van Essen, David C. 2012. "Cortical cartography and Caret software." *NeuroImage* 62:757–764.

van Horn, John Darrel, John Wolfe, Autumn Agnoli, Jeffrey Woodward, Michael Schmitt, James Dobson, Sarene Schumacher, and Bennet Vance. 2005. "Neuroimaging databases as a resource for scientific discovery." *International Review of Neurobiology* 66:55–87.

von Hippel, Eric, and Georg von Krogh. 2003. "Open source software and the 'private-collective' innovation model: Issues for organization science." *Organization Science* 14 (2):209–223.

Wagstrom, Patrick Adam. 2009. "Vertical interaction in open software engineering communities." PhD, Carnegie Insitute of Technology/School of Computer Science, Carnegie Mellon University.

Waldby, Catherine. 2000. "The Visible Human Project: Data into flesh, flesh into data." In *Wild science: Reading feminism, medicine and the media*, edited by J. Marchessault and K. Sawchuk, 24–38. New York: Routledge.

Warach, Steven, Jochen Gaa, Bettina Siewert, Piotr Wielopolski, and Robert R. Edelman. 1995. "Acute human stroke studied by whole brain echo planar diffusion-weighted magnetic resonance imaging." *Annals of Neurology* 37 (2):231–241.

Weber, Steven. 2004. *The success of open source*. Cambridge: Harvard University Press.

Yoxen, Edward. 1987. "Seeing with sound: A study of the development of medical images." In *The social construction of technological systems: New directions in the sociology and history of technology*, edited by Wiebe E. Bijker, Thomas P. Hughes, and Trevor J. Pinch, 281–303. Cambridge: MIT Press.

Ziewitz, Malte. 2016. "Governing algorithms: Myth, mess, and methods." *Science, Technology & Human Values* 41 (1):3.

Zimmer, Carl. 2004. *Soul made flesh: The discovery of the brain—And how it changed the world*. London: William Heinemann.

1stPart

Socializing Code

The goal of this part is to analyse the collaborative relations, structures, and values present in neuroimaging software development. Emphasis is given to personal relations (chapterTwo Sharing Code: Social Mediations in Software Development) and the notions and ideologies held by programmers (chapterThree Writing Code: Software Development and Communication).

```
// social code
// source code
```

chapterTwo (Sharing Code: Social Mediations in Software Development) {

```
660    // Instantiate image interpolator
661          UniquePtr<InterpolateImageFunction> interpolator;
662          interpolator.reset(InterpolateImageFunction::
                 New(interpolation));
663          if (!IsNaN(target_padding)) {
664                  interpolator->DefaultValue(target_padding);
665          }
666
667    // Initialize output image
668    // Note: Always use floating point for intermediate
              interpolated image values!
669    UniquePtr<RealImage> target;
670    if (target_name) {
671          UniquePtr<ImageReader> reader(ImageReader::
                 New(target_name));
672          UniquePtr<BaseImage> image(reader->Run());
673          if (image->T() == source->T()) {
674                  target.reset(dynamic_cast<RealImage
                        *>(image.get()));
675          }
676          if (target) {
677                  image.release();
```

© The Author(s) 2019
E. Bicudo, *Neuroimaging, Software, and Communication,*
https://doi.org/10.1007/978-981-13-7060-1_2

```
678                } else {
679                        target.reset(new RealImage(image->Attributes(),
                              source->T()));
680                        target->PutTSize(source->GetTSize());
681                        for (int l = 0; l < target->T(); ++l)
682                        for (int k = 0; k < target->Z(); ++k)
683                        for (int j = 0; j < target->Y(); ++j)
684                        for (int i = 0; i < target->X(); ++i) {
685                                target->PutAsDouble(i, j, k, l, image->
                                    GetAsDouble(i, j, k, 0));
686                        }
687                }
688        } else {
689                target.reset(new RealImage(source->Attributes()));
690                if (!IsNaN(target_padding)) {
691                        const int nvox = source->NumberOfVoxels();
692                        for (int vox = 0; vox < nvox; ++vox) {
693                                target->PutAsDouble(vox, source->
                                    GetAsDouble(vox));
694                        }
695                }
696        }
697        if (IsNaN(target_padding)) target_padding = -inf;
```

The piece of computer code presented above belongs to a software package called MIRTK.[1] Its author is Andreas Schuh, who developed the package in the C++ programming language. I would like to highlight three lines: 660, 667, and 668. These lines are a special component of computer code: they are called comments. The reader will notice that all these lines begin with double slash ("//"). When the computer is "reading" this code and bumps into this symbol, all the characters that follow, up to the end of the line, are ignored. Technically, then, these three lines are not really part of the program, and could be simply deleted from the source code without any impact on the package's performance. Therefore, Andreas Schuh wrote three lines in which he is not "addressing the computer"; he is giving direct explanations to people who might read the code in the future. Hence the fact that comments are written in plain everyday language, no matter if this is done in English, Portuguese, Chinese, or any other natural language. In our example, it is possible to know that the output image of the brain is initialized in line 667, a computing process that finishes in line 696. There is even a warning, ending with an exclamation mark, in line 668.

Comments can provide code readers with some guidance pertaining to what each block of code is doing. Whenever code is shared by many software developers, with different people working on different parts of it, comments prove particularly helpful. By the same token, comments can surely help developers who joined a project at a late moment of its existence. This is very common in software development, as different parts of computer code (or functions in the programmers' jargon) may not result from the same programming effort. The programmer can design a full-fledged, working software package and, years later, add a whole set of new functionalities. For example, a neuroimaging package that shows the brain from always the same angle may receive a new function with algorithms enabling to rotate the brain image.

---

[1]https://biomedia.doc.ic.ac.uk/software/mirtk/.

This is precisely what happened to MIRTK [hist, mirtk]. During several years, this program grew up as an amalgam of code written by several students. Even though its current design results from the notable programming effort of just one person, around 20% of its code is still composed by lines written by different researchers.

> [hist, mirtk]
>
> The MRITK software package, initially called IRTK, began to be produced by Professor Daniel Rueckert in the late 1990s, when he was still a PhD student at Imperial College London, UK. On completion of his degree, Daniel became a professor in the same university and his students took over IRTK's development. For more than ten years, the package was consistently modified by several students who implemented series of new functionalities as part of their PhD studies on the brain. In 2012, Andreas Schuh, a German computer scientist who had had a research stay in the University of Pennsylvania, United States, joined the Dept of Computing of Imperial College London. Doing his PhD there, he quickly came into touch with the IRTK package and decided to finally refactor it, reviewing the whole source code. This is how he summarized, in his interview, what happened thereafter:
>
> *"So the old version [that is, IRTK] no longer exists.*[2]
>
> *Well, it exists [...] That is because, of course, PhD students, during all those years, ten or fifteen years now, contributed, made adjustments, added their own code in that software, and that's all part of what's now available [...] So I think Daniel [Rueckert] has done a great job at this time, but then, during all those years when, of course, many PhD students messed around with the code, added their contributions, but only with the focus to get their things working, without consideration with how the code can work together or be useful. And then there's a lot of duplication of code, like copy and paste from one file to another and so on, instead of making, like, reusable components. So that's why it is a rather messy project [laughter]. And that's was part of the step of going from IRTK to MIRTK, basically, to clean all this up."*

[2]When quoting parts of interviews, I sometimes include my question in bold letters.

> *"For example, If I'm going to write some code for my colleague who's psychologist and knows very little of MatLab, I try to comment it as much as I can, so she understands, for example... let's say... If I say in the code that it has to go to a certain folder to read twenty images, if I comment this for her and one day she wants to change the code to twenty-two images, it will be much easier for her to realize that the number twenty has to be replaced with twenty-two."*
>
> (Researcher, Univ of Coimbra, Portugal)

> *"If we have a complex code, we created the rule... [...] It's a rule or a good practice... It's better to say good practice. As a good practice, we put a comment above the code. 'Oh, it's complex, so I think I'm going to add a comment here because if somebody reads this code, he won't understand, or I suppose he won't understand.' So in those blocks that are more complex, we end up commenting. And then, sometimes, in methods that are not very clear, we also put a comment. But we avoid putting too many comments because, sometimes, it is too polluting, isn't it?"*
>
> (Researcher, Univ of Sao Paulo, Brazil)

The history of MIRTK [hist, mirtk] serves to illustrate the occurrence of collective software development. Code writers frequently engage in collaborations, which requires a whole set of small, albeit most meaningful, technical preoccupations. One of these is precisely the use of comments, additional yardsticks left for the sake of code readers. Writing comments can be either a personal preference or the result of agreements made within a development team. If computer code carries a rationale, as will be seen in chapterThree, then comments may help unveil such rationale, which may be difficult to grasp. For example, a study with developers based in Microsoft showed that 66% of participants declared to have serious difficulties to understand the logic behind other developers' code (LaToza et al. 2006).

> *"[...] I've been here in this project for one year, and before me, there were other people who are no longer here. Yes. So I have to work on code that they made.*
> **And is it easy to understand?**
> *It is very complicated.*

> *Really?*
> Yes, because you have to grasp that person's rationale. Every person has a certain logic to write code. So I have to understand what the rationale was [...]
> **Is the code well commented?**
> Sometimes it's not [...] Some parts are not well commented [laughter].
> **Otherwise, that would help you to understand, wouldn't it?**
> Yes. Because what people think is... I mean... People write some phrase, for example, "Processing by participant." Processing by participant, okay. They think that this says everything, this explains everything that... But sometimes it is not so [laughter].
> **Sometimes it is harder to understand the comment than the code.**
> Yes [laughter]."
>
> (Dr. Margarita Olazar (Dept Computer Science, Univ of Sao Paulo, Brazil)

The efficacy of comments should not be overstated, though. Generally, developers are quite brief, economic, and unspecific, writing comments that may turn out to be rather murky. However, in addition to this and other technical strategies, there is a series of non-technical practices guaranteeing transparency and mutual support among developers. In this sense: "Software development is a highly social process" (LaToza et al. 2006, p. 500). In order to analyse these systems of communicative actions in neuroimaging software, four issues will be explored in this chapterTwo: relations and messages among developers; the modular configuration of software packages; the transmission and reception of pieces of code; and the dynamic negotiations entailed by collaborative code writing.

section1 [Swift Relations, Agile Development] {

1.1. The Technical Worth of Collaboration

Even though code writing is for the most part a lonely task, there is much interaction in developers' working routines. Previous studies have shown the centrality of personal relations between software developers (Galloway 2012; Herbsleb and Grinter 1999; LaToza et al. 2006), stressing the occurrence of much "face-to-face communication" (LaToza et al. 2006, p. 492).

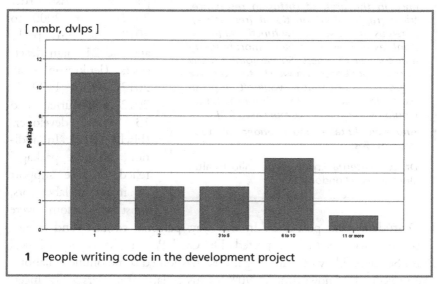

1 People writing code in the development project

Increasingly, software development becomes a collective endeavour. Even though the number of projects carried out by one single developer is still considerable, the number of projects in which two or more people are engaged tends to grow. In neuroimaging software, this trend can also be verified, as showed by some findings from Survey 2.[3] Even though most participants declared that their project is carried out by only one person, the number of projects with collective code writing (two or more people), cannot be neglected [nmbr, dvlps].

---

[3]To know details about the Surveys, see the Methodological Appendix.

"[...] it was a... a team. We were really working in a team throughout the process [...]

Would it be possible to realize... to develop this kind of tool [software] if there was not a team?
I think it would, but it would be very complicated because... In my experience with André [Ribeiro], for example, we were side by side [...] So I think it is easier with a team, because each person has a different experience [...] Because I had experience in the area of diffusion resonance, tractography, and functional [resonance], whereas André also knew functional [diffusion], as well as the area of morphometry and cortical thickness. So we joined many of our experiences and we started to make it grow. It is not that it wouldn't be possible to do something alone, but I think that would be much more complicated [...]

And it would take a much longer time too.
Yes. Yes, yes."

Dr. Luís Lacerda (Institute of Child Health, Univ College London, UK)

Today's most complex and popular neuroimaging packages have been designed by a considerable number of people. For example, from 1991 to 2012, the names of 31 different people figured on the list of main developers of the SPM package (Ashburner 2012). From 2000 to 2012, the FSL package had 21 main developers (Jenkinson et al. 2012). From 1999 to 2012, FreeSurfer had 13 main developers (Fischl 2012). In addition, all those packages relied on the support of many collaborators, most of whom were PhD students and post-docs. In this way, personal contacts and friendships are obviously to be expected. Dr. Cyril Pernet (Centre for Clinical Brain Sciences, Univ of Edinburgh, UK) has joined some online collaborative projects; he knows around 90% of his collaborators personally, meeting them occasionally in conferences and other events. At the Donders Institute for Brain, Cognition and Behaviour, Netherlands, Professor Robert Oostenveld's office is located next to Dr. Marcel Swiers'. They occasionally visit each other to discuss the FieldTrip project, of which Robert is the creator and coordinator, and Marcel an occasional contributor.

At this early point of this book, it is important to highlight an idea that will be paramount: collaborations, interactions, and support happen not only because people are moved by communitarian or altruistic sentiments. Even though such motives may be present, collective software development happens, to a large degree, because there are technical benefits for developers who do not work in isolation. A quite long list of advantages can be given. When the project is collaborative, the resulting package is more robust and sophisticated, because each programmer brings particular skills and knowledge to the project (Fielding 1999; Kitchin 2017; Ashburner 2012); the work is more productive even if each individual programmer can devote little time to the project (Fielding 1999; Ghosh 2005); the resulting code tends to be more conducive to additional modifications and improvements (Schwarz and Takhteyev 2010); and as a consequence, the package is enhanced more speedily (Raymond 2001; Weinberg 1998; Torvalds and Diamond 2001). If one of the main hurdles of software development consists in finding and fixing problems (the so-called bugs), then such task can obviously be attended to more efficiently when there are many people dealing with it. The bigger the group of collaborators, the more efficiency will result. Hence the formulation of a sentence that has become a sort of mantra for software developers: "With a million eyes, all software bugs will vanish" (Torvalds and Diamond 2001, p. 226). As seen in the sequence, there are many ways in which such collaborative work is organized.

1.2. Expressions of Collaboration

It is perhaps noticeable that collaborative development may become a technical trap. For if there are benefits to be gleaned from the participation of large crowds of programmers, then it is crucial to ask the question raised by Weber (2004): how is it possible to realize such complex division of work? Clearly, several kinds of problems can emerge when software development is undertaken by different programmers who become responsible for different parts of code and may be working in different locations (Herbsleb and Grinter 1999). Collaborators may be sticking to divergent rationales, considering different time frames, trying to use the same piece of code to implement different functionalities, trying to solve doubts by addressing the least knowledgeable contributors, among other problems. Because the list of confusions likely to occur is indeed long, some developers sometimes "[...] agree that in an ideal world all programs would be written by a single person [...]" (Galloway 2012, p. 58).

However, those difficulties have never been a reason for coding capitulation, as programmers, and especially those engaged in open source projects, recognize that the solution lies in the search for proper and clear schemes of co-work. "To get past the boundary where the complexity of software would be limited by the work one individual programmer can do on his own, the development process has to implement its worn principles of collaboration" (Weber 2004, p. 82). In his interview, as well as in conversations after the interview, Dr. Matthew Brett (School of Biosciences, Univ of Birmingham, UK) made clear that he is not a professional programmer and therefore lacks the knowledge necessary to coordinate the work of people involved in collaborative development. Being the leader of the NIPY community[4] (which designs neuroimaging software by using the Python language), he feels that such knowledge would greatly facilitate his tasks. Nevertheless, even if those coordinating skills are lacking, Matthew and other programmers devoted to collaborative software development have been able to create and follow some norms and rules pertaining to collective work, tentative, fluid and implicit as they may be. Let us then consider five situations in which such organizational efforts are made.

> *"So the advantage of working with neuro-imaging here is that you're helped. Is that correct?*
> *For neuroimaging, yes. There are both students and old researchers here, people who work in the area [...] So I have much support here in this area.*
> *That is, if you have any doubt...*
> *Yes, there are many people to help.*
> *Does that frequently happen?*
> *Yes. When I arrived here, I didn't know much. I have learnt a lot with people here in the five years I've been here in Rio [de Janeiro] [...]."*
>
> Researcher (D'Or Research Institute, Brazil)

First, in software development teams, each individual possesses a particular set of skills and coding experience. In this way, considering the frequent hurdles of software development, people are very likely to recur to their colleagues' help when it comes to solving a coding problem that proves particularly hard. This likelihood is increased when developers work on pieces of code they do not have deep knowledge of (LaToza et al. 2006). In this sense, a development team represents, from the viewpoint of each of its components, a reservoir of knowledge to be explored whenever necessary.

---

[4]http://nipy.org/.

Second, pieces of code are frequently shared between colleagues. If work duplication seems wrong and futile, this feeling proves particularly strong in software development. Unless a programmer wishes to face some hard exercise to acquire coding skills, nobody will ever wish to spend days or weeks struggling to make a buggy piece of code work only to find that somebody else has already written the perfect, desired lines of code and is willing to provide it for free. In my interviews, I heard countless times the old saying that there is no use in reinventing the wheel. This phenomenon is actually an expansion of the previously described one. For in addition to being a reservoir of knowledge, development groups are, from the viewpoint of each particular programmer, a potential source of pieces of code incorporating both technical knowledge and personal help. In many neuroimaging laboratories, there are even informal archives formed by pieces of code that have been written, modified and perfectionated by different people throughout the years. Newcomers, because they gain access to this coding store, are never completely unassisted.

"So I imagine that there is a kind of code archive that can be used and adjusted in new projects.
Yes, yes, we adjust it, we do that.

Is it an archive of yours or is it an archive of the group?
Ahm, we don't... we don't actually accord it an owner. I adjust some code and if my colleague next to me wants it to adjust it, I simply give it to him [...] A team should work like this, so people advance more quickly. If there is some work done, it shouldn't be repeated."

Researcher (Univ of Coimbra, Portugal)

"Okay, but has somebody here at the Institute, for instance, asked you something: 'Oh, I need to do such a thing.' And you: 'Oh, I have the code.'
Yes, yes, this is very common.

Common.
Yes.

And have you also ever asked...
Yes. Sometimes I ask for help from colleagues who work at the laboratory. We do like this.

There is code exchange. [Laughter.] Code sharing.
Yes, there is [...] When we find some code, we let people know too: 'Look at this code. It's interesting. It solves this problem.' We can adjust it and pass it on. We say: 'I've enhanced it, I've implemented something, you can carry on working on it.'"

Researcher (D'Or Research Institute, Brazil)

Third, software development groups, in addition to forming this collectivist environment, generally possess hierarchical features, which is to be expected from teams based in universities. Heading those teams, there are frequently professors who have sometimes had some experience in companies or other universities, either nationally or abroad. Even though those leaders do not necessarily have deep programming skills, they have the technical and institutional savviness to provide their students and supervisees with much support. At times, they are also the original developers of a software package that has been worked on by successive students. In this way, they have been in touch with a certain codebase for many years, thus acquiring technical legitimacy. According to LaToza et al. (2006, p. 498): "Almost all teams have a *team historian* who is the go-to person for questions about the code." One classic example is Karl Friston, the initiator of the SPM neuroimaging package. As shall be seen in chapterFour, Karl used his notable programming skills to launch SPM, one of the world's first specialized neuroimaging applications, which gained rapid popularity. He attracted many collaborators and obtained funding for the continuation of his brain studies. Nowadays, he continues to be active as head of the SPM development group. In order to provide his supervisees with support, he created a scheme that he names "clinics": every week, there is some time slots reserved for people to visit him and ask general questions about software design or specific questions about the SPM's codebase.

The fourth phenomenon is a consequence of the previous ones. The collaborations verified within a development group can persist even after its members stop sharing a laboratory. This is what happens with Ricardo Magalhães (Institute of Life and Health Sciences-ICVS, Univ of Minho, Portugal) who had a research stay in France, returned to Portugal, and continued to be in touch with a French collaborator. Such persisting bounds seem to be stronger between supervisors and supervisees. Studying a software development company based in California, United States, Mockus (2009, p. 73) found "[…] an indication of continued professional collaborations between a follower and a mentor even in cases where the mentor leaves the organization […]."

Finally, there are examples of personal and moral support that can be verified not only in software development. Hackers frequently receive inspiration, insights, and even personal encouragement from colleagues. This help can be especially important in an activity where the constant occurrence of bugs and technical failures may become discouraging. Programmers know that some tricky bugs are mysteriously solved when one is away from the computer and suddenly has an idea that pops up in one's mind like a revelation. In this way, having the opportunity to socialize or talk about a particular coding problem can turn into an indirect technical support. The story recalled by a researcher based in the University of Sao Paulo is on this point very illustrative [hit, tactic].

[hit, tactic]

"Would you say there is any advantage in working in a group?
Yes, there are many advantages.
What, for instance?
Oh... [Pause.] You have more ideas, right? [...] I think this is the main thing. You have more ideas... ideas, opinions, you solve doubts, sometimes you share probl... Sometimes, just because you share a problem... [laughter.] Sometimes the person doesn't even give you an answer, but just because you share it, you end up thinking differently [...] By the way, in a company I worked for, there was a guy who gave a name for this technique: hit.
Hit?
Hit. He said: 'Evandro, come on here, I'll give you a hit.' And what did he do? He told me the story of what he was developing [...] So we were very close friends, and he called me: 'I have a problem here.' I said: 'I don't know if I can help you.' He said: 'Yes, yes, you can.' He told me the whole story: what he was developing, how we wanted to do it, and so on. And then.... I was hearing his problem for the first time, right? I didn't give my opinion, I didn't hasten to give it. Then, many times, he said: 'Oh, I think I am going to do it in such way.' And I said: 'Explain to me how you want to do it now.' And he explained to me the way he had imagined. This is

> *when I gave my opinion. And I said: 'Oh, it is perfect this way.' Or I proposed some adjustment to what he... So this was my help [...]*
>
> **This technique is interesting. In fact, he solved the problem himself.**
> *Yes [...] This frequently happens here too [...] You're so involved with a problem that you can't think differently. You're too involved there. So you need somebody to give you an opinion, or somebody who draws your attention somewhere else, or to keep at distance for a while so you can look at it with different eyes afterwards [...]."*
>
> Researcher (Dept Computer Science, Univ of Sao Paulo)

> *"How do you imagine your career from now on? Do you think you'll remain in this area, in the academia?*
> *Yes, I... [...] I'm not a physicist, I'm not a physician. I'm a biologist who works in neuroscience. So what I have realized is that I have managed to make people converse. So sometimes there is a physician who can't understand what the physicist is saying, and I, as a biologist working in the area of medical physics, of resonance, and having this biology background, I can sometimes enable this dialogue. So I see myself, from now ow, providing this interface between physicians and physicists or radiologists [...]."*
>
> Dr. Carlo Rondinoni (Department of Physics, Univ of Sao Paulo, Brazil)

There is one characteristic turning this mutual support particularly crucial in brain studies: the field is very much interdisciplinary, as pointed out previously (Dinov et al. 2014; Filler 2009; Bear et al. 2007; Padma 2008). For example, one of the UK sites visited in my fieldwork was the Centre for Neuroimaging Sciences, a large building in one of the university campuses of King's College London, where Dr. Vincent Giampietro is based. According to him, "in this building we have everything, you know, we do have physicists, engineers, mathematicians, but we do have nurses, psychiatrists, neurologists, psychologists, clinical psychologists [...] Everybody works together and there's a lot of overlap." Those scientists have convergent research interests, as they all wish to understand the brain, but their scientific background is different. Furthermore, they frequently produce and share pieces of computer code while having different programming skills. To work in these conditions, much dialogue, negotiation and exchange of information is necessary, especially when it comes to connecting two disparate sides: on the one hand, there are people with medical and biology background and deep knowledge of the brain's anatomy and genetics; on the other hand, there are physicists, radiologists, and engineers with deep knowledge of the operation of magnetic resonance scanners and the initial process-ing of data with which brain images are generated. For neuroimaging studies to be carried out, these two poles need to converse, which is realized by means of personal relations and verbal dialogues, but also by means of computer code. In this sense, code writing assumes the form of what I called *mediational action* (Bicudo 2014), as it incor-porates, at the same time, the requisites of people searching technical success, and the communicative logic of scientists searching for inter-comprehension. To a large degree, then, neuroimaging studies, and more specifically the development of neuroimaging software, open up much leeway for social mediation, as illustrated by the story of Professor Alexandre Franco [alxd, frnc].

[alxd, frnc]

Professor Alexandre Franco has background in electrical engineering. He pursued his engineering studies in the United States, where he completed the Master's Degree in 2005, and the PhD in 2009. Afterwards he worked as a post-doc researcher at the University of New Mexico. Thanks to the knowledge and experience gained, he could have stayed in the United States but was invited to be one of the research leaders of the Brain Institute of Rio Grande do Sul (InsCer), in the south of Brazil. Realizing the soundness of the project, and making sure that proper research infra-structure would be provided, he accepted the challenge and went back to his native country. Nowadays, in addition to being one of the research leaders of InsCer, he is a professor at the Pontifical Catholic Univ of Rio Grande do Sul where he has supervised many students with several back-ground and at different levels.

"I am the guy who is between the resonance and the final image. I am in-between. Because resonance generates very ugly images [...] The cute image showing brain activations... there is a long way to get there [...] So I have to have very good communication with people who interpret

> *those results; they are the clinical staff, the scientists [...] Those people understand what those activations mean, because they have clinical questions [...] And I can also talk to the physicist who is at the development side saying: 'What is the pulse sequence that we need to apply? How can we improve the quality of the image that we're going to collect?'*
>
> **It's mediation.**
> Yes.
> **Then at one side, which is the physics side, a language is spoken. At the other side, which is clinical, a completely different language is spoken.** *And those two sides don't talk to each other. And I have to converse with both of them."*

In face of these considerations, it is possible to ponder the meaning of the relation between the group and its constituent people, an issue that is more than classical in sociology.

1.3. Individuals and Groups

In chapterThree, it will be seen that one of the main features of code writing is the cognitive creations it involves. In this way, collaborative code writing would facilitate the sharing of these cognitive tasks. As claimed by Berezsky and colleagues (2008, p. 21): "[...] computer code enables new communicative processes, and with the increasing social dimension of networked media the possibility of new and exciting forms of collaborative thinking arises. This is [...] the promise of a collective *intellect.*" In this regard, however, the interpretation should not be too radical. Instead of describing software development as a pure thought process, it is preferable to depict it as a task that lies at midpoint between cognitive tasks and material production. On the one hand, programming does not involve only thought, because code gains a certain material reality (as a coding text that can be seen, read, shared, and so on) and its existence is therefore not bound to a particular mind or even to a particular development team. On the other hand, programming does not precisely involve the production of material things, so it is not subjected to the limitations that apply in material production (such as the costs related to the acquisition of raw materials).

This nature of computer code creates a most fruitful interplay between the development group and individual programmers, as illustrated by the Apache group. Apache is an international, web-based community dedicated to the development of different open source software packages, including the HTTP server, the world's most successful internet server software. Today organized as a foundation,[5] the community has over the years formulated ways to organize its collective work. According to Fielding (1999), the Apache experience has been successful because it was possible to find an organization whereby the community constitutes a space for intellectual stimulation and support whereas it is up to individual developers to deal with concrete programming issues and take most technical decisions. Therefore, the Apache community, like many other software development teams, has much in common with society. Individuals perform tasks while being moved by their own interests, but these interests are only meaningful because there is a social context where they are framed as reasonable and where there are recommended, accepted ways to pursue individual goals. In this way, it is not correct to claim that we are dealing with purely purposive, instrumental actions. Neither is it possible to conclude that programmers are completely moved by collective, communicative purposes. The presence of mediational actions can in this sense be pointed to.

Software engineers and programmers have realized the benefits brought about by collaboration, code sharing, and personal support. As a result, development groups have created and fostered working schemes conducive to such collaborative atmosphere. The most famous and most well-succeeded of such initiatives has been the so-called agile software development approach. This is a set of methods for software development created in 2001 by seventeen software developers. They proposed some rules for software development, aiming at reinforcing collaboration and the sharing of resources and ideas. They published the "Manifesto for Agile Software Development,"[6] outlining the movement's twelve principles, two of which are worth mentioning here.

---

[5]https://www.apache.org/.

[6]http://agilemanifesto.org/.

According to the sixth principle: "The most efficient and effective method of conveying information to and within a development team is face-to-face conversation" whereas the eleventh principle goes: "The best architectures, requirements, and designs emerge from self-organizing teams." Therefore, the manifesto highlights the value of personal relationships and stress the development team's capacity to produce an internal collaborative dynamic.

The agile development approach has reinforced the social traits of software development, as its "adherents value individuals and interactions over processes and tools" (Galloway 2012, p. 71). More specifically, the movement created or helped diffuse coding practices that have gained much popularity. Code ownership within the team, the realization of frequent meetings, and incentives to face-to-face interactions are among those practices (LaToza et al. 2006). However, if there is one practice that has come to characterize the agile development approach, this is surely pair programming, in which developers (usually two people) sit at the same computer and work on the same piece of code. Occasionally, this practice is used in development teams, even if people are not really aware that they are incorporating something from the agile development movement. For example, the group responsible for the development of the NES neuroimaging software package at the University of Sao Paulo, Brazil, used pair programming at an initial phase of its work. According to one of the programmers involved in this project, the practice is beneficial because while one person writes

> **"Did you ever write code together?**
> Yes, yes.
>
> **At the same time, at the same computer?**
> Yes.
>
> **Many times?**
> Yes.
>
> **Do you enjoy this kind of simultaneous code writing?**
> Yes, yes. By the way, it's something I wanted to do more frequently during my PhD, for example. Because during my PhD I was... there were not many people working on what I was doing, and I think If I had had other person to work with and write code with more closely, I could have done more things in the PhD."
>
> Dr. Luís Lacerda (Institute of Child Health, Univ College London, UK)

the code, the other becomes a sort of reviewer; in this way, every programming decision requires a quick conversation and an agreement whereby two ideas are compared and the best one picked up. According to the results of Survey 2, pair programming is rather rare in neuroimaging software development, but a considerable number of projects follow other practices of agile development. Of the 23 projects included in the Survey, seven subscribe to the rules of either pure agile development or extreme development, which is a variation of agile development [agl, dvlpmt].

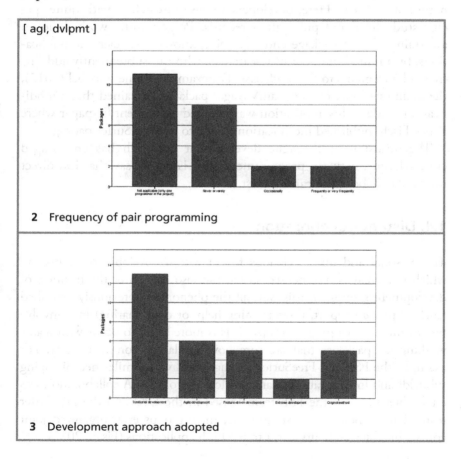

[agl, dvlpmt]

2 Frequency of pair programming

3 Development approach adopted

In software development teams, like in so many other social groups, some people display a stronger tendency to engage in frequent contacts with colleagues. In average, however, each individual tends to have constant interaction with only a few peers. In his survey with over 2700 developers involved in open source projects, Ghosh (2005, p. 37) verified that "more than 50 percent are in contact with one to five others." Therefore, it is possible to say that in software development, there are small communicative islands formed by small groups of programmers. In neuroimaging software, such tendency is accentuated by the scientific nature of the field. Here, developers are also researchers with quite specific study lines and publication records. Programming work done on a certain software package ends up being known, via scientific publications, by a limited number of researchers, who can subsequently add further enhancements to that codebase. For example, Rainer Goebel (2012), the main developer of the BrainVoyager package, explained that a modification made to his application was inspired by a scientific paper where Bruce Fischl published modifications made to the FreeSurfer package.

The diffusion of the agile development approach has encouraged close relations between programmers. This is not to say that less direct forms of collaboration cannot exist.

1.4. Distant Collaboration

The personal and direct relations found within development teams, on which this analysis has focused so far, can also cross over the frontiers of development groups. In this way, all the phenomena previously described (such as moral support, programming help, or code sharing) may involve two or more development groups. This is more likely to occur with teams working on packages that are somehow similar or complementary. For example, the FSL and FreeSurfer packages deal with similar neuroimaging subfields and incorporate similar statistical approaches. A collaboration was established between the FSL team (based in the UK) and the FreeSurfer team (United States). For some years, these groups even organized joint courses in which students learnt to use both applications (Fischl 2012).

If collaborations can involve direct contact, they can surely happen through digital channels as well. Two main phenomena can be pointed to here. First, specialized websites have been created on which developers learn either basic or advanced programming techniques. Some of these websites serve also as a space where people can share doubts pertaining to programming issues, thus becoming able to deal with particular coding problems. One of these websites, named Stack Overflow,[7] is turning into a sort of Wikipedia for software developers. Second, the webpages of neuroimaging software packages generally contain a forum enabling users to send queries and tackle problems faced while using or modifying the package. Frequently, questions are answered by members of the development group, so responses are quick and precise.

"When there is some defect, some problem in the code and it doesn't work, a bug, how do you solve this?
This is a very common thing. So we analyse it, we analyse it. As I told you, there are several online fora that... I always think this way: 'This problem, somebody has already had it.' Yes. And this is generally true. It is very, very rare to come across something that is not on the internet. So when I have some problem and I can't get rid of it, I use the internet.

"Have you ever used those software fora where you go to to solve doubts, ask questions...?
Yes, mainly for AFNI [the AFNI neuroimaging package, developed in the United States]. There was an error [...] I couldn't solve it. [Professor] Alexandre [Franco] [her supervisor] couldn't solve it either.
This kind of problem is very frequent, isn't it?
Yes, so we went to a forum to ask. We showed the problem. 'What is going on?' And it was solved. Those fora are very good. I thought it wouldn't be solved, nobody would read it, but people do read the fora.

---

[7]https://pt.stackoverflow.com/.

You use Google.	**Why did you think nobody would read it?**
Google and it takes me to those sites like Stack Overflow, which I like very much. There are others as well [...] This is my first route. This is my first way, to use the internet. After that, if I can't solve it, then I ask my colleagues who are around. I talk to them: 'I have this problem, I can't solve it', to see if they can help me [...]."	[Pause.] Oh... I thought nobody lost time trying to answer other people's questions. I thought people only wanted to have their own questions answered [...] No, they answered very quickly [...] It took one day. The answer came [...] It was one of the AFNI developers that sent the answer."
Dr. Margarita Olazar (Dept Computer Science, Univ of Sao Paulo, Brazil)	Dr. Nathalia Esper (Laboratory for Images-Labima, Pontifical Catholic Univ of Rio Grande do Sul, Brazil)

Other modes of indirect collaborations are analysed in the following sections of this chapterTwo. Therefore, the realm of programming is not only a space for solitary code writing. In addition to that, there is much leeway for manifold interactions and personal support. The longer the list of human needs and activities that become dependent on software usage, the more intense these relations between software developers will become. What is important to highlight is that these social dimensions of software development are nurtured by two factors. First, mutual support is a communicative need manifested in several human activities. Second, there are also technical requisites making collective development not only possible but also necessary. These technical and instrumental aspects are stressed in the following section.

}

section2 [Globalization, Modularization] {

2.1. Modules and Toolboxes

As claimed in section1, collective software development needs to follow certain rules or patterns. This division of work very frequently means division of code. In other words, it is very usual that different programmers are made responsible for different parts of the codebase, which generally perform very specific tasks. For example, whereas some people implement and refine algorithms

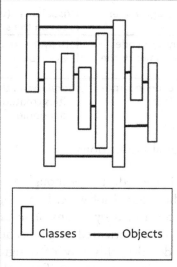

This image is a tentative representation of a C++ program. Generally, new C++ modules are created by introducing new classes into the program. Some classes are bigger than others, as they contain algorithms that are more important. Objects guarantee the connections between classes. Different classes can be worked on by different programmers or teams.

In this way, the package can be consistently modified and expanded with little technical disturbance.

pertaining to the visual representation of the brain, other programmers would deal with the statistical tests performed by the software package. Software engineers and programmers call modules each of these different parts of the program. In fact, people may have different things in mind when using the word "module." Moreover, other words are sometimes used, such as "toolbox" or "plug-in." However, for the sociological analysis looked for here, full precision, in terms of technical terminology, can be relinquished. What needs to be understood is that the principles of modularization have enabled software developers to implement

"[…] the division of a complex task into a series of simple tasks that can be carried out by essentially autonomous modules […]" (Weber 2004, p. 27). In the following sections and chapters of this book, I shall have recourse to a specific usage, as summarized in the following scheme:

Code level (what the programmer work on)	Visual level (what the software user sees on the screen)
Source code: all the lines of code written by the programmer or the programming team	*Software:* all the functionalities made available to the software user
Module: new lines of code incorporated into an old code base	*Toolbox:* all the additional functionalities created through the addition of a module

Classification of Digital Productions

When a module is created, it is as though a "mini-software package" were appended to the original software. In a program's source code, different modules are connected by means of so-called objects, which are used in the so-called object-oriented languages like C++ and Python.[8]

Throughout the decades, this way of designing software has been consolidated. In neuroimaging, a proliferation of toolboxes can also be verified, with many options available today (Ribeiro et al. 2015). Table 1, with data from Survey 1, gives an overview of the historic multiplication of toolboxes.

Table 2.1 Number of software packages and toolboxes cited in the neuroimaging literature

Year	Software packages	Toolboxes	Total
1995	13	2	15
2005	42	27	69
2015	39	69	108
Average growth per year (%)	10.0	167.5	31.0

Source Survey 1 (see Methodological Appendix)

---

[8]I come back to the issue of objects and object-oriented languages in chapterThree.

Because of this unequal expansion in the number of packages and toolboxes, Aguirre (2012, p. 766), in a probably exaggerated vein, pointed to a "friendly Balkanization of software empires into smaller tools." Thanks to modularization, collaborative projects get more feasible, as different programmers or teams, even if based in different laboratories, universities, cities or countries, can work in apparently independent pieces of code.

"Okay, so what I find interesting is that users can develop modules for...
Yes, that's true [...] I developed a model which is the same model for the people in Brain Innovation [the company developing the BrainVoyager neuroimaging package] as well as people outside to write these modules from any other university or institution, even commercial institution. And that's what I call modules or plug-ins [...] And they can then do whatever they want to do with it, right? [...] 'Oh, no, I have this good idea. I want to add this routine.' They can just write a few lines of code and they're done. They don't have to reinvent the wheel. That's what I mean with modules. Modules, you can just focus on your idea and do not have to program simple things again and again and again, you know.*

Okay, but if I'm going to write a module for BrainVoyager... What's a module? Is it a class? Is it a set of functions?
It's like a class. Exactly. You derive your class [...] and then you can add whatever you want and can call routines of the class, you know, like 'Load this file, make a statistical map,' and then you use the map and do something yourself [...] So you can basically do everything.*

So I actually use objects to access the class that is already there.
Exactly. Yes. Yes [...] So if someone sits in some lab and has an idea, they can just program in C++ [...] And they can just link it to BrainVoyager and make a new product out of it [...]."*

Professor Rainer Goebel (Brain Innovation, Netherlands)

Modularization is by no means a recent strategy in software development. In 1945, Hungarian-American computer scientist John von Neumann, one of the fathers of modern computing, summarized the principles that, from his times on, have underpinned the construction of computers. Using modern terms, this is how Grcar (2011, p. 615) describes von Neumann's conception of computing operations: "[...] separate arithmetic, control, and memory repeating a fetch-and-execute cycle instructions alterable by conditional branching [...]." Therefore, since early times, computers were designed to perform minimal,

> *"Do you think that at some point you'll end up developing a toolbox for FSL [the FSL neuroimaging package]?*
> No, I don't think so [...]
>
> *But you would be able to do it.*
> Yeah. If I want to, yes. If I really think there is a need. But I wouldn't do it on... alone [...] I would like to have a... be part of a bigger team.
>
> *Why?*
> Then it's much... There's more people who are hoping... you know, help you develop that tool much faster [...] You cannot be alone in the project [...] You need a team in these software developments. You need teamwork. There's more... You have to divide different modules between different teams. You cannot work on... It's not one module. Developing a software, you have to divide it into, like, say, different modules, smaller modules, and each module is given to one team and that team is responsible for developing only that, and then that's being integrated into the...
>
> *Because it's too complex work.*
> It's not about complexity. It's about making work look precise and good, and doing exactly the right thing. I think working with people don't add complexity, it removes complexity."
>
> Researcher (Donders Institute for Cognitive Neuroimaging, Netherlands)

repeating, and separated functions. However, the principles of modularization were radicalized in the 1970s when the Unix operating system began to be developed at Bell Labs, in the United States. Unix was composed of several modules performing simple tasks, a technical organization which rapidly attracted the interest of many developers based in several countries, thus becoming a successful open source project (Schwarz and Takhteyev 2010). According to Linus Torvalds: "Unix [...] comes with a small-is-beautiful philosophy. It has a small set of simple basic building blocks that can be combined into something that allows for infinite complexity of expression" (Torvalds and Diamond 2001, p. 55). The success of the Unix project (which can be considered as the father of the equally successful Linux project) is to a large degree responsible for the diffusion of the modular design of software, as well as for turning this way of working into common sense among software developers, especially in open source.

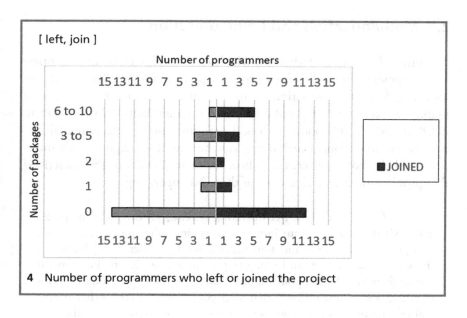

4   Number of programmers who left or joined the project

Obviously, programmers treasure modularization because it brings about many kinds of technical advantages to software development. Parnas (1972) stressed three of such benefits: each development team work on its own, which speeds up the work; profound modifications made to one module do not affect the operations of other modules; and the final codebase becomes quite organized and understandable. In another paper, Parnas highlights another crucial benefit: "The decomposition into and design of submodules is continued until each work assignment is small enough that we could afford to discard it and begin again if the programmer assigned to do it left the project" (Parnas and Clements 1986, p. 254). Indeed, software development projects can be quite fluid in the sense that programmers are constantly joining and leaving programming teams. In Survey 2, it was seen that even though the majority of projects maintained all the developers, some leavers and newcomers can be detected. In two projects, three developers quit whereas in over 6 projects, five developers arrived [left, join]. Modularization is important because when some developers leave the project, other modules can be further developed while some specific modules' progress halts for a while.

2.2. Modularization and Communication

The principles of modularization, which cut complex tasks up into operating fragments, lie at the core of computing technologies. "For computer scientists, it is the translation of the continuum into the discrete that marks the condition of possibility for computationality" (Berezsky et al. 2008, p. 14). At this point, technical requisites meet social potentialities, because computational partitioning can be smoothly aligned with interpersonal help. For example, this is how Stallman (2002a, p. 153) described the spirit behind his development of the GNU operating system:

> [...] I didn't write a whole free operating system [...] I wrote some pieces and invited other people to join me by writing other pieces [...] I said, 'I'm going in this direction. Join me and we'll get there' [...] So if you think in terms of, how am I going to get this whole gigantic job done, it can be daunting. So the point is, don't look at it that way. Think in terms of taking a step and realizing that after you've taken a step, other people will take more steps and, together, it will get the job done eventually.

Software modules can be seen as steps taken by different individuals engaged in a project. Nowadays, long programming walks often begin with the assumption that several steps will be taken by different people designing different modules. However, the number of these modules, as well as their internal configuration, are not defined beforehand: they are possibilities envisaged along the development pathway. This collective coding work is possible because there are both social and technical factors enabling it. On the social side, there are manifestations of personal support (as showed in section1) and established norms to solve coding conflicts (as showed in section4). As for technical factors, there are largely known rationales pertaining to software engineering and programming languages (as shown in chapterThree). In this way, a development team can be composed by developers who share the physical location, but can also be the "implicit team" described by Mockus (2009, p. 68) and characterized by the cooperation between people separated in place and or time. Current software developers often bear in mind the existence of those implicit teams, as illustrated by the story of Fabrício Simozo [fabr, smz].

[fabr, smz]

Fabrício Simozo has background in medical physics. In the Master's Degree, he studied the brain by using electroencephalogram techniques, which do not involve the production of images. Initially, he used JBios, a software package that had been developed at the Computational Group for Medical Signals and Images, Univ of Sao Paulo, Brazil, where he was based. As the package did not provide him with sufficient analytical versatility, he decided to program his own package. However, he considered only his research needs and, as a result, the code became very intricated and obscure. For his PhD, he decided to do neuroimaging in the same research group, focusing on dysplasias, which are brain malformations generally associated with resistant epilepsy. At the beginning of this degree, Professor Luiz Murta Junior, his supervisor, introduced him to 3D Slicer, open source software that can be used for neuroimaging, and taught him the scientific value of open source and collaborative projects. Contrary to what was done in the Master's Degree, he is intent on helping other scientists in his PhD, in two ways. First, he will publish the module he is developing for the identification of dysplasias. Second, he will produce clear code which can be taken on and modified by other programmers.

"Your idea is to install a plug-in, a module, something like that.
Right. It is to develop a module that carries out the type of analysis that we want and integrate this into 3D Slicer, so that a physician, for example, can upload his image, execute this analysis and see the outcome within 3D Slicer. He'll obtain some feedback on what had been classified as a possible dysplasia, a possible brain abnormality [...]

And is your idea, in the end, to upload your modification or not?
Yes, yes. When it's ready, the idea is to make it available.

It's a goal.
It's a goal, right [...] I'm making the code a bit cleaner, a bit more commented, so that, when everything is ready, it is easier to make it available for other people."

According to Gilbert Simondon (1969), a difference must be drawn between "abstract technical objects" and "concrete technical objects." Whereas the former are constituted by closed, self-sufficient and self-sustaining parts, the latter have open, mutually dependent and mutually cooperating parts. In this sense, software packages can be described as concrete technical objects, because their evolution also "[...] tends towards a state that would turn the technical object into a system which is completely coherent with itself, completely unified" (Simondon 1969, p. 23). This unification of software happens at the technical level but has also a social dimension, as coding work undertaken by different people can be consistently put to overlap. In neuroimaging, this other-centred perspective may be reinforced when the programmer works on an application that will be directly used by physicians trying to identify brain diseases.

Logical and modular code is not only a condition for collaborative software development: it is also its consequence. Many times, when a software package is developed by one single person, the resulting code is quite murky and intricated, with large blocks of code that can barely be separated into smaller and clear lines. In its turn, code written in collaborative projects tends to be organized in small reusable and clearly readable blocks. Therefore, collaboration becomes, at the same time, a source and a product of modularization.

The notion that software development is facilitated by the presence of large numbers of programmers working simultaneously received a big blow in the mid-1970s when Frederick Brooks, an American computer scientist based in IBM, published a series of essays on software development. According to him, when software is produced by many programmers, the project tends to progress very slowly because of two problems. First, each new developer joining the project needs to be trained so as to understand what has been done. Second, much time is needed for communication among developers. In this way, "[...] the sheer number of minds to be coordinated affects the cost of the effort, for a major part of the cost is communication and correcting the ill effects of miscommunication (system debugging). This, too, suggests that one wants the system to be built by as few minds as possible" (Brooks 1995, p. 30). Therefore, Brooks' ideas contradict the benefits of largely collaborative projects previously pointed out. However, as noted by both Raymond (2001) and Schwarz and Takhteyev (2010), Brooks ignored the central fact that many projects follow a modular organization whereby most communication happens within small groups responsible for each module.

These technical objections to Brooks' arguments can be complemented by some considerations made from a communicative point of view. Although Brooks did not provide a direct definition, it is clear that when he denounces the "added burden of communication" (Brooks 1995, p. 18), he makes the common confusion between communication and transmission of information. He takes into account the instrumental failures that may occur when information is added to information, or computer code is added to computer code. He was not considering people's communicative capacity to deal with inconsistencies and create mechanisms to reach "states of intercomprehension" (Habermas 1987). Brooks' main (instrumental) concern was the elaboration of a model with which software development could be speeded up. Nevertheless, in order to be viable and consistent, every human activity needs to be not only quick and efficient but also endowed with sense. Software developers would be glad to finalize a project in one day only if such fast production could make sense in the light of some normative or personal motivations. Nowadays, it is for them frequently motivating to consider that their work belongs to a coding effort joined by many other programmers, even if such coding crowd takes long to finalize every development phase. If communication and collaboration were to be framed as burdens, then there would be high risks of extinction not only for software development but for social life and reality itself, because "[...] objective reality is constituted for an observer only together with the intersubjecitivity of possible communication concerning his cognition of events in the world [...]" (Habermas 2008, pp. 168–169).

Throughout the years, Brooks' ideas have been contradicted by analysts and software developers alike, as testified by the consolidation of modular development, and the growth of developer communities. Collective software design is indeed a common practice but its outcome, in terms of the nature of packages developed, is not always the same. At this point, a classification needs to be proposed which will underpin many of the subsequent analyses of this book.

2.3. Software Flexibility

It is not sufficient to point to the presence and relevance of communication and collaboration in software development. In order for a precise description to be given, it is necessary to consider the different degrees collaboration is taken to in different projects. Once again, let us focus on the configuration of computer code and claim that coding projects can be of three kinds, which I propose to name "rigid," "semi-flexible," and "flexible." Rigid software packages (those which result from rigid codebases and rigid schemes of collaboration) are the ones whose code is not made open and cannot therefore be downloaded and modified by individuals or groups other than the original developers. Here collaboration is taken to a very low degree, as the final product depends on the coding effort of a small group of people, or even one single programmer, with some space left open for only occasional inputs from external contributors. This type of software development corresponds to what analysts generally call "closed source software" or "traditional development." Frequently, such packages are produced by companies which due to proprietary concerns, keep the source code as a protected asset. Even though they are generally acknowledged as excellent products, their use is considerably restricted by both the high prices of their licences and the closed nature of their source code, this last feature being frequently rejected by scientific researchers. One example is the LCModel neuroimaging package, produced by L.A. Systems, a company located in Tokyo, Japan.

In their turn, flexible packages are the ones whose development encourages collaboration to the highest level. The source code is published on the internet so any programmer can download, analyse and modify it. However, one additional feature is key to characterize flexibility: there must be efficient and swift schemes enabling modifications to be incorporated into the old codebase. This additional feature has been greatly facilitated by the advent of Concurrent Versioning Systems (von Hippel and von Krogh 2003) and online version control platforms, two technologies that solve a tricky challenge. Let us imagine that the source code of a certain package is published on the internet. Simultaneously, five programmers download it and modify different parts of it. At this point, there is no longer just one source code, but six versions: the one originally published and the ones modified by the five contributors. So the question is: how is it possible to turn these six products into one single product, keeping the lines of code that have not been modified and incorporating the new ones? Version control platforms are applications realizing this task, which would be very slow and tedious if conducted by a human programmer. The application "reads" the different versions of source code, identifies modifications, protects non-modified lines, and eventually generates the resulting source code. Launched in 1999, SourceForge[9] was one of the first of such online services. Over the last years, its popularity has been overshadowed by GitHub, another platform launched in 2008 (see the Empirical Example at the end of this chapterTwo). Flexible neuroimaging packages exist in a high number. Although most of them are used by few researchers, some are gaining popularity such as the ANTS and the MRIcro packages.

This definition of software flexibility is close, but not completely similar, to that given by Naur (2001, pp. 232–233) when he claims: "In including flexibility in a program we build into the program certain operational facilities that are not immediately demanded, but which are likely to turn out to be useful." In the categorization proposed here, these facilities can be designed by the original developers or external contributors, and can be implemented as original features or later modifications. In flexible software packages, collaboration has the potential to reach very high degrees, because they can be consistently expanded and improved through the work of programmers who can be in high numbers and send contributions from several countries.

However, it is crucial to explain that such possibility of collaboration is sometimes not made real. In fact, some codebases are published and not modified for many years. One of the points made in chapterFour is precisely that some software packages do receive more attention because of the hierarchies typical of the academic universe. However, two points need to be made here, with the support of some results from Survey 2. First, what characterizes flexible packages is not their popularity but the fact that their developers are open to contributions. This openness

---

[9]https://sourceforge.net/.

has become very frequent among software developers, and the neuro-imaging field is not an exception. Second, even though many packages receive a small number of external modifications, it is rare to find packages that remain completely ignored [ext, modif]. As a consequence, software development becomes a sort of language game of questions and responses, because publications constitute coding provocations that inspire coding reactions.

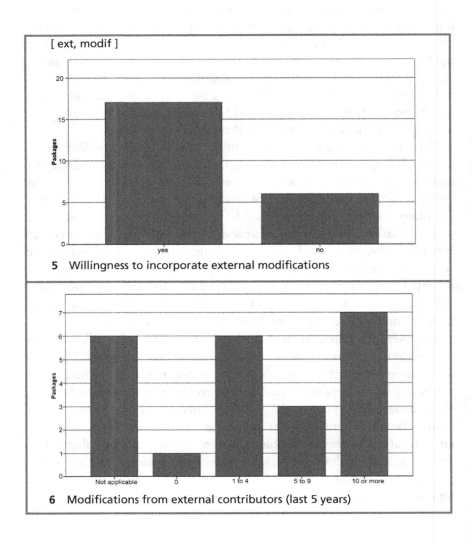

5   Willingness to incorporate external modifications

6   Modifications from external contributors (last 5 years)

A third category of software needs to be considered which tends to be ignored in most analysis of software development. This ignorance is due to the fact that most authors, while being at least indirectly aware of technical and social factors, are generally oblivious to situations in which technical and social factors mingle together, generating hybrid phenomena. In other words, instrumental and communicative logics are indirectly acknowledged, while mediational logics tend to be neglected. Semi-flexible packages are those which combine the features of rigid and flexible packages. On the one hand, they are produced by a small group of developers who generally frame their production as a sort of scientific propriety in need of tight control. On the other hand, the source code is made largely available, so any programmer can download it. As a result of this seemingly contradictory situation, the source code is indeed accessed by many people who come to analyse and modify it, but there is no efficient and swift schemes for people to publish these modifications. Resources like version control platforms are not used, the source code being generally published in the highly controlled premises of institutional websites. In this way, local modifications are allowed to happen but are very likely to remain local.

Semi-flexible packages, which are expressions of mediational actions, have gained much importance over the last decades, as they guarantee, at the same time, collaboration and hierarchization. In chapterFour, it will be seen that they are today's most widely used packages. The FSL software, produced by the Centre for Functional Magnetic Resonance Imaging of the Brain (FMRIB), University of Oxford, UK, is a classic example. The package gained international reputation and its code is frequently downloaded, modified, and provided with new modules. However, FMRIB's researchers prefer to publicize only their own modifications, a situation that restricts the diffusion of external contributions quite severely. Modularization, collaboration, and academic prestige have been decisive factors in the definition of the current configuration of the FSL package, as well as its development model [hist, fsl].

[hist, fsl]

FSL is nowadays one of the world's most popular neuroimaging software packages. In its acronym, "F" comes from the centre which develops it (FMRIB), "S" stands for software, and "L" means library. This is precisely the nature of the package: it is a library, or a set of digital books used to analyse neuroimaging data. In the late 1990s, when FMRIB was founded, its researchers used a package called MEDx, which was selected because, in spite of being commercial software, was in line with the Unix rationale whereby the available range of toolboxes can be replaced with new code. As explained by Mark Jenkinson and colleagues (2012, p. 784): "Part of the Unix philosophy that we liked was the fact that it is built from small components (tools/programs/executables/scripts – call them what you like) that do individual jobs but can be put together in a very large variety of ways to accomplish a huge range of tasks. This was exactly what we wanted to be able to do with our tools, and so this was the approach we took." As the centre's analyses became increasingly sophisticated, researchers did begin to replace the MEDx's modules with original code. At a certain point, it was realized that the centre's researchers were using a small proportion of MEDX's toolboxes while the most substantial part consisted of toolboxes developed in the centre. When it was seen that MEDX's had become only a support for the centre's computer code, the decision was made to build up a completely independent package, replacing all the MEDX's code with new code. In his interview, this is how Christian Beckmann (now based in the Donders Institute for Brain, Cognition and Behaviour, Netherlands) recalled the process that took place in the beginning of the twenty-first century:

"So initially Steve Smith started developing tools and techniques as plug-ins into MEDx that would start kind of allowing you to look at time series and so on and so forth. And different PhD projects then got initiated that would look at different aspects of data processing. A friend of mine at the time [...] started developing tools for automatic segmentation. Mark Woolrich started to develop tools for better GLM [general linear model] estimation, working with the stats department in Oxford. And I started looking into exploratory data analysis and independent component analysis. Mark Jenkinson concentrated on registration at the time. And a lot of that was initially implemented as plug-in tools into MEDx. But very quickly the set of plug-ins and the tools that we developed became, in volume, really kind of quite substantial. And initially we got asked by the other labs that were MEDx-based whether we could provide the tools and techniques. Then I think at some point MEDx even decided to ship FSL as part of MEDx [...]."

> *"Would it be interesting for you to integrate the modules that are produced by somebody into BrainVoyager officially?*
> *That has happened already, from time to time.*
>
> *Really?*
> Yes. There's, for example, a flexible way to segment [brain] maps, so find clusters in maps, where we got a mathematical institute from Berlin who said: 'We can add this to BrainVoyager.' And then they wrote such a module and now we can distribute it. I paid them something, one time, but now I can just distribute that to all other users of BrainVoyager and they can use this tool."
>
> Professor Rainer Goebel (Brain Innovation, Netherlands)

It was claimed that companies generally produce rigid packages. However, the social and technical dimensions of semi-flexible development have become so strong that even some companies have preferred this kind of design rationale. One telling example is the BrainVoyager neuroimaging package, designed by a Dutch company called Brain Innovation. Professor Rainer Goebel, its initiator and main developer, has made it possible for people to add toolboxes (or plug-ins in the BrainVoyager jargon) to the package. Interestingly, even though the package is commercial, people are even let free to decide if the plug-in's source code will be open or not.

The consolidation of the modular design of software has been turning modularity into "hyper-modularity." By this, I mean that software developers, and especially those with limited programming skills, instead of designing a whole new software, have preferred to select a consolidated package to which they append very specific modules performing very specific tasks. For Master's Degree and PhD students, this scheme is advantageous because they can have, by the end of their programmes, a working application enabling the conduct of the analyses needed in their study. For example, Dr. Nathalia Esper (Laboratory for Images-Labima, Pontifical Catholic Univ of Rio Grande do Sul, Brazil) developed, for her Master's Degree, a small application that analyses data taken from children with dyslexia. She added two mathematical models found in the literature to the AFNI neuroimaging package. Inspired by some pieces of code circulated in her research group, she implemented what she describes as a "slight modification" to AFNI's code. Therefore, the time has come when the principles of modularization can be taken to extreme degrees, enabling the conduct of many PhD thesis and Master's Degree dissertations. In this way, hyper-modularity reinforces trends of massification of science, making it possible to integrate large numbers of students into an international system in which new pieces of code can be consistently produced without disturbing the stability of available software packages.

Thanks to hyper-modularity, collaboration in software development becomes not only a technical and social phenomenon but also a geographical event. For the proliferation of semi-flexible and flexible packages has constituted a kind of technical invitation to programmers of several countries who can now participate in a global endeavour of software development. Indeed, programmers who have joined collaborative projects over the last decades, especially in scientifically less dynamic countries, have preferred to work on either semi-flexible or flexible packages, as illustrated by Maps 2.1, 2.2, and 2.3 resulting from Survey 2.[10]

---

[10]The methods used in the preparation of maps is explained in the Methodological Appendix.

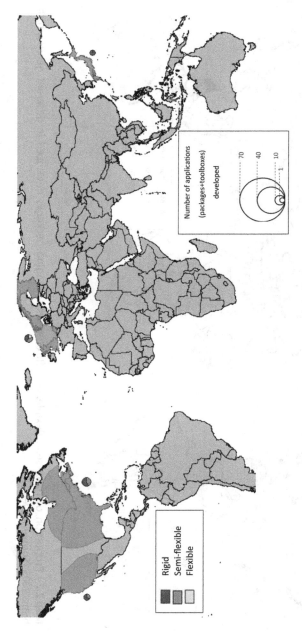

Map 2.1 Proportion of rigid, semi-flexible, and flexible neuroimaging packages developed in different geographical hubs: 1995 (*Source Survey 1*)

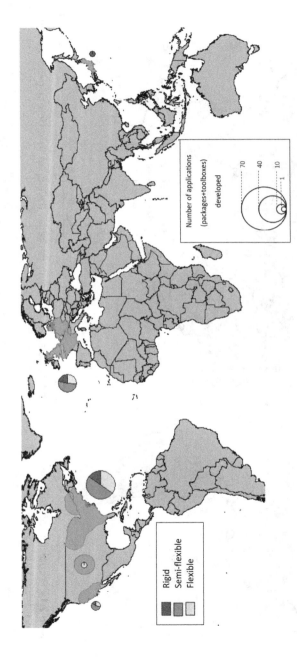

Map 2.2 Proportion of rigid, semi-flexible, and flexible neuroimaging packages developed in different geographical hubs: 2005 (*Source* Survey 1)

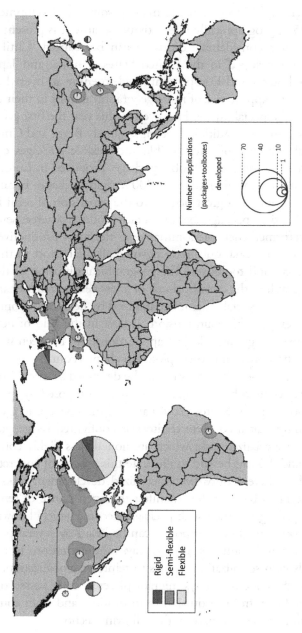

Map 2.3 Proportion of rigid, semi-flexible, and flexible neuroimaging packages developed in different geographical hubs: 2015 (*Source* Survey 1)

These maps clearly show that neuroimaging software development has been, since 1995, a geographically rare activity, because it is present in few world regions, but a considerable diffusion can be realized. Until 2005, applications were developed in the United States, Europe, and Japan. In 2015, the United States and Europe continued to enjoy an overwhelming dominance, with the appearance of more geographical hubs in their territories, which indicates specialization. However, in this year the first geographical hubs appeared in non-traditional settings, namely Brazil and China.

The historical relevance of semi-flexible and flexible packages can also be perceived. In 1995, only one flexible package, produced in Europe, was cited in the neuroimaging literature. In 2005, rigid packages became less central, which was due, to a great degree, to the expansion of semi-flexible packages. In 2015, rigid packages became a rare product, whereas semi-flexible packages reaffirmed their predominance in the most productive hubs. Interestingly, non-traditional hubs have devoted all their effort to the work on flexible packages and toolboxes. The geographical diffusion of the coding enterprise, as well as the constitutional of a less centralized scenario, has been dependent on the social schemes guaranteed by flexibility. Contrary to what one might expect, the main tension, at least in the field of neuroimaging, is not between rigid and flexible applications, but between flexibility and semi-flexibility, a phenomenon explored in chapterFour.

Previously, it was argued that software can be described as what Simondon (1969) named concrete technical objects, which are composed by specialized but highly connected parts. Santos (2000) applied this concept to geographical space, claiming that a concrete territory is configured by geographical units undertaking specialized tasks which are nonetheless highly connected. Maps 2.1, 2.2, and 2.3 can be said to illustrate this process of concretization of space whereby different hubs host specialized development tasks which come to be integrated because they manifest the open stance of semi-flexible and flexible packages. When such packages become largely known, then the coding provocations they incorporate can be heard by several people in several institutions and countries. In this way, modularization, in addition to joining hands with socialization, comes to reinforce globalization. This is possible because there are nowadays migrant pieces of code passed on from developer to developer, from institution to institution, and from country to country, a phenomenon scrutinized in the following section.

}

section3 [Inherited Knowledge, Inherited Code] {

3.1. Inheritance

When designing software, programmers can use a functionality called inheritance. To understand what this means, let us imagine a basic program composed by only three classes called "Brain," "Cerebellum," and "Neurons." By means of some simple commands, the programmer can link those classes together in a hierarchical mode. In this way, "Brain" becomes the parent class (or base class, depending on the terminology of the programming language at use) whereas "Cerebellum" and "Neurons" will be child classes (or derived classes). Since the moment when the classes are thus linked, all the computer code written on the "Brain" class becomes available in the other classes, with no need to rewrite pieces of code. This is useful because the base class presumably contains fundamental operations which are useful in other classes as well. In our example, the usefulness is clear, for the "Cerebellum" class depicts not only the cerebellum itself but the cerebellum and its place in relation to the brain, whereas the "Neurons" class represents not only neuronal networks but also their location inside the brain. Inheritance is a meaningful coding strategy because it helps turn software packages into concrete technical objects, that is products with high levels of internal coherence. Inheritance, in addition to being realized within the technical fabric of software, has found many expressions in relations between developers. It is as though one could point to the existence of parent developers and child developers, almost in the same way as there are parent and child classes.

Software developers do not always need to have recourse to sophisticated coding strategies. Sometimes they can also mobilize more mundane tactics, one of them being arguably very familiar to most readers: the copy-and-paste approach. As shown in section1, research teams constitute repositories of code available to their members. Very frequently, students and researchers move to new groups and places, leaving behind them pieces of code that may continue to be used or modified for many years.

"When I arrived here, four years ago, a small development took very long. Today there are many things that are copy-and-paste, you know, we've done it for other projects, for other tasks. It's in practice integrating things, combining some things with others, and it eventually becomes relatively quick [...] If I do something from scratch, it takes longer, but most of things we now do are not... nothing is done from scratch, because we've already code from other situations, and it eventually becomes relatively quick.

That is, nowadays you already have access to, so to say, an archive of [...] pieces of code that can be reutilized.
Exactly. This includes people in the group [...] If I want to integrate some eye-tracking data, I know that somebody has done it, then [...] I talk to the person. If there's a new person in the group who has never seen that, we can share the code we already have, so the person can work on that code [...]."

Researcher (Institute for Nuclear Sciences Applied to Health-ICNAS, Univ of Coimbra, Portugal)

Researchers are very much aware of such code sharing practices. As claimed by a researcher based in the Donders Institute for Cognitive Neuroimaging, Netherlands, "[...] if I want to reconstruct [a particular dataset] using some reconstruction method, and if one of my post-doc [colleagues] has some code, then I will ask him and use it." Therefore, code is meaningful because it is written by somebody but also because it is received by somebody else (Mackenzie 2006). In open source development, code sharing tactics are particularly abundant, enabling developers "to build on the work done by others" (Torvalds and Diamond 2001, p. 238). In this regard, the history of the Draw-EM package is very telling [draw-em, hist].

[draw-em, hist]

The Draw-EM neuroimaging package was completely developed by Dr. Antonios Makropoulos. Formed a computer scientist in Greece, Antonios moved to the UK where he did the Master's Degree in artificial intelligence and the PhD in neuroimaging. Draw-EM's development was initiated in his PhD and completed by the end of his post-doc position at the Biomedical Image Analysis Group (BiomedIA), Imperial College London, UK. In spite of working alone, Antonios relied on decisive work previously done at BiomedIA. A set of functions had been previously designed which were fully integrated into Draw-EM. Before the beginning of his coding work, Antonios had discussions with one of the PhD students who had produced those sets of functions.

"Did you start from scratch?
Well, not from scratch. So we had, in my... in the BiomedIA group of Imperial [...], this team had already developed a library of software, written in C++, that had a lot of commonly used functions [...] for doing some basic processing, and it [Draw-EM] was based on this library. It is using this library and then it is extended [...]
Without this previous library, it would have been impossible for you to write Draw-EM.
Err... yeah, it would be very difficult, because starting from completely scratch... I mean, nowadays it is not that difficult because of Python and all of the packages that have now been published, but when I started my PhD there were no such packages, so starting from scratch would be quite difficult."

Reusing pieces of code is an old practice in programming. According to Friedman (1992), in the late 1950s developers used to manually copy pieces of binary code onto tapes that were introduced into computers so they performed the desired tasks. Nowadays, this copy-and-paste approach has obviously been made much simpler. So this is the first phenomenon that has to be pointed out: code is passed on from person to person, something that has become very usual within research groups, as showed in section1. Now another phenomenon has to be noticed: code is passed on from moment to moment. On this point, the account given by Aguirre (2012, p. 767) can be considered:

With all the distance that neuroimaging software has travelled, I am amused at how archaic bits of code still linger within modern packages. One of the first pre-processing routines that we wrote for 'Tenure Maker' [the package that preceded the neuroimaging VoxBo package] was a slice-acquisition correction routine [...] I wrote a routine [...], and Eric and I described our efforts in a posting to the SPM mailing list. Darren

Gitelman at Northwestern University modified the code to work in MATLAB, which was subsequently tweaked by Rik Henson and his colleagues at the FIL [Functional Imaging Laboratory at the Wellcome Trust Centre for Neuroimaging, UK], and incorporated into the SPM package. In this way, the comment header I wrote describing the rationale for slice timing correction hopped from one software package to another [...] to survive in SPM [...].

"[...] they [people based in the Centre for Medical Image Computing, Univ College London, UK] have implemented Nifty Reg, a software, which is one of the best, I think, free-form deformation implementations. So Marc Modat, he's been the one who basically took the algorithm of Daniel [Rueckert] that IRTK [the IRTK neuroimaging package developed at Imperial College London, UK] implements and made major improvements which made it a lot faster [...] And the MIRTK [package, an enhanced version of IRTK developed by Andreas Schuh], basically, the new registration... some modules are based on Nifty Reg, in a way [...] I studied also their source code and I also learnt about the improvements, like... The original IRTK had not an analytic computation of derivatives of the objective function [...] And that was something that Marc has done, which he published in Nifty Reg [...] And that is, in a way, replicated in MRITK but within a... within a framework that's more modular and like more object-oriented than it is the case in Nifty Reg [...]."

Dr. Andreas Schuh (Biomedical Image Analysis Group, Department of Computing, Imperial College London, UK)

Many of the currently available software packages, whether they are open or closed source code, are full of small, and sometimes not so small, pieces of code written in different moments. In this way, software packages also contain those "wrinkles" which according to Santos (2000), are dispersed throughout the geographical space: creations that come from previous periods, being preserved and integrated into the present territory. This is so because software developers are very much willing to "take pieces out of some existing program" and even "cannibalize large pieces from some existing free software package" (Stallman 2002c, p. 177). Even if no codebase is available to do exactly what programmers need to do, they will be likely to look for a similar package from which some lines of code can be taken and adjusted

(Raymond 2001). Obviously, this behaviour has nothing to do with coding laziness; it is actually provoked by two concerns. First, most software packages, and particularly those aimed to perform complex tasks like neuroimaging, contain several layers of specialized functionalities, making it difficult, if not impossible, for any programmer to master such large range of notions and concepts. Second, people would not feel scientifically and personally motivated to work on already existing algorithms; generally, they would prefer to sail in uncharted waters, which is even more true for scientists constantly urged to be innovative. In this way, software packages end up becoming a virtual patchwork of lines of code written in different historical moments, sometimes kept in their original form, sometimes adjusted to new programming needs, sometimes translated into another programming language.

In addition to being passed on from person to person, and from moment to moment, code can also travel from place to place, something that most analysts have been oblivious to. On this point, three phenomena deserve to be highlighted. First, there is a transnationalization of code. Some toolboxes result from intense collaboration between two or more teams located in different countries, making it eventually difficult to point to something like a main developer or a project leader. One example is the history of the collaboration between one of my Brazilian interviewees and the Martinos Centre [mrtn, cntr].

[mrtn, cntr]

During his PhD, my interviewee had a research stay in the United States. Subsequently, still as a PhD student, he had another stay in the UK. In this way, he started to build up an international network of academic collaborations on which he continues to capitalize. Thanks to these relations, he has managed to collaborate with researchers of the Martinos Center for Biomedical Imaging, Harvard Univ, United States, where the worldly famous FreeSurfer neuroimaging package is designed.

"I think with Harvard, what we have now... it's a large project that involves modifying a software package that the community uses, which is FreeSurfer, you know. We have written some code, part of the code of

the software, with a student of mine, [...] who did the PhD. We've written an application [...] which is available for download [...] we did it with the Martinos Centre. Now we... we have a bigger challenge, which is modifying the software for high-resolution images and prepare a high-resolution atlas [...]."

"I had even to take a toolbox for statistical analysis which is called Network Statistics and adjusted... and I wrote code that I found online to do some more complex statistical tests, and I inserted that code into that toolbox [...] So it was quite difficult but the result was quite good.

That is, it was code fusion.
Yes.

You mixed code with code.
Yes. Yes. On the one hand, I had a toolbox that was good to use with neuroimaging data and didn't do want I wanted, and on the other hand I had a function that did what I wanted but couldn't be used with neuroimaging data. So I blended both codes so the thing could work.

Did you take long to do that?
One month [laughter]."

Dr. Ricardo Magalhães (Institute of Life and Health Sciences-ICVS, Univ of Minho, Portugal)

The second geographical phenomenon to be stressed here is something that in one of my interviews, I suddenly named "code fusion." This happens when programmers solve coding needs by taking pieces of code written by different hackers, combining them, and blending them together in creative ways. For example, the FieldTrip package was initiated by Robert Oostenveld (MR Techniques in Brain Function group, Donders Institute for Brain, Cognition and Behaviour, Netherlands) who continues to be its main developer. Pieces of code from this package have been incorporated into the EEG Lab project (San Diego, United States), as well as in the SPM neuroimaging package (London, UK).

Finally, there is another geographical phenomenon that can be called naturalization of code. When a certain piece of code is taken from one location to another, it may be subjected to changes so it is adjusted to the institutional and technical conditions of its new place. In this regard, Luiz Fernando Dresch (Laboratory for Images-Labima, Pontifical Catholic Univ of Rio Grande do Sul, Brazil) told me an interesting story. His research group has the AFNI package as preferred analysis resource. He was made responsible for learning to use another software, SPM, in order to realize some specific tasks. At a certain point, he was struggling to carry out some data processing with SPM. At that moment, an international event would take place at the university, and he heard that a professor from the United States, an SPM specialist, would be present. Luiz asked professor Alexandre Franco, his supervisor, to approach the American professor and request some technical help. Luiz and Alexandre were then provided with a piece of code. However, the code contained some technical assumptions that only applied to the American laboratory where it had been originally written. Luiz then proceeded to analyse the code and see how it could be made to work in his laboratory. The problem was finally solved when he modified the ways in which the code sends data to different locations within the computer. In this way, code can surely travel and find new homes but for so doing, it frequently needs to be worked on and tinkered with before being able to comfortably settle down.

These three geographical phenomena (code personal transfer, code fusion, and naturalization of code) show us that when programmers use the copy-and-paste approach, they can be realizing quite complex social, temporal or spatial connections. While the "control C" phase is made possible by an open attitude which pours lines of code into the international coding ocean, the "control V" phase does not go without the willingness to sip at those digital waters and mix them with local resources. Talking about a "coding ocean" may sound somewhat exaggerated, but let us consider that, in fact, the currently available store of computer code, largely published on the internet, compose big waves of toolboxes, intermediate streams of functions, and profound waters of algorithms. As for those more profound waters, the existence of libraries represents a powerful example.

3.2. Libraries

In the software engineering jargon, a library is a set of very basic operations necessary for implementing the complex tasks performed by software packages. For example, the C++ programming language contains the cmath library, which is composed by several mathematical operations like exponential functions and logarithms, while the CString library enables many operations with words, like searching particular characters in a text or replacing a certain word with another one. If a programmer needs to display a circle on the screen, it is not necessary to teach the computer to realize that operation: one just needs to locate the proper library. Therefore, if a software package was to be described as a huge building, then libraries would be collections of bricks. This is why the centrality of libraries in software development has been highlighted (Kitchin 2017; Skog 2003). "To implement any new idea of real substance, the developer must make his selection of supporting component libraries that at the final stage merge with his own programs" (Skog 2003). This is also why, in our times, software development continues to be a solitary task but, strictly speaking, nobody really builds up an application without some basic help from other programmers. Every developer always relies on work done by people who construct and improve fundamental libraries.

> **"Is there a language that you consider as your favourite language?**
> [Pause.] Favourite... Er... I think I like Java the most [...]
>
> **Why?**
> Java. Because Java... Er... Java is a language that is open, isn't it? And there are many, many people who develop... they create new libraries. So it is a language in constant growth, isn't it? It is like you don't have limits there. And other languages like... [...] C# [pronounced "C sharp"], C# is a proprietary language, of Microsoft. So I need Microsoft to add new libraries.

In his interview, Dr. Andreas Schuh (Biomedical Image Analysis Group, Imperial College London, UK) explained that when it comes to learning a new programming language, "[...] the main part is about getting familiar with standard libraries, like standard functions that are available within

> *Not with Java. With Java I can create new libraries, can't I? So the software's architecture grows much more dynamically with Java [...]."*
>
> Dr. Margarita Olazar (Department of Computer Science, Institute of Mathematics and Statistics, Univ of Sao Paulo, Brazil)

the language [...] so you don't need to rewrite all the functions [...]." In Python, a language that has gained much popularity, especially in open source projects, two libraries, named SciPy and NumPy, are particularly relevant for scientific analysis. Moreover, they have been taken as building blocks for the development of other libraries. A researcher who participates in a neuroimaging software development project at the University of Sao Paulo, Brazil, explained to me that he uses a library which depends on NumPy. "So I'm not really using NumPy but the library that I'm using requires NumPy. So I end up having to include NumPy into our project [...] It's because there is a dependency [...]." Nowadays, these kinds of dependencies become increasingly frequent, suggesting the image of a building work that puts bricks on top of bricks.

As claimed before, code frequently is passed on from person to person. However, such transfer does not need to happen in a direct fashion. Nowadays, computer code is frequently offered and received on the internet (von Hippel and von Krogh 2003), a crucial phenomenon that has not been sufficiently stressed by most analysts. It is indeed becoming easier and easier to find whole pieces of code on the internet. For example, somebody could find an efficient way to colour certain areas of a brain image and publish the code on some specialized websites like Stack Overflow. This is particularly helpful for programmers who use several programming languages. For example, Andreas Schuh (Biomedical Image Analysis Group, Imperial College London, UK) declared: "If Stack Overflow didn't exist, my job would be a hell lot harder [...] Because I swap between languages, sometimes I forget how to do the basics, and I if you just type that into Google, a Stack Overflow question comes up [...]."

"Do you often reutilize code?
[...] Oh, yes. Yeah. So... If I don't know how to do something, you can Google the problem [...] Programmers, in general, could not function without control C and control V [laughter]. It's vital. Yeah.

And is it easy to find things on the internet?
Yeah. Err... I mean, the vast majority of problems with big software packages like MatLab have been experienced by other people [...] It's very unusual that I would get stuck on a programming issue [...]."

Researcher (Univ College London, UK)

"[...] was there any time when you went on the internet, saw a code suggestion or an example, took it, adapted it, and...
Yes, yes. This is frequent.

It is frequent, isn't it? Do you remember an example?
[...] For example, when you do, for instance, a loop, you know. I want to do a loop and I have doubts on how to do it. In the beginning, we didn't know what the best way was to do a basic loop to store data [...] So: 'How can I do it?' It was not so obvious. On the internet, there was a guy who gave some examples. I tried out two of his examples. I tried it out. I liked one of his examples and adapted it to our usage.

Was this guy Brazilian?
I don't think so. I don't even know who the guy was. It was a blog. It was the blog of a guy who developed a tool similar to ours.

Wasn't it in Portuguese?
No, it wasn't. So he posted it...

He posted the code.
He posted the code [...]."

Researcher (Univ of Sao Paulo, Brazil)

Increasingly, programmers use the internet as a source of pieces of code which can be copied and modified. As a result, similar pieces of code are generated here and there; they are aimed to perform similar things, but not exactly the same things. LaToza and colleagues (2006, p. 498) gave the name of "example clones" to these creations that in a way belong to the same family. In addition, they described another kind of clone, which are formed when a certain piece of code is translated into a different programming language.

The large circulation of code on the internet belongs to a historic period when programming has become a rich field for communicative processes to flourish. Nofre and colleagues (2014) stress that the idea of language marks the early history of computing. However, most authors employing this idea (including Nofre and his collaborators themselves) tend to have a limited, technical approach in which language is framed as a carrier of the concepts of a discipline (in this case, computer sciences). As a result, language is put, for example, on a par with "the universal language of mathematics" (Nofre et al. 2014, p. 50). The examples reviewed in this section3 enable us to have a more assertive approach. Programming languages can be thought of as proto-languages, because they have an expressive dimension (as claimed in chapterThree) but also because they have a material form, serving as mediations between people, historical moments, and places. In this sense, computer code enables a frequently distant and wordless dialogue between programmers, creating a space where they can make sense of each other's actions.

However, if it is true that hackers enjoy much liberty when implementing coding ideas, it is also true that such creative tasks are increasingly dependent on, and sometimes restrained by, the work done by other hackers, at previous moments, and in distant locations. Much negotiation, and sometimes much tension, is involved when it comes to realizing such combinations of coding work, as shown in section4.

}

section4 [Social Connections, Technical Connections] {

The social dimensions of software have long intrigued and even confused many analysts. In their search for a plausible interpretation, those people have proposed two main approaches which are sometimes blended in the scope of the same analysis. I will claim that such interpretations can be described as: "software development as state of nature"; and "software development as culture."

4.1. Software Development as State of Nature

For some analysts, software development constitutes a field where rules and norms are tepid or even absent. As a result: "A diverse crowd of people, using a wide array of techniques and understandings, produce the 'algorithm' in a loosely coordinated confusion" (Seaver 2017, p. 3). This tendency to lack logic, analysts say, becomes stronger for open source software. In this regard, the most classic analysis was proposed by Raymond (2001) for whom software development can follow two clearly distinguishable models. On the one hand, there would be a "cathedral" model, generally enforced in software companies, in which development happens under strict coordination and hierarchization. On the other hand, there would be the open source, "bazaar" model where development relations tend to be fluid and messy, with no clear hierarchies and divisions of work between developers whose accuracy derives more from their sheer number than any kind of logic coordination. According to Raymond, the creation of the internet, as well as its increasing use in software development, has provoked a rapid diffusion of the bazaar model.

In this second model, software development would be largely motivated by basic human wishes and instincts, which run freely thanks to the absence of clear norms and rules. Hence my claim that this approach assumes the existence of a state of nature. This claim will probably sound less artificial when it is considered that Raymond does recur to models taken from philosophers for whom the idea of state of nature was key. For example, he argues that programmers fight for the property of development projects like people would fight for land property in a state of nature. In this vein, Raymond (2001) proposed that project property, in open source software, can be likened to the Lockean concept of property, in which people turn into owners by homesteading the land, that is by exploring it and mixing their work with it. For him, a software project has similarities with the primitive functions of territoriality because in both cases, the definition of people's territories prevents violent acts from occurring. "These things remain true even when the 'property claim' is much more abstract than a fence or a dog's bark, even when it's just the statement of the [software development] project maintainer's name [...] It's [...] based in territorial instincts evolved to assist conflict resolution" (Raymond 2001, p. 98).

This approach gives serious attention to issues of code ownership (Galloway 2012; LaToza et al. 2006; Raymond 2001). "Clearly there is a tacit etiquette to the ownership of program code in team projects, which programmers transgress at their social peril: you may not alter the code of a colleague without his or her permission, just as it is not acceptable to discipline another person's child" (Galloway 2012, p. 68). Coming back to Raymond (2001), he affirms that programmers have three ways of taking possession of a development project: by starting it; by receiving it from its founder; or by taking over a project that has been relinquished by the original developer. Apart from these situations, Raymond claims, taking possession of a project will trigger unfriendly reactions from programming peers.

Because of this emphasis put on issues of property, much attention is given to the practice of forking. In most projects, there is a group or a person taking the main decisions about the software package. In this way, it is decided what modules will be implemented, what type of code organization will be used, what classes will be created, and so on. If source code is open, individuals who are not happy with the package's evolution can take the code, use it to open a separate development line, and invite people to join the dissenting project. This is called forking the project. Studies that analyse this practice (LaToza et al. 2006; Raymond 2001; Weber 2004) recognize that programmers avoid forking as much as possible, because of the conflicts that are thus generated.

Raymond (2001) goes as far as describing forking as one of the main taboos in software development. Therefore, this approach gives interpretive priority to potential conflicts, thus diminishing the relevance of collaboration.

This approach (software development as state of nature) is not completely false. Indeed, programmers sometimes need to deal with disagreements and conflicts. In my fieldwork, three types of conflicts were frequently mentioned or suggested. First, there are indirect conflicts generated when people criticize the coding work done in development projects they do not participate in. When source code is open, developers sometimes analyse the code and check the strategies adopted, so as to get inspiration or maybe learn with bad examples. In this way, when a developer or development team publishes source code, it is important to be prepared to be evaluated and criticized. When the internet began to be largely used in software development, this potential for criticisms and tensions was increased. "The same bandwidth that enables collaboration on the Internet just as readily enables conflict [...]" (Weber 2004, pp. 88–89).

> *"Has anybody ever criticized the solutions you used in the code?*
> Well, in the natural sciences, yeah, everything is criticized. You know, we expect to be criticized.
> *But does that happen from time to time? Like people emailing you to say: 'You could have done things differently here in this part of the code; this is not the proper, best approach...'*
> Ah... That doesn't happen so much in the mailing list because the people on the mailing list tend not to be the people that read the code. There's a lot of general criticism because we write in MatLab [a proprietary programming language] rather than Python [an open source language]. So the young hipster developers are very critical. In terms of how the algorithms work and whether or not they work? Yeah, there are often criticisms but I think that's to be expected."
>
> Professor John Ashburner (Wellcome Trust Centre for Neuroscience, Univ College London, UK)

> *"[...] SPM, AFNI [neuroimaging packages], it's all usually based on parametric statistics [...] And then, every five, six years, there usually was a couple of papers saying people should be using non-parametric statistics because the assumptions behind parametric statistics are violated in parts of the brain [...] But these papers were largely ignored. Then there was a bit stronger movement like six, seven years ago, about trying to push non-parametric stats again, by some big people this time. So more people took notice. FSL as well, about ten years ago, which was non-parametric as well, just went up. And now, suddenly, over the past two, four years, everybody is realizing... There's been papers showing that non-parametric statistics are the way to go, much better, much more reliable, give much better, reliable results. SPM is still, you know, all the old way, is still all parametric statistics [...] initially people would just, you know, say 'Why would we bother?' you know. And suddenly now people are realizing: 'Oh, yeah, maybe these guys were right twenty years ago.'"*
>
> Dr. Vincent Giampietro (Centre for Neuroimaging Sciences, King's College London, UK)

Second, there are scientific conflicts about the pathways that neuroimaging and neuroimaging software should take. For example, the determination of sample sizes, the choice of significant thresholds, and the selection of statistical tests have always been major issues in the analysis of data produced by magnetic resonance scanners (Gold et al. 1998; Savoy 2001). In the early years of the twenty-first century, there was much debate about the statistical approaches that should be adopted in the analysis of neuroscience data. In the SPM neuroimaging package, parametrical tests (which assume a normal distribution of data) were incorporated. In the equally popular FSL package, initial analyses (the so-called pre-processing phase) also follow parametrical routines but in the final analysis steps, non-parametrical tests are performed. After FSL, other packages were launched which were partially or completely based in non-parametrical tests. These packages were released in a context of hot academic disputes over the "most precise" or "most correct" way of analysing neuroimaging data. For decades, parametric statistics were the preferred approach of most researchers but over the last years, non-parametric strategies have gained increasing scientific endorsement.

In my fieldwork in London, it was not possible to meet a programmer who develops a quite old package completely based in non-parametric statistics, because he was traveling abroad. We could only exchange some emails, and I was allowed to publish parts of our written conversation. In one message sent to me, it was said: "Given the number of disputes I have been involved in over the last 20 years [...], I have no desire whatsoever to become involved in any more of the narcissism and power struggles that have characterised fMRI [functional magnetic resonance imaging] analysis over the last two decades." This researcher went on to write: "[...] despite the current fashion for asserting that [a particular non-parametric test] is the 'Gold standard' for fMRI analysis, this opinion has only been consistently held for twenty years or so by [a colleague] and myself. Many of the new 'converts' to the method have spent most of their careers trying to deny that it was needed and attempting to suppress publications using it."

The third kind of potential conflict is represented by direct disagreements that may emerge between programmers engaged in the same project, whether those people are based in the same institution or have online collaboration. Modifications made to a program by one developer may be at odds with the goals of the original developers (Naur 2001). What is more, such disturbing modifications can abruptly clash with the intentions of many people who have previously worked on the software, triggering "cascading effects throughout the project" (Wagstrom 2009, p. 134).

As put forward by talented developers (Stallman 2002b; Torvalds and Diamond 2001), progress of programming generally depends more on cooperation and small contributions than on massive solitary work done by talented and brilliant people. However, when it comes to generating breakthroughs, what Raymond (2001, p. 54) claimed frequently applies: "[...] the cutting edge of open-source software will belong to people who start from individual vision and brilliance, then amplify it through the effective construction of voluntary communities of interest." The point is that some of these new ideas may turn out be considered, by most developers, as too visionary and brilliant. "The iconoclastic, the maverick and the marginal may find a highly collaborative world a difficult place to flourish" (Adams 2012, p. 336). As a result, an innovative developer, and even one who comes up with useful ideas, may be temporarily or permanently ostracized.

> *"So my final point is: in spite of all these powerful tools available today, is there still any advantage in developing software alone like you did in MarsBar [neuroimaging toolbox], or is collaborative development always more effective?*
>
> [Pause.] Hmm... I think what happens or seems to happen is that somebody decides that a problem needs to be fixed and nobody agrees with them, they spend three or four years working very hard on their own with no help, often complaining that nobody is listening to them, and then somebody somewhere realizes that what they're doing is really important, and then they start helping, and then other people start helping, and then the stuff at the same time is becoming useful to doing something. So an example is a package called Pandas. It is... You know, the guy who did that was a statistician who thought that Python didn't have the right tools for doing the kind of analysis that he'd like to do and he developed those tools. And so he kept on telling everybody that Python needed those tools. And he says that people ignored him [...] He went on and did it himself and now his package is one of the most popular used packages in all the Python ecosystem, because he was right. So err... I remember there's another guy who wrote packages for doing symbolic mathematics [...] and he was just doing a lot of work by himself and nobody seemed to agree with him that it was important, and then suddenly people started to notice [...]
>
> **Hm. But sometimes technical development still needs this kind of visionary people who can see something that other people are not seeing.**
>
> Yeah. But I think, in that case, it's not that they want to work on their own, it's just that nobody wants to work with them, because nobody understands what they're... their vision [...] The person who wrote the plotting library in Python was fond of... (he died) he was fond of saying that everybody told him not to do that because there was already plotting libraries in Python. He was kind of: 'I know I can do it better.' And he did. He made his difference [...]."
>
> Dr. Matthew Brett (School of Biosciences, Univ of Birmingham, UK)

However, the presence of these three kinds of conflicts (criticisms to one's code; scientific disputes; and direct disagreements) does not allow us to conclude that people are prisoners in a permanent state of confusion and always threatened by violent reactions from peers. It was already seen that modularity provides programmers with a sort of technical shield, allowing people to modify one module without messing up with code written by other people. In addition, programmers have displayed a remarkable capacity to formulate schemes to make decisions without provoking disturbances. An interesting example comes from the Apache development community, which managed to grow very large and find effective rules of coordination. The range of responsibilities accorded to each programmer depends on the previous amount and quality of coding work done, and new members join the Apache community by means of an online voting system (Fielding 1999). In other projects, decisions follow more intuitive schemes, as exemplified by the Linux development project. According to its leader, in such type of huge collaborative endeavour there is a moment when "[…] the process for maintaining all the subsystems becomes organic. People know who has been active and who they can trust, and it just happens. No voting. No orders. No recounts" (Torvalds and Diamond 2001, p. 121). Therefore, even if nothing like formalized and rigid rules can be verified, it is not possible to point to a sort of anarchical nature of software development. In this sense, Weber's (2004, p. 89) conclusion sounds very precise: "What needs to be explained is not the absence of conflict but the management of conflict […]." At every moment, programmers, and especially those engaged in large projects, are looking for ways to coordinate actions. This is so not because programmers fear violent reactions from peers, but because coordination and the search for inter-comprehension are presuppositions of any collective endeavour.

4.2. Software Development as Culture

In parallel to this approach (software development as state of nature), there is an interpretation in which software development is framed as culture. Quite often, analysts oscillate between these two approaches.

As a result, they can foreground the wars of software development and, in the very next moment, speak of the peaceful and altruistic stance of programmers. The most famous version of this second approach mobilizes the concept of gift culture, a description that is said to be particularly valid for open source software. Once again, Raymond (2001, p. 110) provides us with a clear example, claiming that developers of open source software participate in a culture that can be depicted as "a gift culture," "giving time, energy, and creativity away." The internet is pointed to as a resource fostering knowledge and data sharing. "One of the most striking features of the net is the ubiquity of its hi-tech version of the gift economy" (Barbrook 2003, p. 91).

The concept of gift culture (or gift economy) belongs to anthropology's tradition. It began to be moulded in the 1920s by classical anthropologists such as Bronislaw Malinowski and Marcel Mauss, who studied the patterns of exchange of material goods in primitive social groups. The concept was subsequently refined by sociologists who transformed it into the so-called social exchange theory (Blau 2006; Clark and Mills 1979), according to which social life is composed by the interdependence created when social actors exchange resources, whether these latter are material goods or abstract ones such as knowledge. Inspired by these ideas, analysts went on to say that hackers and internet users live in a real gift culture (Barbrook 2003; Raymond 2001). Therefore, if the previous interpretation (software development as state of nature) emphasizes the impending conflicts of a domain marked by weak or absent rules and shaped by fundamental human drives, in this second approach (software development as culture) analysts highlight the role of morality and cooperation. If the first approach frames the fear of violent reactions as a promoter of social integration, for the second approach social integration is guaranteed by a voluntary commitment to the group. In this vein, Lakhani and Wolf identified, for open source software, "[...] a strong sense of community identification and adherence to norms of behavior" (Lakhani and Wolf 2005, p. 5).

In the same way that the first approach contains some truth, the second approach is not completely false. Indeed, programmers have to look for collaborative schemes in order for their coding products to be minimally coherent. In Dr. Matthew Brett's (School of Biosciences,

Univ of Birmingham, UK) words: "You can't put code in that people can't take care of. It can't be a mess. That's a dangerous thing for a project." Hackers have elaborated some formalized ways to organize things and take collective decisions, which eventually potentializes their sense of belongness. In the Linux project, for example, there is an "ordered and methodical" way of taking collective decisions via discussion lists (Weber 2004, pp. 63–64). Such collective schemes can become so effective that for some hackers, the centrality of collaboration may turn into a value or personal motivation. In this vein, Brooks (1995, p. 7) affirmed that one of the joys of programming "is the pleasure of making things that are useful to other people." For some other hackers, collaboration may turn into a political and moral project. The clearest example, here, comes from the movement headed by Stallman (2002d, p. 62), whose adherents encourage programmers to "adopt an ethical perspective, as we do in the Free Software movement."

In spite of its contributions, the software-as-culture approach brings about interpretive complications that manifest themselves from both a technical and social viewpoint. On the technical side, the gift culture approach forces us to think that software development is mainly, or even purely, dominated by altruistic motives. Software developers, in the many situations when code is shared, would be like people of primitive groups exchanging goods to express hope, faith, respect or fidelity. In this way, one tends to overlook the fact that the search for harmonic relations may be secondary, or even meaningless, for some developers, because many of them are moved, to a great degree, by the simple willingness to produce an efficient, well-written, and coherent program. In many development groups, the major concern is maintenance of the package's technical integrity. An interesting example comes from the Computational Group for Medical Signals and Images, Univ of Sao Paulo, Brazil. This group is responsible for the open source JBioS package, which analyses biomedical signals. On the package's website,[11]

---

[11]http://dcm.ffclrp.usp.br/csim/jbios/.

some recommendations are given to external contributors: "When writing plugins, some rules must be followed: (1) Compiled classes must be placed in folders signals, tools or methods; (2) Those folders must be placed in the same folder as 'lib' and JBioS.jar.; (3) Classes names must end with underline char ('_')." Other recommendations are given, none of which are anyhow aimed to foster solidarity or shape contributors' ideology. Therefore, one should not forget that even though developers do engage in personal relations and have created collaborative schemes, they are often primarily concerned with their technical product. Let us remember that packages derive, to a large degree, from the coding work done by PhD and Master's students moved by the self-centred goal of completing their programmes and acquiring their degrees.

On the social side, it is worth mentioning the two objections that Ghosh (2005) raised to the gift culture approach: one, programmers frequently share code with an abstract community, whereas in traditional groups permeated by a gift culture, the giver and the receiver generally know each other personally; two, it sounds inadequate to speak of programmers as people closely attached to communitarian bounds, as many of them are rather willing to be creative, provoke changes, and disrupt the field. It is frequently possible to claim that "the politics of developers in particular tend towards the libertarian rather than the communitarian" (Ghosh 2005, p. 32).

The development-as-state-of-nature and the development-as-culture approaches are not completely denied by the communicative approach adopted in this book. Rather, they are taken in a balanced way. For communication is precisely the process through which people can manage conflicts without succumbing to disintegrating forces, and express communal bounds without sacrificing personal goals. According to Kitchin (2017, p. 18), "[...] creating an algorithm unfolds in context through processes such as trial and error, play, collaboration, discussion and negotiation." Indeed, the right balance between disruption and maintenance, self-centrism and altruism, war and peace, can be struck because of constant communicative efforts. Hence the occurrence of reactions typical of communicative contexts, as seen in the sequence.

4.3. Trust and Leadership

As a field marked by communicative processes, software development opens up some leeway for phenomena that shape interpersonal relations, one of them being trust. In most open source projects, it is possible to identify one person who plays a key role of leader (Pavlicek 2000). Such leadership, instead of stemming from bureaucratic or institutional criteria (as would be expected, for example, in a company), is generated by the leader's coding experience or achievements, which are largely recognized as sources of trust. In the Linux project, this becomes clear. According to Linus Torvalds, its leader, "[…] everybody connected with Linux trusts me more than they trust anyone else" (Torvalds and Diamond 2001, p. 168). Whereas some leading programmers, like Richard Stallman, the leader of the Free Software movement, have quite assertive and strong stances, Linus has preferred to realize a sort of soft leadership. One might even say that in parallel to his programming successes, Linus' main achievement is the kind of leadership he has come to intuitively formulate whereby, instead of imposing his view, he leaves contributors explore all kinds of coding insights, knowing that bad ideas will eventually die away.

Strong or soft, leadership can only be built up in groups where different actors know what can be expected from each member. Software developers have learnt that in some projects, things may go amiss if a leader is lacking. For example, the NIPY community,[12] which develops neuroimaging packages, started out with no leader. However, this situation proved unproductive because, as explained by Matthew Brett (School of Biosciences, Univ of Birmingham, UK), "[…] there were three different teams and it wasn't clear who was kind of running the project. So the fight broke out. So there was […] sort of problems, because, you know, group A wanted to sort of do something and group B didn't want to do something, and it wasn't clear who was going to make the decision. So it became very chaotic." As a result, Matthew decided to assume the role of leader, which imparted more productivity and clarity to the community.

---

[12]http://nipy.org/index.html.

For the FieldTrip project,[13] leadership was established more naturally, because Professor Robert Oostenveld (MR Techniques in Brain Function group, Donders Institute for Brain, Cognition and Behaviour, Netherlands), its main developer, was obviously to assume this role because he initiated the project and because of his scientific expertise. In this case, leadership is so largely recognized that Robert needs at times to recede so his figure does not overwhelm other key participants. For example, he realized that he should play a minor role in the project's online discussion list, as his presence there can intimidate participants. "So that's why I'm keeping low in the discussion list. Because if I keep low, then the community is willing to… to chat along, and people that are perhaps not so secure of their own skills are still willing to share their expertise whereas if I start to talk, then everyone is going to wait for me to give the next answer." Therefore, leadership needs to be not only gained, acknowledged and legitimized; it also needs to be managed and adjusted to each particular context.

> *"And when you finish your PhD, what's going to happen with MIRTK?*
> [...] I hope that also other people will find more interest to actually contribute but, realistically, it will be me, still, maintaining it for the future [...] Most people are not really familiar with the source code, but of course I am [...] So when people obviously would send emails to the mailing list or report issues, maybe bugs, I would of course try to take care of it, just like I am doing for other projects (open source projects) that I'm maintaining that have nothing to do with Imperial [College] or my current position."
>
> Dr. Andreas Schuh (Biomedical Image Analysis Group, Imperial College London, UK)

The relevance of leadership, as a communicative phenomenon, can be detected not only in huge projects such as Linux and middle-sized projects like FieldTrip. It is also valid in small projects which are yet to acquire external contributors. This is why the main developers of small software packages acknowledge their future responsibilities pertaining to the package's advancement, including the possibility of becoming the leader of a grown-up project in the future.

---

[13]http://www.fieldtriptoolbox.org/.

This is not to say, however, that the presence of a clear leader is always in need. Weber (2004) explains that the viability of a project, and also the possibility of quickly solving conflicts, can be defined by the presence of a leader but also by the decision-making structure chosen for the project. It makes no interpretive difference if the final solution involves some formal rules and the presence of a leader or if a more fluid and intuitive arrangement is in place. Sometimes, coordination is obtained via schemes that appear somewhat anarchic, which happened, for example, with the Apache development community. "There was no Apache CEO, president, or manager to turn to for making decisions. Instead, we needed to determine group consensus, without using synchronous communication, and in a way that would interfere as little as possible with the project progress" (Fielding 1999, p. 42). Coordination, not leadership, is the main goal. More than communicative actions (which are dependent on individual actors), systems of communicative actions (which gain some independence from individuals) are needed.

Anyway, if trust in the leader is absent, then it is necessary to trust the group and its organizing rules. Because of the trust reserved to the group, a communitarian feeling is more likely to emerge in flexible packages, as illustrated by the findings of Survey 1 [fidel, soft].

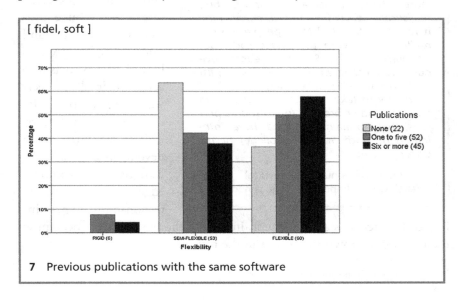

7 Previous publications with the same software

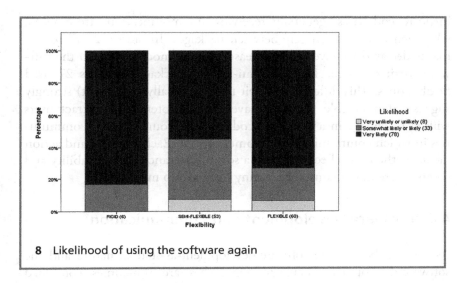

8   Likelihood of using the software again

Considering the publications that have been made by the survey participants, an interesting difference was verified. No firm statements can be made about rigid packages, because only six questionnaires pertaining to them were received. However, it is possible to point to an important difference between users of semi-flexible and flexible packages. For semi-flexible ones, a (statistically significant) difference was verified, most researchers using them for the first time, and the minority of them having had six or more publications with the same package in the past. For flexible packages, the trend is opposite, with most researchers having had six or more publications (although this difference was not statistically significant). When one looks at the likelihood of using the package in the future, the proportion of users of flexible packages declaring to be very likely to use them is again bigger than the same proportion of users of semi-flexible packages.

The findings summarized in these two charts show that flexible applications are more likely to be used repeatedly. This is so because flexible packages can be modified more easily, so their users are more likely to apply some coding modifications, enhance the software or design a new module, thus establishing personal bounds with the software. In their turn, semi-flexible packages are more user-friendly but have more intricate codebases, generally receiving no modifications from users, who then move more freely between packages. In Survey 2, 47 participants declared to have made at least a minor modification to the software used; of them, 18 used semi-flexible packages whereas 27 used flexible ones. This difference (albeit not statistically significant) strongly suggests that flexible packages have greater potential to attract users willing to engage in a collective coding endeavour, creating communities in which communication becomes central. Each decision and action taken by the group becomes then a sort of utterance whose viability and meaning are under constant scrutiny from group members.

4.4. Software Development and Communication

As claimed before, the software-development-as-state-of-nature and the software-development-as-culture approaches are sometimes combined by some authors. This analytical oscillation is understandable, for we are in fact dealing with a field marked by twofold phenomena. Thus what seems inappropriate is the terms in which the paradox is presented, not the paradox itself. By the way, as I claimed elsewhere (Bicudo 2014), precise descriptions of contemporary social life demand an interpretive stance whereby one faces paradoxes instead of fleeing them. In software development, the paradox assumes precisely the shape of a tension between, on the one hand, the instrumental requisites of an activity where most crucial decisions are generally taken by a single developer concerned with the performance of a particular package, and, on the other hand, the collaborative schemes that renders software development more robust and speed up coding work. Because of this constant search for the right balance, software development is a process marked by several discussions and agreements, as exemplified by the history of the NES package [hist, nes].

[hist, nes]

The open source NES (Neuroscience Experiments System) package, whose development began in 2014, is produced within Neuromat, a neuroscience research network involving several institutions of the Brazilian state of Sao Paulo. Its developers are all based in the Dept of Computer

Science, Univ of Sao Paulo. The package is composed of different modules dealing with different types of brain data. The design of a module for neuroimaging data analysis is supposed to be initiated soon. The development group uses some techniques of the agile development approach, including the organization of two meetings per month. In addition, there is constant dialogue with a neuroscience research group based in another state (Rio de Janeiro). Because this group has become an intense user of NES, its members are in frequent touch with the developers, so the evolution of the package can be constantly discussed. Frequently, the Rio de Janeiro group comes up with ideas and feature requests whose viability is then discussed with the package's developers. Professor Kelly Braghetto coordinates the project since its beginning. In her interview, in addition to speaking of these ongoing negotiations, she anticipated new kinds of negotiations in the future.

"One day, if somebody wants to do it, the person can download the code, implement, for instance, a new module...
Yes.
It has not happened so far, has it?
No.

So if one day... When this happens, how will the process be? The person will apply for the modification. And who will decide whether this is integrated or not into...
[...] When we use software to control versions, the software controls it... We can create what we call branches. These are divisions in the development line [...] Does somebody want to create a new module? The person can create a new development line and in this new line, you have all the code you had before this new module. After testing it, checking it and seeing that everything is okay, we can mix the lines again, we can merge the result. We could integrate this module that has been built in a parallel line into the main line corresponding to the software that we disseminate for download and call the official version.
This is the ideal thing, by the way.
This is the ideal thing [...] I hope this will happen. As long as we have a dedicated team for this, our team can deal with this integration, right? [...] Because I think that for this to happen, there will have to be some previous communication between the group that is willing to add the new module and the group that develops the tool today. Because I think it is not such a trivial thing if somebody just looks at the code and says: 'Oh, I will sit here and develop a new thing.' Because it is software that's been developed for some two, three years [...] So if there is somebody interested in developing it, it would be nice if the person talked to the

> *developers. I think this frequently happens in open software, doesn't it? When there is a new collaborator, the guy doesn't just work on his own and starts developing a new module...*
> **Without revealing it to anybody.**
> *He interacts with the development team, he scans for important issues... So I think that in the ideal situation, there should be this interaction. If there is no interaction in the development phase, this will fatally have to happen at the integration phase. I mean, otherwise this will never be considered as part of the official version [...]."*

A communicative approach makes it possible to consider the importance of conflicts and collaborations in a balanced way. From this viewpoint, software is not only what results from different pieces of code coherently combined, because this is only the technical side of the story. In addition to this, software is what derives from a communicative process in which projects and intentions get to be discussed and negotiated, and, at the same time, some rules or potocols are produced to steer those interactions.

On the one hand, there is no need to overstate possible conflicts between programmers. The confusions and disagreements that may emerge between them can be solved by means of the (technical and natural) language they master. "The confusions which occupy us arise when language is like an engine idling, not when it is doing work" (Wittgenstein, 1963, p. 51). Hence, the emergence of schemes such as the ones reviewed in this chapterTwo (control version platforms, online voting systems, agile development techniques, code sharing, face-to-face conversations, and so forth). On the other hand, there is no reason to get too excited about the collaborative deeds of open source and the internet. Technical justifications, practical interests, and even code hoarding, have much leeway to emerge, as they also make sense in the scope of the mediational actions of software development.

}

section5 Empirical Example: GitHub

In section2, it was explained that collaborative schemes in software development were greatly facilitated by the appearance of version control platforms. SourceForge was the first of these services to gain popularity, losing much of its space for the benefit of more recently released platforms such as BitBucket, created in 2008. In the same year, another online platform was launched which would rapidly become the preferred means of collaborative code writing: GitHub.[14] The main feature of GitHub is similar to that of SourceFourge: we are dealing with version control systems allowing large number of developers to contribute code to open source projects. Every new piece of code is incorporated into the old codebase and an updated version is constantly produced. Across the years, GitHub assimilated a series of additional tools such as repositories for documentation, space for bug reports and feature requests, an automatic subsystem that tests every new piece of code and chase bugs, an internal mailing service allowing developers to exchange messages, among others. These additional features were among the reasons why the platform has become very popular among developers. Dr. Matthew Brett (School of Biosciences, Univ of Birmingham, UK), who has been very much involved in open source software and is the leader of the NIPY neuroimaging software community, showed me some features of GitHub on his computer screen. According to him, "[...] this interface you're looking at here is vastly superior to anything that we previously had. And, you know, it's universal. I mean, like, everybody is now on GitHub and they're all using this." In 2018 the number of users with active accounts was approaching 30 million.

This is not to say that GitHub is free from the ambiguities present in contemporary social life. On the one hand, GitHub is today's most popular platform for collective development, holding "the world's biggest collection of open source software" (Finley 2015). On the other hand, the service used to be provided by a company (headquartered in San Francisco, United States) whose structure became increasingly

---

[14]https://github.com/.

rigid. If in the beginning, one could barely identify hierarchical relations, in 2014 the company adopted a more stratified structure, with management layers. In 2018, when this book was being written, the company was acquired by Microsoft, which triggered unfavourable reactions. Some developers, concerned with the commercial rationale that Microsoft could impart to GitHub, transferred their projects to other code sharing platforms (Tung, 2018). From the users' perspective, ambiguities are also to be found. On the one hand, GitHub can be used for code sharing and collaborative development, but, on the other, it can also be used as a personal code store, as users are given the possibility of having closed accounts so their pieces of code remain hidden from other users. One of my interviewees, a researcher based in the D'Or Research Institute, Brazil, has taken some of his development projects to GitHub. Even though his account is not closed, one of his main purposes is to create a personal code archive. In spite of allowing some space for those personal projects, GitHub encourages code sharing and collaboration. For so doing, one of the strategies adopted is the limitation of the memory space available to users with closed accounts, while open accounts have no limits.

In addition to being used for the control of personal projects, GitHub has proved useful for companies. Nowadays, over 110 million companies have active accounts on the platform. Large companies such as Facebook or Google pay fees for using the private version of the service, so they can use all the available features, block access to external users, and work with no limitation in terms of memory space. Brain Innovation, the most successful company developing a neuroimaging package, is also becoming more and more interested in GitHub. According to Michael Luehrs, one of the company's programmers, the platform is used as a code repository. "So we have a lot of routines, for example, for file operations, which is a pain if you have to rewrite these features [...] We just put these [into GitHub] as [...] little projects and then we can reuse them right away. Also, if there's external people coming in [joining the company], they can just use them." Even though the company has created a closed account for code storage and version control of its Brain Voyager package, Professor Rainer Goebel, its CEO and main software developer, has decided to open the access to some pieces of code. He is intent on enlarging the proportion of open code in the years to come.

In spite of these personal and commercial kinds of use, GitHub's greatest impact has certainly been on open source and collaborative development. For example, it has served as an incentive for programmers to produce clear and organized code, a concern which is associated with other-centred stances, as seen in chapterThree. Antonio Senra (Computational Group for Medical Signals and Images, Univ of Sao Paulo, Brazil) declared that his research group is becoming increasingly interested in strategies of collaborative development. "GitHub is an example of this. There we can make the whole code from our work become available and in this way we are able to receive some help from other people who perhaps got interested in the project."

GitHub is used by developers of several types of software, including neuroimaging packages. In Survey 2, of 23 respondents, 11 declared to use GitHub to publish the software and its code, as well as to realize version control, the second most used means being either an institutional website or a website created to host the package [dvl, platf].

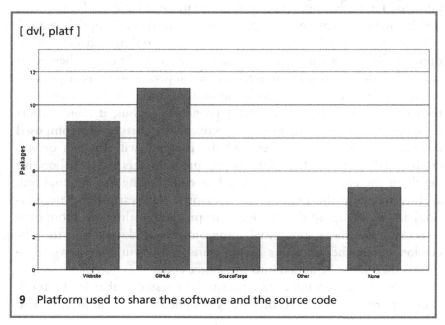

[dvl, platf]

9 Platform used to share the software and the source code

In my fieldwork, I came across many examples of programmers using GitHub. To give only three examples: the MIRTK package, developed by Andreas Schuh (Biomedical Image Analysis Group, Imperial College London, UK) was taken to GitHub in 2015; Rodrigo Basílio (Laboratory of Image Processing, D'Or Research Institute, Brazil), uses the platform for managing the Friend package; and Makis Marimpis (Brain Innovation, Netherlands) uses GitHub for controlling his personal project called Dyfunconn (Dynamic Functional Connectivity), having his former Master's Degree supervisor as his main collaborator. These and other projects show that a whole set of resources (such as computer systems, internet connections, collaborative development platforms, and so on) need to be in place so communication can effectively happen. This is why I claim that, more than communicative actions, it is possible to identify the presence of systems of communicative actions, which involve not only motivations and personal initiatives but also material and digital infrastructures.

The growing popularity of GitHub brings about communicative consequences to software development. Let us consider three pivotal phenomena. First, platforms such as GitHub make it easier to form online development groups, enlarging the scope of those "implicit teams" referred to by Mockus (2009). In this regard, online software development displays an important advantage over traditional schemes. It is largely acknowledged that software developers who share a physical location must establish a smooth and swift exchange of information, so their collaborative project can make progresses. "Thus, if their authors have not succeeded in interacting effectively, the parts of a decomposed program will not do so either, and the program will not run or will crash" (Galloway 2012, p. 59). When programmers use collaborative development platforms, some tasks like code testing are automatized, slightly reducing the need for interpersonal clarification and negotiation, and speeding up the development process. In this way, more time is available for more substantial communicative tasks, such as the discussion about the software's architecture and about innovative algorithms that might be added into the source code.

The second communicative aspect to be stressed is that in the traditional approach, in which software is designed by a small group based in a certain institution or company, packages can eventually "die away" when the group disintegrates and no external programmer is able to fully grasp the rationale behind the codebase, taking over the project. In this sense, Naur (2001, p. 234) claimed: "The death of a program happens when the programmer team possessing its theory is dissolved." GitHub stores not only computer code (with its comments) but also technical information and supporting documents. In this way, the increasing use of such collaborative platforms reduces packages' likelihood of dying away as their codebase and all the associated information is taken from institutional storage systems to the internet. As a result, the rigid schemes of software ownership which prevailed in the past come to be slightly softened.

Finally, GitHub and its brother systems have reinforced a new logic, namely a non-commercial logic, in software development. Without overstating the weight of this phenomenon (especially after Microsoft's acquisition), it is possible to point to a "decommodification of information" (Barbrook 2003, p. 92) promoted by the internet. Describing what he called "control revolution," Beniger (1986, p. 39) stressed the impacts of information technologies on the possibilities for control, this latter being defined as "[...] purposive influence toward a predetermined goal." However, it is also paramount to recognize that informational technologies and the internet, in addition to having the instrumental capacities highlighted by Beniger, have enabled many kinds of collaboration and negotiation, such as the ones made possible by GitHub. In this way, one should only admit the existence of a control revolution if, at the same time, one points to the occurrence of a parallel coordination revolution.

}

References

Adams, Jonathan. 2012. "The Rise of research networks." *Nature* 490: 335–336.

Aguirre, Geofrey K. 2012. "FIASCO, VoxBo, and MEDx: Behind the code." *NeuroImage* 62:765–767.

Ashburner, John. 2012. "SPM: A history." *NeuroImage* 62:791–800.

Barbrook, Richard. 2003. "Giving is receiving." *Digital Creativity* 14 (2): 91–94.

Bear, Mark F., Barry W. Connors, and Michael A. Paradiso. 2007. *Neuroscience: Exploring the brain.* 3rd. ed. Philadelphia: Lippincott Williams & Wilkins.

Beniger, James R. 1986. *The control revolution: Technological and economic origins of the information society.* Cambridge: Harvard University Press.

Berezsky, Oleh, Grigoriy Melnyk, and Yuriy Batko. 2008. "Modern trends in biomedical image analysis system design." In *Biomedical engineering: Trends in electronics, communications and software,* edited by Anthony N. Laskovski, 461–480. Rijeka: InTech.

Bicudo, Edison. 2014. *Pharmaceutical research, democracy and conspiracy: International clinical trials in local medical institutions.* London: Gower/Routledge.

Blau, Peter Michael. 2006. *Exchange and power in social life.* New Brunswick: Transaction.

Brooks, Frederick P. 1995. *The mythical man-month.* Reading: Addison-Wesley.

Clark, Margaret, and Judson Mills. 1979. "Interpersonal attraction in exchange and communal relationships." *Journal of Personality and Social Psychology* 37 (1):12–24.

Dinov, Ivo D., Petros Petrosyan, Zhizhong Liu, Paul Eggert, Sam Hobel, Paul Vespa, Seok Woo Moon, John D. va Horn, Joseph Franco, and Arthur W. Toga. 2014. "High-throughput neuroimaging-genetics computational infrastructure." *Frontiers in Neuroinformatics* 8:1–11.

Fielding, Roy T. 1999. "Shared leadership in the Apache project." *Communications of the ACM* 42 (4):42–43.

Filler, Aaron G. 2009. "The history, development and impact of computed imaging in neurological diagnosis and neurosurgery: CT, MRI, and DTI." *Nature Precedings*: 1–76. Available at: http://dx.doi.org/10.1038/npre.2009.3267.5.

Finley, Klint. 2015. The problem with putting all the world's code in GitHub. *Wired.* Available at: https://www.wired.com/2015/06/problem-putting-worlds-code-github/.

Fischl, Bruce. 2012. "FreeSurfer." *NeuroImage* 62:774–781.

Friedman, Linda Weiser. 1992. "From Babbage to Babel and beyond: A brief history of programming languages." *Computer Languages* 17 (1):1–17.

Galloway, Patricia. 2012. "Playpens for mind children: Continuities in the practice of programming." *Information & Culture* 47 (1):38–78.

Ghosh, Rishab Aiyer. 2005. "Understanding free software developers: Findings from the FLOSS study." In *Perspectives on free and open source software*, edited by Joseph Feller, Brian Fitzgerald, Scott A. Hissam, and Karim R. Lakhani, 23–46. Cambridge: MIT Press.

Goebel, Rainer. 2012. "BrainVoyager: Past, present, future." *NeuroImage* 62:748–756.

Gold, Sherri, Brad Christian, Stephan Arndt, Gene Zeien, Ted Cizadlo, Debra L. Johnson, Michael Flaum, and Nancy C. Andreasen. 1998. "Functional MRI statistical software packages: A Comparative analysis." *Human Brain Mapping* 6:73–84.

Grcar, Joseph R. 2011. "John von Neumann's analysis of Gaussian elimination and the origins of modern numerical analysis." *SIAM Review* 53 (4):607–682.

Habermas, Jürgen. 1987. *The theory of communicative action, vol. 2: Lifeworld and system*. Cambridge: Polity.

Habermas, Jürgen. 2008. *Between naturalism and religion*. Cambridge: Polity Press.

Herbsleb, James D., and Rebecca E. Grinter. 1999. "Archictectures, coordination, and distance: Conway's Law and beyond." *IEEE Software* 16 (5):63–70.

Jenkinson, Mark, Christian F. Beckmann, Timothy E. J. Berens, Mark W. Woolrich, and Stephen M. Smith. 2012. "FSL." *NeuroImage* 62:782–790.

Kitchin, Rob. 2017. "Thinking critically about and researching algorithms." *Information, Communication & Society* 20 (1):14–29.

Lakhani, Karim R., and Robert G. Wolf. 2005. "Why hackers do what they do: Understanding motivation and effort in free/open source software." In *Perspectives on free and open source software*, edited by Joseph Feller, Brian Fitzgerald, Scott A. Hissam, and Karim R. Lakhani, 3–22. Cambridge: MIT Press.

LaToza, Thomas D., Gina Venolia, and Robert DeLine. 2006. "Maintaining mental models: A study of developer work habits." Proceedings of the 28th International Conference on Software Engineering, New York, USA.

Mackenzie, Adrian. 2006. *Cutting code: Software and sociality*. New York: Peter Lang.

Mockus, Audris. 2009. "Succession: Measuring transfer of code and developer productivity." International Conference on Software Engineering (ICSE 2009), Vancouver, Canada.

Naur, Peter. 2001. "Programming as theory building." In *Agile software development*, edited by Alistair Cockburn, 227–239. Boston: Addison-Wesley.

Nofre, David, Mark Priestley, and Gerard Alberts. 2014. "When technology became language: The origins of the linguistic conception of computer programming, 1950–1960." *Technology and Culture* 55 (1):40–75.

Padma, T. V. 2008. "India plans for interdisciplinary neuroscience research centre." *Nature Medicine* 14 (11):1133.

Parnas, David Lorge. 1972. "On the criteria to be used in decomposing systems into modules." *Communications of the ACM* 15 (12):1053–1058.

Parnas, David Lorge, and Paul C. Clements. 1986. "A rational design process: How and why to fake it." *IEEE Transactions on Software Engineering SE* 12 (2):251–257.

Pavlicek, Russell C. 2000. *Embracing insanity: Open source software development*. Indiana: Sams.

Raymond, Eric S. 2001. *The cathedral & the bazaar: Musings on Linux and open source by an accidental revolutionary*. Sebastopol: O'Reilly.

Ribeiro, Andre Santos, Luis Miguel Lacerda, and Hugo Alexandre Ferreira. 2015. "Multimodal Imaging Brain Connectivity Analysis (MIBCA) toolbox." *PEERJ* 3 (e1078):1–28.

Santos, Milton. 2000. *La nature de l'espace: technique et temps, raison et émotion*. Paris: L'Harmattan.

Savoy, Robert L. 2001. "History and future directions of human brain mapping and functional neuroimaging." *Acta Psychologica* 107:9–42.

Schwarz, Michael, and Yuri Takhteyev. 2010. "Half a century of public software institutions: Open source as a solution to hold-up problem." *Journal of Public Economic Theory* 12 (4):609–639.

Seaver, Nick. 2017. "Algorithms as culture: Some tactics for the ethnography of algorithmic systems." *Big Data & Society* 4 (2):1–12.

Simondon, Gilbert. 1969. *Du mode d'existence des objets techniques, Analyses et Raisons 1*. Paries: Aubier.

Skog, Knut. 2003. "From binary strings to visual programming." In *History of Nordic computing*, edited by Janis Bubenko Jr., John Impagliazzo, and Arne Solvberg, 297–310. Boston: Springer.

Stallman, Richard M. 2002a. "Copyright and globalization in the age of computer networks." In *Free software, free society: Selected essays of Richard M. Stallman*, edited by Joshua Gay, 133–154. Boston: GNU Press.

Stallman, Richard M. 2002b. "The danger of software patents." In *Free software, free society: Selected essays of Richard M. Stallman*, edited by Joshua Gay, 95–112. Boston: GNU Press.

Stallman, Richard M. 2002c. "Free software: Freedom and cooperation." In *Free software, free society: Selected essays of Richard M. Stallman*, edited by Joshua Gay, 155–186. Boston: GNU Press.

Stallman, Richard M. 2002d. "Releasing free software if you work at a university." In *Free software, free society: Selected Essays of Richard M. Stallman*, edited by Joshua Gay, 61–62. Boston: GNU Press.

Torvalds, Linus, and David Diamond. 2001. *Just for fun: The story of an accidental revolutionary*. New York: HarperCollins.

Tung, Liam. 2018. GitHub rivals gain from Microsoft acquisition but it's no mass exodus, yet. Availalbe at: https://www.zdnet.com/article/github-rivals-gain-from-microsoft-acquisition-but-its-no-mass-exodus-yet/.

von Hippel, Eric, and Georg von Krogh. 2003. "Open source software and the 'private-collective' innovation model: Issues for organization science." *Organization Science* 14 (2):209–223.

Wagstrom, Patrick Adam. 2009. "Vertical interaction in open software engineering communities." PhD, Carnegie Institute of Technology/School of Computer Science, Carnegie Mellon University.

Weber, Steven. 2004. *The success of open source*. Cambridge: Harvard University Press.

Weinberg, Gerald M. 1998. *The psychology of computer programming*. New York: Dorset House.

Wittgenstein, Ludwig. 1963. *Philosophische Untersuchungen / Philosophical Investigations*. Oxford: Basil Blackwell.

```
# social code
# source code
```

chapterThree
(Writing Code: Software Development
and Communication) {

```
92      def PolhemusTracker(tracker_id):
93          try:
94              trck_init = PlhWrapperConnection(tracker_id)
95              lib_mode = 'wrapper'
96              if not trck_init:
97                  print 'Could not connect with Polhemus
                        wrapper, trying USB connection...'
98                  trck_init = PlhUSBConnection(tracker_id)
99                  lib_mode = 'usb'
100                 if not trck_init:
101                     print 'Could not connect with Polhemus
                            USB, trying serial connection...'
102                     trck_init = PlhSerialConnection
                            (tracker_id)
103                     lib_mode = 'serial'
104         except:
105             trck_init = None
106             lib_mode = 'error'
107             print 'Could not connect to Polhemus.'
108
109         return trck_init, lib_mode
```

© The Author(s) 2019
E. Bicudo, *Neuroimaging, Software, and Communication*,
https://doi.org/10.1007/978-981-13-7060-1_3

The small piece of computer code presented above is part of a software package called InVesalius.[1] Dr. Victor Hugo Souza wrote it in Python, a programming language that allows developers to spread the code into several sections called classes. These latter, as explained in chapterTwo, play specific roles. Whereas a certain class would, for example, store all the numerical data pertaining to a brain scan, another class would be responsible for converting such data into visual outputs. The class from which the previous piece of code was taken is named "PolhemusTracker," as defined in line 92. I would like to call the reader's attention to line 94 where an object was created with the name of "trck_init." Python is one of the so-called object-oriented languages, enabling programmers to create these kinds of objects. Roughly speaking, objects are virtual bridges connecting the different classes of a program. They were created in the 1960s, firstly appearing in a programming language called Simula (Skog 2003). In our case here, the "trck_init" object gives access to the coding elements of the "PlhWrapperConnection" class. If the association fails, then the program tries to use the same object to access another class in line 98, and then another class in line 102. In this way, a Python program consists of many separate parts which are nevertheless connected by means of objects. Connections and attempts, not only technical ones but also personal ones, are an integral part of the story of the Invesalius software [str, invls].

---

[1]https://softwarepublico.gov.br/social/invesalius/download.

[str, invls]

Dr. André Peres was trained as a medical physicist at the Department of Physics, Univ of Sao Paulo, where he also did his PhD. During his Master's Degree, he started to develop a neuronavigator, which is a software package allowing to precisely detect the brain areas being stimulated at any given time. The work carried on in his PhD. At a scientific congress, André met the people responsible for the development of InVesalius, a software whose design relies on the support of the Brazilian Ministry of Science. At that time, InVesalius was exclusively used for visualizing technical devices. An opportunity to join the two projects (the visualization software and André's neuronavigator) was then identified. André began to work on this combination, a work that was shortly after taken over by Dr. Victor Hugo Souza, also a medical physicist. In Victor's words:

"We took to InVesalius a series of functions, including the work with resonance images (which the software did not do), support for functional resonance images, and the main thing, which is neuronavigation. So we incorporated neuronavigation into InVesalius [...]."

Objects are particularly crucial when different software packages are combined, as in the case of InVesalius [str, invls]. Thus this software's story constitutes a good illustration of a taken-for-granted idea among software developers: in order for a program to properly run, a coherent and effective chain of computational events must be created. Indeed, bugs frequently happen not because of problems in the internal design of algorithms but simply because the computations have not been put to work in the correct order, or simply because a specific object is pointing to the incorrect class. In this sense, it can be assumed that programmers effect, in the digital realm, what human beings have always effected in the material world: the production of "states of affairs," to use an expression by Wittgenstein. This concept makes reference to the contexts created by a combination of objects and events, in accordance with a certain rationale. In the following passage, Wittgenstein (1922, p. 26) speaks of phenomenological objects observed in the world, but he could be speaking of objects created in Python: "[...] we cannot think of *any* object apart from the possibility of its connexion with other things."

In a sense, a Python program can be described as an overall idea (the whole source code) composed by specks of ideas (classes). It is then possible to understand the words of Richard Stallman (2002a, p. 105), a prominent name in the history of computer programming: "[…] software packages are usually very big. They use many different ideas in combination.

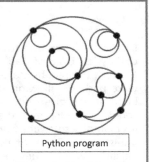

This image is a tentative visual representation of a program written in Python. The biggest circle represents the whole program whereas the smaller circles are classes, each one responsible for a specific task. The black dots are the objects used to connect different classes and make it possible to produce a fully coordinated program. Thanks to its dynamic features, Python (which is very similar to R) enables programmers to express technical solutions in very creative ways.

Python program

If the program is new and not just copied, then it is probably using a different combination of ideas – embodied, of course, in newly written code […]." From a sociological standpoint, it is possible to explore Stallman's description by asking: what does it mean to combine ideas in software development? Is such creative work laden with only technical dimensions or does it also impact the communicative flows that link people and constitute society? To address these issues, let us consider four phenomena: the ways in which human thought can be externalized, gaining a material expression; the generation of coordination by means of computer code; the discourses formulated by social agents to make sense of their activities; and the emergence of developer communities as a motivating idea. After having analysed the relations between programmers in chapterTwo, this chapterThree aims to scrutinize the expressive roots of such relations, showing that, since its early moments, code writing presupposes negotiation and communication.

section6 [Open Talk, Open Code] {

The thousands of lines of code written by hackers eventually produce software, making the computer behave in anticipated fashions. One might then consider that these highly formalized and complicated lines constitute the most basic pillars of software. However, this conception is limited, as it disregards aspects that are crucial not only for social scientists but also for computer scientists. It was by the way computer scientist Peter Naur (2001, p. 227) who proposed the so-called "theory building view" of software, affirming that "[...] programming properly should be regarded as an activity by which the programmers form or achieve a certain kind of insight, a theory, of the matters at hand." In this way, the actual pillar of any software is not really the coding text but the ideas that, woven together, enable the program to function in a predetermined manner. Hence Brook's (1995, p. 185) statement that software "is pure thought-stuff," as well as Moran and Carroll's (1997) claim that computer code always incorporate a particular "design rationale."

By stressing the theoretical dimensions of software development while giving less attention to its syntactic aspects, Naur promoted a profound interpretive shift. The same change can be verified among social scientists who instead of seeing software development as a technical phenomenon merely influenced by cultural aspects, describe it as one of the activities that constitute culture. In this vein, it was anthropologist Nick Seaver (2017, p. 5) who claimed: "We can call this the *algorithms as culture* position: algorithms are not technical rocks in a cultural stream, but are rather just more water."

If software can be described with reference to its creative dimension, then it should be explained what particular features these creative tasks take. Moreover, it should be made clear in which ways this creative process holds interest in sociological inquiry. At first sight, such interest would be lacking, as we are dealing with an effort that is usually made in isolation, away from the collective relations focused on by social scientists. As seen in chapterTwo, the majority of neuroimaging software packages are in fact developed by one single person.

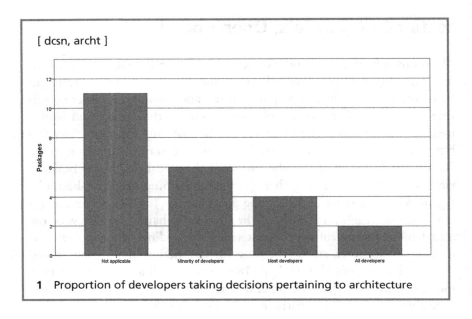

[dcsn, archt]

1 Proportion of developers taking decisions pertaining to architecture

Even when there are two or more developers in the project, the main decisions, such as those pertaining to the software's architecture, are taken by the minority of them [dcsn, archt]. In the software engineering jargon, architecture is the overall structure of a software package, the ways in which different computing operations are separated into classes, as well as the ways in which those classes are linked together. If software is primordially composed of cognitive solutions, architecture is *the* major solution presiding over its design. Therefore, findings from Survey 2[2] show that the cognitive dimension of software development seems to lack firm connections with social life.

---

[2]For information about the two surveys conducted, see the Methodological Appendix.

However, even solitary activities such as thinking carry some imprints of society. As I showed elsewhere (Bicudo 2012), people's viewpoints are frequently shaped by unspecific ideas that gain social diffusion and become ideological tenets for large (and even international) groups of social agents. According to Arendt (1978, p. 45), men are endowed with an "ability to think" which "[...] [...] permits the mind to withdraw from the world without ever being able to leave it or transcend it." If the thinker never ceases to be a man of this world, there is no reason why the programmer could ever acquire such transcendence.

At this point, Galloway's (2012, p. 59) daring claim can be invoked: "[...] programmers have at their disposal an especially powerful means of externalizing their thinking: they write program code to embody it." If as Kant proposed, thought never appears, not even to the person who thinks it, then computer code constitutes a sort of inside-out thought which dares to have a physical appearance. It is again Arendt (1978) who claims that thought is swift, as it is free from many restraints typical of the physical body. If computer code is an externalization of human thought, as it were, then an operating computer code (that is, code being "read" by the computer) is thought being thought (with no thinking mind) in an especially swift way, freed from even the biological limitations of the human nervous system. In times when the processing capacities of computers escalate (Hood 1990; Stein 2010), promises of swift thought become increasingly grand.

In the light of these considerations, the field of neuroimaging software is particularly interesting, as it is largely dominated by open source software (Aguirre 2012). Survey 2 provides us with good indications, showing that most neuroimaging researchers do publish the source code of their applications. Of the 23 packages involved in the Survey, only five are not open source. The most largely used neuroimaging packages, such as FreeSurfer, AFNI, and FSL, have their source code published, in addition to being completely free of cost. A basic reason why software development can be the object of sociological analysis is that even though programmers are to a great degree engaged in a solitary and cognitive task, their ideas must comply with socially established (programming) rules, in addition to gaining a material form (the source code) which is frequently seen, studied, and scrutinized by other programmers.

The externalizing act of open source software is not similar, for example, to the externalizing act of a philosopher who discloses ideas by means of words. While the philosopher can skilfully hide away some prejudices or political biases underpinning the theory, an open source software developer can hide nothing. When computer code is published, every line (that is, every step of the logical trajectory followed by the developer) must be disclosed. Considering that a software package may have its operations compromised, if not spoiled, by the removal of just one character from its codebase, it is easy to realize that publishing computer code becomes a futile act if full disclosure does not happen.

Because it is a sort of thought whose premises are not hidden, open source software becomes an invaluable resource in the academic realm. In his interview, Professor Carlos Garrido (Dept Physics, Univ of Sao Paulo, Brazil) pointed out that with open software, it is possible to exactly know what happens with data during data analysis. This clarity is in line with some tenets of scientific activity, as "[…] science is a socially shared and socially validated body of knowledge" (Merton 1968, p. 59). Every developer engaged in an open source project is in a way subjected to public scrutiny. The "readers" of open source code may evaluate the different aspects of the package, one of which is surely its architecture. If computer code contains a logic that comes to be revealed, it is worth exploring the components of such logic, as well as its social dimensions. This is the goal of the following sections of this chapterThree.

}

section7 [Social Morphology, Code Syntax] {

7.1. The Logics of Programming

Computer code writing requires the possession of some mathematical skills but it is far from being synonymous to mathematical reasoning. Frequently, programming demands an ability to create and improvise that is often absent from mathematical operations. One could even argue that a software developer is a craftsman, in the sense proposed by Sennet (2008), because code writing requires intuition, engagement, everyday practice, and, as shown in chapterTwo, much social cooperation.

In order to realize their creative deeds, programmers need, to put it simply, to think, as far as possible, in tune with the computer. Every programming student quickly realizes that computer code frequently does not work when it is written in accordance with the practical rationales of everyday life. A new rationale is needed, which must be acquired through some sort of training.

For example, one of my interviewees is a student based in the University of Campinas, Brazil. This person learnt to program during the Bachelor Degree, by joining a course on programming syntax, in which students do not learn any specific language; instead, they are taught basic commands and operations common to several languages.

> *"[...] learning languages gets more and more rapid.*
>
> **Oh, really?**
> *Yes.*
>
> **Why?**
> *I think... I already know the logic, which you took from other languages. The logic doesn't change. What changes is the syntax. So over time, you get the syntax of a new language much more easily [...]."*
>
> Dr. Nathalia Esper (Laboratory for Images, Pontifical Catholic Univ of Rio Grande do Sul, Brazil)

> *"[...] you're learning, you know, compu-tational thinking, you're learning about problem solving. And then whatever language you use after that doesn't really matter [...] After that, it's just about the syntax."*
>
> Dr. Vincent Giampietro (Centre for Neuroimaging Sciences, King's College London, UK)

With this knowledge, my interviewee could subsequently learn two languages, C++ and MatLab, on his own. What was acquired in that course is the basic rationale of code writing, which enables the programmer to "instruct" the computer appropriately. This is why programmers frequently claim that even though they cannot write in a certain language, they can still read that language.

In this respect, some languages, more than others, impart programmers with deeper knowledge of the computing rationale. This is surely the case of C and its child language, C++. While some languages (like Java) have been designed so as to allow programmers to produce more "economic" code, with a small number of code lines, in C and C++ the programmer has to take care of many computing steps. For example, if a considerably large memory space is necessary for storing some data or graphical element, the programmer has to write some code to allocate and deallocate that memory space, a procedure that is unnecessary in many other languages. On the one hand, writing C or C++ code becomes then a more arduous task. On the other, however, the programmer comes to understand the computing logic more comprehensively.

> *"Is there any [programming] language that is your favourite one?*
> *Err... C, maybe [...]*
>
> *Why?*
> *[...] With C I worked mainly during the Master's [Degree] [...] most of other languages work very much on the basis of so-called functions that are available, and you fit together things that already exist. With C you have to do things much more from the root. You take the basic blocks and you build up complex things from there [...] during the Master's we did some things that I really enjoyed to do. It was good to realize how things work at different levels of abstraction [...]*

> So if I'm not wrong, C is considered as a low-level language, because it's very close to... to the machine itself.[3]
> Yes. Yes.
>
> Do you believe that, because you've worked with C, that gives you more tools for you to work with other high-level languages afterwards?
> Yes [...] For example, when we're going to process neuroimaging data and we have n cases and we have a computer with 32 gigabytes of memory and with 24 core [processor], for example [...], it's worth doing things at lower level so we can later manage to build up multi-framing scripts. I mean, we can start several processes at the same time so we can process various data at the same time [...] I can speed up things or realize how high-level things work to speed them up [...]."
>
> Dr. Ricardo Magalhāes (Institute of Life and Health Sciences-ICVS, Univ of Minho, Portugal)

Whenever a program is being written, the programmer needs to anticipate how the computer will react to the instructions it receives, so "miscommunication" is avoided. Without assimilating such computational logic to some degree, and without complying with those technical constraints, the programmer will surely fail to produce useful applications. Because of such technical requisites of code writing, some analysts have displayed a strong tendency to overstate the rigid contents of programming, formulating an approach with problematic consequences for social analysis. This is the focus of the following section.

7.2. Control and Coordination

The rigid rules of code writing have a twofold nature. On the one hand, the computer is an obedient worker. When coherent code is produced, the computer will precisely comply with the instructions, always repeating prescribed operations with no surprises or deviations. On the other hand, the rules of programming languages (the syntax that programmers have to obey) are much more stringent than those of natural languages. Somebody who travels to a foreign country

---

[3]In chapterFive, it will be seen that C++ is actually a middle-level language.

without having complete mastery of the local language may still be understood. Even if pronunciation is not very accurate, and even if grammatical errors are made, local people may grasp the meaning of what is imperfectly said. No such linguistic tolerance is observed when one tries to "converse" with the computer by writing code. If the syntax is not strictly stuck to, the machine will simply not "understand," and the code will just not run. These two technical dimensions of computer programming allow us to have recourse to the concept of *facticity*. Phenomena that impose their rules to social actors, acquiring an unavoidable binding force, are expressions of facticity in social life (Habermas 1996). In the realm of facticity, there is little space, if any, for negotiation and improvisation, as divergent actions get subjected to sanctions like moral disapproval, practical failure, interpersonal misunderstanding... or the production of a buggy code.

Because of these technical and rigid aspects, some analysts have associated programming with rule-making, establishment of criteria or, in a word, control. The clearest example is the argument defended by Beniger, who speaks of a "control revolution," a process that is said to mark human history. "All control is [...] *programmed*: it depends on physically encoded information, which must include both the goals toward which a process is to be influenced and the procedures for processing additional information toward that end" (Beniger 1986, p. 40). In this view, the need for control would be already engraved on the basic biological structures of life, because "[...] everything living processes information to effect control; nothing that is not alive can do so – nothing, that is, except certain artifacts of our own invention, artifacts that proliferated with the Control Revolution" (p. 35). The computer, as one of such artefacts, would be reinforcing this process.

The danger of this approach is that it makes believe that, eventually, the logic behind computer programming is the same logic behind the laws of matter and energy. Every form of programming would be the manifestation of some physical force, biological need, or unavoidable trend, and would therefore be related to the need for control. This interpretation is in line with a current revival of determinism which is manifested in the genetical or neuroscientific explanation of human behaviour, as well as in the tendency to believe that human communication can be equated with the exchange of molecules between plants or the exchange of signals between animals. In this "neodeterminism," the import of norms, values, and choices is replaced with a biological explanation. We are dealing with "[...] a kind of naturalization of the human mind that places our practical self-understanding as responsibly acting persons in question [...]" (Habermas 2008, p. 141).

This is not to say that contents of control are completely absent in computer programming. It was already seen that, in a sense, programmers control the computer, shaping the machine's behaviour in the scope of a particular software package, and determining the range of possible tasks to be performed by the future users of that package. At the same time, programmers are in a sense controlled by the computer, because they must abide by the syntactic rules of programming languages. In the 1970s, when the initially intuitive and experiential activity of code writing gave rise to software engineering, a discipline with scientific aspirations, the notions of control, precision, and strictness became particularly heavy. However, as explained by Ensmenger (2010, p. 32), the laws of software engineering failed to abolish "the essentially craftlike nature of early programming practice."

From those early times up to our times, the *instrumental* side of programming has been coupled with a *communicative* side whereby the programmer creates connection and coordination in at least three ways. First, functioning code is the one resulting from a successful dialogue between the programmer and the computer. From this perspective: "A computer program is a message from a man to a machine. The rigidly marshalled syntax and the scrupulous definitions all exist to make intention clear to the dumb engine" (Brooks 1995, p. 164). Second, there must be coding elements guaranteeing the program's internal coordination, a role that in languages like Python or C++, is played by objects. Finally, coding strategies and social relations, many of which were analysed in chapterTwo, must also be in place to ensure coordination between pieces of code written by different programmers.

Therefore, computation does not completely foreclose the emergence of contingency, indeterminacy, and aesthetic purposes (Fazi 2018). If the work with computers were the exclusive realm of rules, control, and obedience, there would be no leeway for tricks and improvisation. By talking to programmers, one can rapidly see that this is not the case. For example, it is possible for Dr. Ricardo Magalhães to implement some "little tricks" in order to "fool the computer" and make applications designed for human beings work with brain images taken from animals [rcrd, mglh].

[rcrd, mglh]

Dr. Ricardo Magalhães (Institute of Life and Health Sciences-ICVS, Univ of Minho, Portugal) is a biomedical engineer. After doing the Master's Degree in medical informatics, he initiated a PhD in neuroimaging, studying the effects of stress by using images of the brains of rodents. At the same time, due to his experience in neuroimaging, he helps students and researchers carry out several neuroimaging studies at ICVS. In this way, he works with brain images of both animals and humans.

"So in terms of data processing, is there too big a difference between working with the human brain and the animal brain?
Err, no. The majority of tools is sufficiently generic to be used with the two, but there is some extra care that is needed with animals. There is, for example, a very basic problem, which is... which has to do with resolution, with the fact that, for example, animals have a 0.3mm resolution while in humans it is 3mm [...] And the majority of software packages... For example, SPM and FSL cannot interpret data that... with such high resolutions, like 0.3. So we had to... To reconstruct those data we had to go to the headers and modify the headers to fool the computer, so the computer thinks this is not at 0.3mm but 3mm. We have to implement an increase, a mock resolution, ten times higher [...]

Okay. So there was one word that I didn't understand: the headers. What are the headers? I think I know it with a different word, perhaps [laughter].
For example, the images, we generally use them in the DICOM format. So this is a part containing information about the resonance, about the resolution, information about TR [repetition time], TE [echo time] and all those things. This is what we call the DICOM headers. They store all the information about the [image] acquisition [...] And we have to edit this, so we can fool the software [laugther] [...]

That is, it's possible to use all those software packages, FSL, SPM, AFNI, as long as it's possible to fool the software...
Yes, yes. You need some little tricks.

Okay. So I imagine that a good part of your study is implementing those little tricks to make it work [laughter].
Yes, yes, yes."

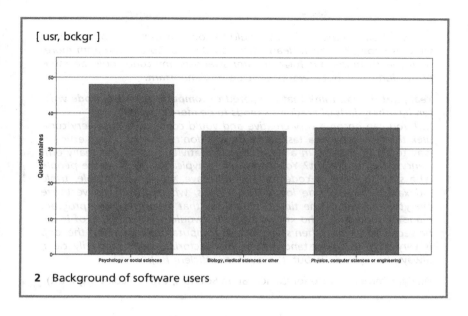

[usr, bckgr]

2   Background of software users

For most researchers developing neuroimaging packages, the acquisition of programming skills is not the result of formal and consistent training. In her interview, a Brazilian researcher said that computer engineering students, at the beginning of their degree, have difficulties to assimilate concepts because they are used to everyday decimal logic; therefore, one of the main goals of their studies is the assimilation of computers' binary logic. However, let us consider that people doing neuroimaging are not always computer scientists who have assimilated the binary logic. Many of them come from very different areas such as medicine and psychology. Interestingly, Survey 2 showed that most users of neuroimaging software are psychologists or social scientists [usr, bckgr]. In chapterFive, it will be seen that in neuroimaging, the figures of developer and user are frequently mixed up. In this way, there is a considerable number of people with background in human sciences who have written at least some basic code for neuroimaging software. Writing code without having had proper training, those people cannot help recurring to some surprising or unorthodox ways to engage in a "conversation" with the computer, whether this involves "fooling" the computer, "instructing" it, or "persuading" it.

> *"I come from psychology. How would I know the standards of computer sciences people? I had to learn them on the job. So as I program more, then I learn more, but for computer scientists, my code could be more professional [...], but it works, which is very important.*
>
> **Yeah, but do you think that compared to computer scientists, code written by other kinds of academics tends to be more creative?**
>
> *[...] You can become very creative and you'd come up with a very complex thing to achieve a task, because you don't know of a very efficient way to do it. And so, in a sense, that's creativity, but it's not really efficiency [laughter], right? You know, like a typical thing from the people who start to learn to code: they would have, you know, double, triple indexation, loops inside loops inside loops, whereas you can vectorize everything. Most of the time, I would say that if you read the program made by somebody who knows to code, I would say two thirds of it can be vectorized, which then saves 90% of computational time. But the loop is very easy to understand. When you vectorize, not necessarily clear, always, how you do it, but it's way more efficient [...]."*
>
> Dr. Cyril Pernet (Centre for Clinical Brain Sciences, Univ of Edinburgh, UK)

Thus it does not seem correct to suppose that programmers aim to only conceive of strategies to realize control. Neither is their activity standardized and rigid to the point of being the object of controlling powers exclusively. As Galloway points out, the diffusion of computers and software heightened the relevance of the intuitive skills possessed by programmers, who are far from performing automatic tasks. "The intellectual work of computer programming therefore remains intractable to routinized control [...]" (Galloway 2012, p. 43). There must be a constant effort to foster coordination, whether this is realized via the technical coordination made possible by computer code or via concrete exchanges between different programmers working on the same project. If, as claimed before, it is possible to identify some contents of *facticity* in programming, it is also possible to point to the presence of *validity*. Phenomena that are subjected to verification, being likely to be either refused or accepted, reiterated or invalidated, in a particular context, are expressions of validity in social life (Habermas 1996). This is what happens when neuroimaging software developers tinker with code and seek to implement their needed functionalities, adopting tactics that can be "refused" by the computer or criticized by their peers.

Therefore, the "tension between facticity and validity" (Habermas 1996, p. 20) can be found in computer programming.

It is not claimed here that neuroimaging software designers are just jerry-rigging solutions and coming up with clumsy kludges. They are in fact experienced scientists whose work (including the software packages they design) is evaluated by their peers and has enabled scientific and clinical achievements. Improvisations and adaptations are stressed here to make the point that even though programming has some aspects of control, these two concepts should not be blended together. If this is done, then the *communicative* dimension of programming is overlooked for the sake of its *instrumental* dimension. Using Dummet's (1993, p. 129) words, let us claim that code writing contains both "the intended interpretation of a formal language" (in this case, all the rigid rules of programming syntax) and "the received interpretation of a natural one" (in this case, the creative solutions that somehow compensate for technical restrictions).

In this sense, computer programming emerges as a set of *mediational actions*, which, as I claimed elsewhere (Bicudo 2014), are characterized by their having, at the same time, traits of the communicative and instrumental rationalities. If this is not understood, the analysis fails to capture, at the same time, the collaborative endeavours of software development (as shown in chapterTwo) and its hierarchical structures (highlighted in chapterFour). Eventually, then, computer code comes to express divergent rationales. So it is important to verify what is actually manifested on the computer screen. More than that: what, from a communicative perspective, is actually contained in programming code?

}

section8 [Discursive Interface, Graphical Interface] {

8.1. The Expressive Dimension of Code

The neuroimaging software package called SPM underwent a huge renovation work from 1991 to 1994. At that time, professor Karl Friston was his main designer (which he still is). In a 2012 publication, this is how John Ashburner (2012, p. 793) described the redesigning process: "The 33,500 lines of MATLAB code from SPM91 had been reduced to a mere 5700. Karl unified many of his earlier ideas into much more elegant formulations […]." The idea of elegance might sound displaced when applied to a technical field such as programming. Is it possible to claim that a piece of programming code can be either elegant or disharmonious? In software development, are aesthetic concerns ever relevant?

The idea that excellent developers have an ability to design programs with almost artistic or magic skill was already voiced in the 1950s and 1960s (Ensmenger 2010). According to Berezsky and colleagues (2008, p. 13), "computational rationality" is made of different moments, including "communicative moments, aesthetic moments and expressive moments." And according to Mackenzie (2003, p. 13), the time has come when "code becomes an aesthetic object." Some analysts have even equated programming with poetry. Weber (2004, p. 58), for instance, based the comparison on the multi-layered nature of code's fabric: "Software is conceptually […] like a complex poem or great novel in which different kinds of flows coexist across different dimensions." These and similar claims have been voiced not only by social scientists but also by software developers. In this vein, Brooks (1995, p. 7) affirmed: "The programmer, like the poet […], builds his castles in the air, from

> "And how is your code? Is your code organized and clear?
> Yes. I can reasonably, confidently say yes […] I'm quite attached to just elegance in the code in general. So that's why I try to keep things clean and that's why I can reasonably, confidently say yes."
>
> Dr. Tim van Mourik (MR Physics Group, Donders Institute for Cognitive Neuroimaging, Netherlands)

> **"So you were talking about the beauty of code. Is your code beautiful?**
> I try to make it beautiful. I'm a strong believer in beauty of code. Code should be beautiful. And... and you can see it straight-away, like when you see a piece of code and you can calmly and logically work through what it does. And... and you can contrast that against a piece of code that when you see it, then it's like spaghetti and... the symbols are not defined, the variables are not defined, there's no documentation, and the difference in emotion between those two situations is significant [...]."
>
> Researcher (Univ College London, UK)

air, creating by exertion of the imagination." And Linus Torvalds, who had the technical savviness to initiate the development of the Linux operating system, claimed: "This is one of the reasons programming can be so captivating and rewarding. The functionality often is second to being interesting, being pretty, or being shocking" (Torvalds and Diamond 2001, p. 74).

The concept of beauty has been studied, for many centuries, by sociologists, anthropologists, philosophers, and others. Without delving too deeply into the conundrums of this debate, let us quickly ask what programmers mean by beauty. Two aspects can be pointed to. First, one conception present in the speech of most programmers is in line with the Platonic view: beauty is the expression of perfect unity. The lines written by Neoplatonist Plotinus (3rd century, p. 48) clearly summarize this view: "Only a compound can be beautiful, never anything devoid of parts; and only a whole; the several parts will have beauty, not in themselves,

> **"And how is your code? Is it organized and clear?**
> No, I wouldn't say that... I'm a really clumsy coder.
>
> **Really?**
> Yeah [...] So I would say it... it depends. If I have to share the code, I would comment it, try to make it beautiful but if I have to... let's say, if I have a meeting on Monday or on Tuesday or another day and I have to finish it, I won't bother much about how beautiful it is; I will bother about 'I have to finish this problem' [...]
>
> **I know. Okay, so when you have to share it, the code is organized.**
> Yeah, I would try to make it [...]."
>
> Researcher (Donders Institute for Cognitive Neuroimaging, Netherlands)

but only as working together to give a comely total." In this sense, beautiful code is fully coherent code. Second, beauty has to do with the "'readability' of code" pointed to by Mackenzie (2006, p. 15). In this sense, there is a concern with "the other of code," the person who may be willing to read it in the future. On the one hand, programmers take into account issues of harmony and coordination, something that also makes material objects be framed as beautiful. It is in this sense that somebody would say: "The table you made is beautiful." On the other hand, one considers the clarity with which code is expressed, something that makes ideas sound beautiful. It is in this sense that somebody would say: "The way you think is beautiful." As for this second notion of code beauty, the vast majority of programmers are aware of it. This is why code is frequently made to appear nicer when it is supposed to be shared with somebody else, hence gauged by somebody else.

To be sure, many programmers, at least in neuroimaging, do not convey such appreciation for the aesthetic dimension of code. Moreover, the production of truly beautiful code requires the mastery of a wide range of software engineering concepts, something that is lacking for many neuroimaging researchers, as explained in section7. However, there is indeed one conception that, in spite of not being always made explicit, is shared by all computer programmers: the notion that code is the expression of personal capacities, skills, and talents. In this sense, code is not only a highly formalized text aiming at certain functionalities; it is also the backstage that can at any time become the main stage on which the programmer is surprised.

The issue of public appearance has for centuries concerned scientific researchers. According to Shapin and Schaffer (1985, p. 78), the first practitioners of experimental philosophy, in the mid-seventeenth century, were already interested in realizing "[…] a crucially important *move towards* the public constitution and validation of knowledge." For today's producers of open source software, this move is very likely to come true, as computer code individually produced can be easily published, downloaded, scrutinized, and evaluated. Therefore, by deciding which of their pieces of code will be published or not, software designers are also choosing what aspects of their work will be revealed or not.

> In addition to the urge toward self-display by which living things fit themselves into a world of appearances, men also *present* themselves in deed and word and this indicate how they *wish* to appear, what in their opinion is fit to be seen and what is not. This element of deliberate choice in what to show and what to hide seems specifically human. (Arendt 1978, p. 34)

In studies that explore the reasons why programmers opt for open source software (Atal and Shankar 2015; Lakhani and Wolf 2005; Schwarz and Takhteyev 2010; Amabile 1996; Deci and Ryan 1985; Frey 1997; Ryan and Deci 2000; von Hippel and von Krogh 2003), the acquisition of reputation has been pointed to as one central motivation. Good and elegant code, when published, can ensure "a reputational payoff" (Atal and Shankar 2015, p. 1390) for its creator. When the software package is particularly well-designed, its creators may restrict membership of the development group in order to increase these reputational gains (von Hippel and von Krogh 2003). Such concerns are particularly pressing for people based in academic settings, as shown in the sequence.

8.2. Scientific Code Writing

Considering personal skills and work settings, Skog (2003, p. 304) divided programming into "[...] non-professional programming and commercial professional programming." If the programmers' purposes were taken into account, a different terminology would be produced. Then one of the categories identified could be something like "scientific coding." This expression makes reference to the writing of code (and consequently software development) by people moved by scientific interests, seeking scientific targets and rewards. The academic environment, with its internal dynamics, disputes, collaborations, and temporalities, makes code writing assume particular features. If programmers wish to use their code so to as present themselves in certain ways, not others, such construction of the self, in academic settings, is definitely informed by their willingness to appear as good, responsible, and competent scientist-programmers.

> *"Okay, so just to close this interview, I would like to ask what are the major trends for the development of Fieldtrip in the near future. What is going to happen with Fieldtrip?*
> *Ahm... Like... There's ongoing work. It's always like maintaining and improving the code, so adding new algorithms, also improving on existing algorithms, fixing bugs, making the code faster, making sure that it runs with the latest versions of software or making sure that it reads all the data or the data from the new systems. That's the ongoing work, that's always going on."*
>
> Professor Robert Oostenveld (MR Techniques in Brain Function group, Donders Institute for Brain, Cognition and Behaviour, Netherlands)

As claimed by Arendt (1978, p. 54), the idea of "*unlimited progress*" has always been one of the major elements in the discourse voiced by scientists. "When the experience of constant correction in scientific research is generalized, it leads into the curious 'better and better,' 'truer and truer' [...]" (Arendt 1978, pp. 54–55). Software development is an ideal realm for the manifestation of such discourse, because as Dr. Cyril Pernet (Centre for Clinical Brain Sciences, Univ of Edinburgh, UK) claimed, "by definition, there isn't completed software." Indeed, software packages are constantly being enhanced, modified, corrected, deprived of bugs, a circumstance that helps reinforce that image of a consistent search for perfection. In the realm of open source software, such modifications tend to be even more frequent, suggesting the image of a continuous research project (Weber 2004).

> *"But do you consider those projects as completed or are they still evolving?*
> *Well, the thing is, a completed software project is a kind of contradiction in terms. So... they are completed in the sense that they at the moment have a clear goal that they can achieve and... But the question is: do you want to add more goals to that? And, well, the answer is: yes, probably. So, in that sense, no, it's never completed. "*
>
> Dr. Tim van Mourik (MR Physics Group, Donders Institute for Cognitive Neuroimaging, Netherlands)

> *"And do you think that these most popular software packages like AFNI, SPM, FSL, will continue to dominate the field for many years or will they be replaced by new software packages?*
> *Well, I don't... It's hard to answer that question because I don't see those packages as just static entities. I mean, they're also continuously developing [...]. "*
>
> Professor Christian Beckmann (Statistical Imaging Neuroscience group, Donders Institute for Brain, Cognition and Behaviour, Netherlands)

The possibility of modifying the codebase is always present. When a software package is at its initial design phase, programmers often anticipate "[...] the areas that are considered likely to change" (Parnas and Clements 1986, p. 253). And when the package is made available to users, developers frequently continue to modify it, releasing new versions from time to time (Galloway 2012). Survey 2 in fact showed that the release of new versions of neuroimaging packages is a very common event [nw, vrsns].

Programmers can stress the continuous improvement of algorithms because these latter can indeed be frequently refined. In this way, "[...] algorithms are rarely fixed in form and their work in practice unfolds in multifarious ways [...] that is, they are never fixed in nature, but are emergent and constantly unfolding" (Kitchin 2017, p. 21). This fluent nature of algorithms, code, and software derives from the personal interest of programmers but also from the technical needs of a dynamic domain where "[...] the product over which one has labored so long appears to be obsolete upon (or before) completion" (Brooks 1995, p. 9).

[nw, vrsns]

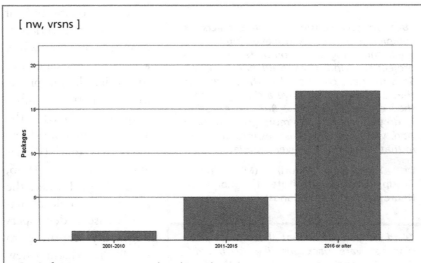

3 Software's most recently released version

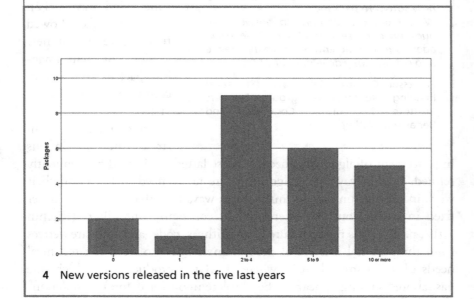

4 New versions released in the five last years

In the same way that programmers can describe software as an ever-evolving entity, they can indirectly describe themselves as code writers in constant improvement. As noted in previous studies (von Hippel and von Krogh 2003; Himanen 2001; Kohanski 1998; Brooks 1995), learning is one of the main motivations of software developers. As the assimilation of new concepts and approaches is a frequent by-product of code writing, programmers can build up a discourse in which they are implicitly framed as people whose knowledge never ceases to grow.

In the software engineering jargon, the graphical interface is a set of visual elements (such as buttons, windows, and colours) displayed on the computer screen to make software intuitive and user-friendly. The graphical interface is the part of code that gains visual manifestation and is perceived by the software user. From a communicative point of view, code has also a discursive dimension (its "discursive interface," so to say), because software developers at times create a socially significant discourse by speaking of their activities, describing their motivations, and stressing the relevance of what they do. This is to be expected, as every kind of social activity, in addition to being performed, needs to be *made sense of.* In this way, choices taken and acts performed by social agents come to have meaning and can be claried in terms of language. "Just as it is self-evident to me [...] that I can, up to a certain point, obtain knowledge of the lived experiences of my fellow-men— for example, the motives of their acts—so, too, I also assume that the same holds reciprocally for them with respect to me" (Schutz 1974, p. 4). In software development, this need for sense underpins the production of a sort of "discursive interface" of code, which is the dimension of programming that comes to be selected and highlighted in the programmers' discourses. Therefore, the externalization of programming thought, pointed out in section6, is coupled with an expressive effort of self-display and self-explanation. Expressive contents (which manifest individual skills, preferences or emotions) are a necessary component of communication, along with normative contents and cognitive-practical contents (Habermas 1984).

> *"Hm. So if I understand, when you begin the project, you don't actually know that it's going to be a software, a complete software.*
> *[...] You're simply talking with some friends, working on a project, on this project, and you were using some data analysis technique and: 'Oh, why don't you use that type of data analysis technique? It would be so much better. You would get much better results and you'd understand better the questions that you want to address.' And so you start with one function, then another, then another, then another. 'The data visualization is not very good so you should start to do some graphical interface.' And so people say: 'Oh, it's very cool. You could distribute that.' And then... [laugther]."*
>
> Dr. Cyril Pernet (Centre for Clinical Brain Sciences, Univ of Edinburgh, UK)

This is not to say that programmers try to deceive people through a discursive embellishment of their work. They simply produce discourses or "fictions"; "[...] fictions, in the sense that they are 'something made,' 'something fashioned' [...] not that they are false, unfactual, or merely 'as if' thought experiments" (Geertz 1973, p. 15). Furthermore, programmers are not pretending when they stress the fluent nature of their work. Software development projects are in fact generally never-ending activities with fluid traits. Many software packages that are today wide, complex, and fully-fledged applications, initiated their stories as unambitious projects, most narrow in scope. Accounts of modest projects which eventually generated sophisticated packages abound in the neuroimaging domain.

> *"When I told you that I spent one year knocking my head against the wall to make things work, all I did during this period is error. And it is not error in terms of results. It is something like: 'The image doesn't appear. The image doesn't appear. The image appears upside down. The image appears upside down. Oh, now the image appears properly [...]'."*
>
> Dr. Victor Hugo Souza (Laboratory for Biomagnetism and Neuronavigation, Univ of Sao Paulo, Brazil)

The image of laborious developers struggling to take their products to perfection can be mobilized because in computer programming, probably more than in many other human activities, the emergence of problems is a recurrent fact. Code is very unstable, in the sense that even slight modifications to the codebase may create huge bugs in a program that, seconds before, was running perfectly. The seemingly perfect program is the one which results from a code that can be perfectly "read" by the computer. This search for technical perfection is, according to Brooks (1995), one of the major hurdles in programming, other important hurdles being: the cognitive efforts and amount of time needed to identify what in the code is generating a bug; and the increasing complexity of the programming work due to the growth of the codebase. Studying the work of software developers based in Microsoft, Latoza and colleagues (2006, p. 500) said: "Developers reported spending nearly half of their time fixing bugs." In this way, the discursive interface produced by software developers makes reference to some technical features of their activities (such as the constant improvement of packages and the hard elimination of bugs) to create an image whereby they appear as serious, competent, and hard-working people. The insistence on such aspects is certainly downplaying other aspects. Indeed, the analysis of programmers' coding products can deny such discursive efforts, installing a fragile balance between pride and shame.

Code is not supposed to be "read" by the computer only. There may always be a group of people intent on having access to the code, analysing its components, and grasping the rationale behind it. In the realm of open source software, this group of readers and critics may be very big indeed. At that moment, code, without ceasing to be the carrier of technical instructions, turns into a sort of proto-language, as it begins to acquire some social meaning. For example, as previously shown, code can be considered as an expression of competence and knowledge, which happens when it is clearly organized.

According to findings from Survey 2, the number of developers declaring that the code of the software project they work on is not organized is slightly smaller than that of programmers declaring that the resulting code is "sufficiently organized" [cd, orgn]. Therefore, a considerable number of neuroimaging software developers seem to reserve some consideration for the coherent appearance of their code.

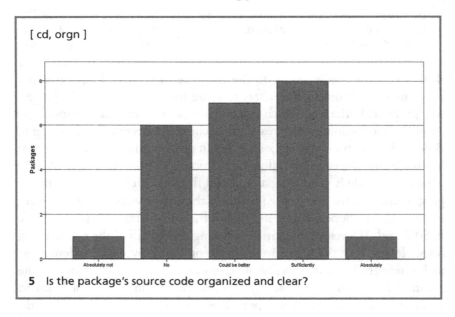

[cd, orgn]

5   Is the package's source code organized and clear?

Because of this growing social dimension of code, open source programming has some aspects of a "reputation-game" (Raymond 2001, p. 92). One of my Brazilian interviewees, who leads an academic team developing a neuroimaging package, declared that even though the development is in line with the spirit of open source, the code has not been published because of its very precarious organization. According to this interviewee, the group would be "ashamed" of publishing the code. Thus researchers and research teams, especially those based in scientific institutions, know that by publishing messy or clumsy code, they may be putting their reputation in jeopardy.

"Well, you are working on open source software. The software is available for everybody and its code is open, isn't it?
Yes.

Do you think that if the code was not open... Would your work be somehow different?
Oh, I've never thought about it, to be honest. I've never thought of it being different. Because I'm very used to this... to collective code. For instance... What is collective code? [...] our code is collective, I mean, we have our system and everybody contributes to it. So the difference is that in open source software, what you did today will be available to everyone who has access to the internet [laughter]. So... I've never had any fear of doing... I don't do it differently because it's open. I don't. Because, actually, even in a company, it is also available to other members of the group, or even to other groups, so it doesn't make much difference.

So collective code ends up being almost open source for those who write the code.
Yeah. Yes, yes. Because the fear of it being open is that sometimes somebody may say that it's ugly and so on. 'It's very ugly, it's ugly. How come you don't know such things?' So you may become ashamed. Err... no, I don't have this fear [laughter]. Because, probably, there is ugly code in there. Because technology progresses very quickly. There must be some methods, for example, that I may use in the old way. But that's science. I don't know all the new methods, so sometimes I'm not updated [laughter]. That may be a reason to become ashamed, because it could be more...

But in a company, you would be equally ashamed.
Yeah, I would. Yeah... The difference is that here you're open to a wider universe. There may be people who are keeping an eye on you, and you don't know who they are [...]."

Researcher (Univ of Sao Paulo, Brazil)

From a source of pride, code may turn into a source of "shame," as long as it does not express the programmer's inventiveness and knowledge of the most modern programming approaches. Helping to build up the programmer's self-image, code is integrated into the communicative realm, thus becoming a discursive tool. If, on the one hand, the lack of code's elegance may represent a sort of

Achilles heel for software developers, they count, on the other hand, with a powerful discursive tool: even when code is not beautiful and technically powerful, it can be effective and useful. Moreover, as shown in the sequence, that useful code can be said to enable the performance of socially beneficial activities.

8.3. Vital Software, Residual Software

When it comes to stressing the social benefits of code writing, scientific programmers can draw attention to knowledge advancement, which is said to be promoted when code is published. As an example, let us consider the point made by Stallman (2002b, p. 61): "Universities should [...] encourage free software for the sake of advancing human knowledge, just as they should encourage scientists and scholars to publish their work." In the field of neuroimaging software, more specifically, the discursive tactics seem obvious: software development is described as desirable because it facilitates the comprehension of the brain, thus helping discover therapies. In this way, programming is discursively connected to clinical purposes, in a discourse aimed to heighten the social legitimacy of code writing. Whenever the relevance of neuroimaging software development is contested, those powerful discursive weapons are available for protecting the field. "A person acting freely always moves within an intersubjective space where other people can challenge them to explain themselves [...] An actor would not feel free if the actor could not give an account of the motive on which he acted if called upon to do so" (Habermas 2008, p. 186).

"*Open source software tends to be better than closed software. But do you do open source software for any ideological reasons too?*
[Paulse] Yeah. Because I work in the academia and I do science [pause] because I like science and because it's supposed to benefit everybody [...] So by definition, for me, science is to benefit everybody and software is part of science, so if I make a software, I want it to be open [...] You know, you're from Brazil. People in Brazil don't always have the resources to acquire data and also may have difficulties to develop the software, but they can plug into my software, they can use my data and they can surely come up with plenty of ideas that I wouldn't be able to have [...] So to me, that's the reason why I do it, because, you know, I think I benefit the largest amount of people possible [...]."

Dr. Cyril Pernet (Centre for Clinical Brain Sciences, Univ of Edinburgh, UK)

In neuroimaging, such protective resources are paramount because of two main reasons. On the one hand, they help justify the activity with a normative language that can make sense to several social agents. On the other, they can be used whenever it is necessary to apply for research funding, as the potential clinical effects of code writing can convince decision makers based on funding agencies. In these ways, the justification provided by software developers leaves the realm of purely technical and pragmatic motives to acquire traits of moral reasoning, which are present when people consider not only their own interests and preferences but also include other people's needs in their justifications (Habermas 1993).

"*The standard model of what researchers should do is that they should do some work, write up some papers, and show that they are advancing the... state of the art in some area [...] There's very much emphasis on novelty in that [...] To make tools which are useful in the community is not necessarily to be novel all the time [...] So I've certainly done things in the past where the main focus is being on getting [computational] tools to work and to try to make something which is as useful as possible,*

The discursive manoeuvres of scientific developers do not cease here. If programmers generally justify their activities by praising the socially beneficial effects of their products, they often, at the same time, downplay the importance of their software

> *which is in a new way more of an engineering problem and less of a scientific, sort of, research problem. I think there's a grey area between those two."*
>
> Professor Mark Jenkinson (Oxford Centre for Functional Magnetic Resonance Imaging of the Brain-FMRIB, Univ of Oxford, UK)

packages. This paradox emerges because within the academic domain, those who gain the highest scientific reputation are the ones who offer explanations, models, concepts, theories. In this environment, the production of a tool like a software package is frequently despised as a minor scientific achievement. This very same phenomenon is observed in the field of bioinformatics; according to Lewis and Bartlett (2013), the "disciplinary coherence" of bioinformatics has been compromised by its being sometimes considered as a mere provider of tools, or even a service.

By the same token, pure software development is sometimes considered, in brain studies, a less prestigious task. In the late 1990s, Beltrame and Koslow (1999) argued that one key challenge confronting the neuroscience community was the formulation of strategies for recognizing and rewarding software development. Nowadays, those strategies are still lacking. As a consequence, even those people who came to release very popular packages talk about their product as ancillary achievements in their careers. This is what happens, for instance, with the developers of largely used

> *"Err... Do you consider SPM as a software development project or a neuroscience project?*
> *[...] I suppose I see it as software development, with the aim of adding to... you know, helping the neuroscience field. But else I see it as just a software development project with the aim to deal with the medical field as well. So I have collaborators in Queen's Square [Univ College London] who are not doing kind of something that is strictly neuroscience or not the same kind of neuroscience that's being done here. So the stuff that I develop is useful for neuroscientists as well as clinicians. Then that's an additional bonus [...]."*
>
> Professor John Ashburner (Wellcome Trust Centre for Neuroscience, Univ College London, UK)

packages like FSL. For Mark Jenkinson (Oxford Centre for Functional Magnetic Resonance Imaging of the Brain-FMRIB, Univ of Oxford, UK): "FSL started as a consequence, just an outgrowing of what we were doing as a group." And Professor Christian Beckmann (Statistical Imaging Neuroscience group, Donders Institute for Brain, Cognition and Behaviour, the Netherlands) claims: "I don't want to be dismissive but, in a way, FSL is a by-product of doing [...] neuroscience, right? If people find it a useful by-product, then that's absolutely fantastic, right? But it's not the primary focus for what I'm doing." The development team of SPM, an equally successful project, echoes these claims. For example, Guillaume Flandin (Wellcome Trust Centre for Neuroscience, Univ College London, UK), claims that "the software is a natural consequence of the research that is based here [...] SPM is a software but, in a sense, it's just a tool for our research. This is a neuroscience lab, it's not a computer science lab [...]."

Once again, such denials belong to an expressive effort of self-display. If the software development facet of those researchers' career were discursively put forward, their scientific reputation would be jeopardized. They understand the rationale of their institutions, knowing that they should foreground the image of scientists formulating innovative models and theories, not the image of technicians dealing with pure methodological issues. In neuroimaging, the most prestigious journals are not the ones devoted to software development. So researchers are officially urged to create new theories and models, not new computer code, thus getting access to key journals in this period when the "publish or perish" gospel is so widely proclaimed.

> "This kind of [software] development is very beautiful, visually. But in computational terms (which is something very relevant for publications and so on), it's rather simple [...] What is difficult to publish and has little impact, so to say, is all these user interfaces. All these tricks that makes things easy for the user are not novel. This is not interpreted as novel. What do you want in a publication? Novelty, right? And this is interpreted as something beautiful [...] It is not like, for example, a guy who is developing an algorithm that detects brain lesions automatically [...] Perhaps this algorithm is as complex as mine, but it has novelty [...]
>
> **But does that mean that your work... that it will be more difficult for you to make your work be recognized as scientific?**
> Right. Right. Right. Right.
>
> ***People will see it as something artistic or aesthetical, perhaps.***
> *Right [...]."*
>
> Dr. Victor Hugo Souza (Laboratory for Biomagnetism and Neuronavigation, Univ of Sao Paulo, Brazil)

In the academic logic, being productive is frequently different from producing beautiful things. From this viewpoint, a different value is attributed to beauty, whether we deal with the beautiful coherence of code or the beautiful graphical interfaces designed by programmers. From a scientific and programming achievement, beauty threatens to become a curse, as it is devalued in academic settings marked by pragmatic and bibliometric rationales. Arguably, it is by considering these issues, among other things, that Professor Christian Beckmann claimed: "I'm not claiming [laughter] that the code I'm writing is particularly nice [...] It was written to do a job. So I'm not spending much time or effort on prettifying my coding."

The discursive interface produced by neuroimaging software developers is efficient because it shrewdly explores the tension between facticity and validity. If deep software engineering knowledge is available, then the person would proudly point to the elegance of code produced and the ability to write code in line with the facticity of programming syntax. However, if such knowledge is lacking, then the person would stress the social contributions of the clumsy code, the scientific novelty it incorporates, and the creativity of somebody who plays with the validity enabled by programming.

In this way, the discourses voiced by programmers are underpinned by a series of implicit values and norms. They can draw attention to their capacity of producing a codebase that is beautiful and easy to be read by interested people, thus disseminating useful knowledge. In parallel, they can stress their capacity of producing efficient software, which can be used to understand the human brain, the human behaviour, and associated diseases. Therefore, it is not possible to claim that those discourses focus on technical matters only. "This conception is based on the unrealistic assumption that one can separate the professional knowledge of specialists from values and moral points of view. As soon as specialized knowledge is brought to politically relevant problems, its unavoidably normative character becomes apparent [...]" (Habermas 1996, p. 361). If expressive self-display through computer code is possible, then it is worth asking how programmers frame the "other person," the one who is going to see their code. If issues of identity can be identified, what happens to issues of alterity?

section9 [Human Community, Developer Community] {

9.1. The "Other Face" of Programming

This chapterThree focuses on the creative and expressive contents of software development, hence its emphasis on the solitary work of programmers. Nevertheless, those who design software need to be aware of other agents, in addition to comprehending the contexts where they work. This includes the knowledge of: how software design and computer languages evolve; the conditions in which research is funded, conducted, and published in the academic environment; opportunities and practices of collaborative software design; the role played by companies; and so on. Eventually, all these notions are combined to form the ideological framework of each particular programmer. For those who adhere to the spirit of open source, for example, a strong preference for all things open is manifested. "Such a desire often leads to distrust of commercial firms in the community [...]" (Wagstrom 2009, p. 180). The most significant historical example is surely David Stallman, founder of the Free Software Foundation, who firmly attacks the existence of copyrights for software and claims that boundless distribution of software has to be fostered because "[...] above all society needs to encourage the spirit of voluntary cooperation in its citizens" (Stallman 2002c, p. 48).

Ideological motives have at least a considerable relevance for people developing software in universities, where the so-called "public" nature of the institution tends to be taken into account. However, when the whole context is carefully analysed, it is possible to realize that public and altruistic motives are sometimes rhetorical tools used to build the image of responsible programmers, as explained in section8. As previously noted (von Hippel and von Krogh 2003), participants of open source communities are sometimes attracted by only the projects' technical features, not by ideological motives. On this point, it is interesting to consider the account given by Linus Torvalds, the founder of the most successful open source project ever seen: the Linux operating system. Torvalds recognizes that "[...] one of the reasons that the open source philosophy and Linux both have major followings in universities is simple: the antiestablishment sentiment" (Torvalds and Diamond 2001, p. 161). Indeed, many academics are willing to support a project which challenges the dominance of large corporations such as Microsoft. However, if this ideology is efficient to secure some diffusion to Linux, the motives leading to its development in the first place had little to do with ideology. According to Torvalds, he initially published the code of Linux just because by doing so, he would attract collaborators to help him design a technically strong package. "I didn't open source Linux for [...] lofty reasons. I wanted feedback" (p. 194). Ideological champions of open source software would probably feel disappointed in hearing Torvalds claim that "[...] open source wasn't conceived in order to detonate the software establishment. It is there to produce the best technology, and to see where it goes" (p. 228).

Torvalds' words have some relation with the concept of instrumental action, which is precisely "[...] an action oriented towards success whenever we consider it from the point of view of the pursuit of technical rules of action and we assess the degree of efficiency of a given intervention in a context of states of things and events" (Habermas 1987, p. 295). One of the main characteristics of instrumental actions is their practical or purposive nature. In software development, what precisely indicates the presence of such actions?

Lakhani and Wolf (2005, pp. 13–14) conducted a survey with over 600 developers involved in open source projects. The majority of participants (58.7%) declared that they participate in open source because they need code for either work or non-work reasons; 44.9% said that they feel personally stimulated; 41.3% declared to be willing to enhance their programming skills; and 33.1% "believe that source code should be open." Only this last reason could be categorized as altruistic or communitarian. According to the authors, the literature shows that programmers are generally motivated by pragmatic or even material reasons, such as career advancement or the acquisition of knowledge so as to find a better job in the future. In their study, the prevalence of such practical motives was not confirmed, but the motivation they point to as the main one, "namely, how creative a person feels when working on the project" (Lakhani and Wolf 2005, p. 3), continues to lack a truly communitarian content. When I speak of instrumental actions in software development, I am referring, among other things, to the fact that many programmers, at many occasions, are mainly moved by technical purposes, instead of being committed to collective causes.

As explained in section8, beauty of code is frequently associated with its clarity. Elegant code is the one whose governing ideas show through smoothly, as it can be easily read and understood. If a considerable proportion of developers manifest some concerns with beauty and readability, who is then the reader that will come along to scrutinize the coding job previously done? The immediate and perhaps surprising response to this question is: the future code reader, for the sake of whom nice code is produced, is precisely the present code writer. For example, Dr. Brunno Campos (Laboratory of Neuroimaging, Univ of Campinas, Brazil) claims that in writing clear code, he thinks about "the future Brunno." I asked Dr. Nathalia Esper (Laboratory for Images-Labima, Pontifical Catholic Univ of Rio Grande do Sul, Brazil) if she produces an organized code for herself to easily read it in the future. "Yes, honestly, I think of me, because I won't remember afterwards." And a researcher based in King's College London, UK, asked why he tries to always produce clear code, responded: "For my own benefit."

In fact, computer code can be so intricate and large that its original writer may struggle to understand its logic after spending some days, or even some hours, away from that coding production. Therefore, the self-centred concerns described above are understandable. What needs to be pointed out here is a very important shift. At an initial moment, as explained above, programmers are to a great degree concerned with current and practical issues like funding, career advancement, or "the race to win the reputational prize" (Atal and Shankar 2015, p. 1395). At a later moment, when code is actually being produced, the focus of concerns begins to fall onto virtual and imagined issues like the possible troubles one might face when reading the code in the future, or the future code's efficiency. From concrete concerns, one goes towards the consideration of more abstract ones. In this way, in addition to being a concrete production that evolves before the programmer's eyes, code begins to also be the future entity that will be read and analysed by an imagined, abstract self, in the possible conditions of an imagined future.

> **"But is there any specific reason why your code is elegant?**
> Ahm, mainly because [...] I have always considered reusability very important, but also because I tend to look at code at a more conceptual level [...] So I always try to think of a new function as... immediately as something that can be used and reused in a more general sense [...]
>
> **Hm. So when you say reusability, you say that at some point in the future you can use it again.**
> Yes, absolutely.
>
> **Okay. Hm. You're thinking...**
> But also you know at some time in the future someone that uses it, from the previous pipeline that they've got... I want them to be able to just look into that function, what it does... and see what it does. And that someone might just be my future self or it might be you or someone else. It doesn't matter. I want anyone to be able to look at it and just be able to introspect the code without being sent, well, in one big spaghetti mess from one side to the next and find out that you went full circle and didn't progress in any way."
>
> Dr. Tim van Mourik (MR Physics Group, Donders Institute for Cognitive Neuroimaging, Netherlands)

This phenomenon through which the self becomes a virtual entity has much sociological import, because this shift can rapidly encompass other social agents. In this way, besides considering the future presence of an imagined self, the programmer also takes into account the presence of an imagined other, a potential reader of the code being worked on. This new shift goes with no discursive sacrifice. As one is always speaking of an imagined person occupying a future position in time, that person can easily be the speaker or anybody else. Practical issues have now been stripped away for the sake of unspecific reasons (such as the need for understanding) that, precisely because of their non-specificity, can be attributed to anybody. In his interview, a researcher based in the University College London, UK, recalled that, at his undergraduate degree, it was said that code should be written for at least three people: "For yourself; your future self, who's forgotten what you did; and somebody else."

That "somebody else" is the entity that Brooks (1995) described as "the other face" of programming. It is an abstract other that emerges as an extension of an abstract self. The appearance of such abstract figures is one of the main phenomena marking the passage from life in small communities cemented by personal bounds to life in society where solidarity becomes dependent on more abstract creations (Durkheim 1925, 1932; Tönnies 1955). Georg Simmel profoundly examined this issue, explaining that in modern social life, relations cease to be the interplay between concrete people with specific features to become abstract connections between people playing standardized roles. These abstract contents of social life are strengthened by the growth of cities (Simmel 1950 [1903]) and the diffusion of the money economy (Simmel 1997 [1900]). From a communicative perspective, this phenomenon can be grasped with a pair of concepts: the we-orientation (in which the actor is in contact with a specific other) and the they-orientation (in which the actor consider the existence of nonspecific people). "The referenced point of the they-orientation is a type of the conscious processes of typical contemporaries. It is not the factual existence of a concretely and immediately experienced alter ego, not his conscious life together with this subjective, step-by-step, constituted meaning-contexts. The reference point of the they-orientation is inferred from my knowledge and form the social world in general, and is necessarily in an Objective meaning-context" (Schutz 1974, p. 75).

> **"Do you know C++ too?**
> *Ah, I know that it exists. I don't have any program experience in it [...] for scientific programming, you usually don't need that kind of optimization where you work with pointers and all the efficient coding. So you can get away, most of the times, with things like Python, I think [...] So you can quickly develop your code and have a more readable code, and more people... They are not pure computer scientists but they're scientists; they can jump into the project and work with you [...]."*
>
> Dr. Marcel Swiers (MR Physics group, Donders Institute for Brain, Cognition and Behaviour, Cognition and Behaviour, Netherlands)

In software development, that abstract other emerges as a future code reader. Programmers generally express their consideration for that virtual alterity by means of expressions such as "people who are not pure computer scientists" or "somebody who has perhaps never used MatLab." It is by considering that unknown figure that hackers try to make their intentions transparent in their code. At this moment, the "conversation" between the programmer and the computer is made somewhat more complex, as a third character

> *"That is, your code is not a very organized and clear code for other people to read it so. Well, it depends on the case. It depends on the case.*
>
> *Okay, sometimes yes, sometimes not.*
> *Err... I mean, when it comes to the tasks of my studies, my projects and my analyses, it is obvious that the code is more for me. But I do some analyses with data from other projects, and I try to structure the code in such a way that somebody who has perhaps never used MatLab can open the code, the MatLab interface, run the code, and understand what the code is doing."*
>
> Researcher (Univ of Coimbra, Portugal)

(the future code reader) is invited to join in. Thus code writing reinforces the communicative dimensions of science, for science can also be framed as "a system of communication" (Merton 1968, p. 59). In this sense, code surely becomes a new type of scientific publication. "[...] for science to be advanced, it is not enough that fruitful ideas be originated or new experiments developed or new problems formulated or new methods instituted. The innovations must be effectively communicated to others" (Merton 1968, p. 59). At this point, it is possible to be more precise and analyse the coding tactics used by programmers wishing to produce clear code.

9.2. Self-Explanation and Documentation

In her interview, a Brazilian researcher explained that today's computers do not essentially differ from the ones originally built in the 1940s, because the fundamental operations conceived of by Hungarian-American computer scientist John von Neumann are still in use. In this logic, computers realize "the sequential execution of a limited number of operations" (Friedman 1992, p. 14). Hence the organization of code in lines, which are "read" by the machine like a normal text, from top to bottom and from left to right. As a consequence of such logic, the whole codebase of a software package turns out to be very complex and code readability ends up relying, to a very large degree, "[...] on the programmer's ability to completely specify in detail precisely how the computing is to be done" (Friedman 1992, p. 14).

Not many programmers are skilful to the point of being capable to repeat what Linus Torvalds declared: "My code is always, um, perfect" (Torvalds and Diamond 2001, p. 44). Frequently, code results from a messy process through which the programmer struggles to produce the desired visual and numerical outputs. Modifications made in order to tackle a certain bug may generate other sorts of bugs, in a convoluted story where the programmer can even lose track of thoughts and solutions. This work process can be highly confusing, as pointed out by Parnas and Clements (1986), but the final source code will ideally be something reasonable to potential readers. Moreover, it will serve as an explanation, insofar as it will reveal the logic that made it possible to turn confused, buggy code into harmonious, functioning software. This pedagogic nature of properly written code is made possible by documentation. In the software engineering jargon, documentation is a series of steps taken to make software and computer code comprehensible. There are two types of documentation, which I propose to classify as "weak" and "strong."

Programmers usually prepare some texts and files (written in normal everyday language) explaining how the software package can be installed, how it works, what the main ideas are behind its development, how to deal with operational problems, and so on. This weak documentation serves as a manual with instructions to software users, as well as to other developers intent on modifying the source code. In this way, it generates the impression of an orderly and rationally designed package, and that is why Parnas and Clements (1986, p. 254) argue that documentation produces a "'fake' rationality."

In its turn, what I call strong documentation is the explanation incorporated into the source code itself. It is a series of coding strategies aimed to enhance the readability of code. Let us consider only three of such strategies. First, the computer reads only the characters written by the programmer: blank lines and spaces are simply ignored. Thus the programmer can use blank spaces and indentations so as to improve the outlook of the text, making it visually agreeable to the reader. Second, the programmer is free to choose the names of variables. A variable created to hold a numerical value pertaining to the size of the brain will be more quickly understood if it is called "brainSize" than if it is simply called "x." Finally, strong documentation will be even stronger whenever the code is carefully separated into meaningful classes. In this way, the programmer can create a class dealing with statistical calculations, another class for the visual representation of the brain, another class for conversions between different file formats, and so on.

According to programmers, and especially those with more advanced programming skills, strong documentation is what really conducts to the production of self-explicative pieces of code. Anyway, both forms of documentation help source code turn into one of the "variety of communicative forms" pointed to by Galloway and Thacker (2006, p. 26). Coming back to the "theory building view" of software, proposed by Naur (2001, p. 230), documentation manifests the "[...] knowledge a person must have in order not only to do certain things intelligently but also to explain them, to answer queries about them, to argue about them, and so forth." These clarifying tasks are surely performed through code itself, by means of strong documentation. Hence the presence of some explicative concerns like the ones expressed by Matthew Brett [mthw, brtt].

[mthw, brtt]

Matthew Brett's (School of Biosciences, Univ of Birmingham, UK) story begins like that of many hackers who at an early age felt an attraction towards computers. When he was 15 years old, he had some basic classes in programming at school. Thereafter he began to spend all his school breaks writing code at the computer installed in the school, the only one he had access to. At university, he studied medicine with specialization in neurology. At the Master's Degree level, he conducted a neuroscience study and began to mix up his main interests: brain studies and programming. In 2002 he released MarsBar, a neuroimaging application that would become very popular and frequently cited in the literature. In 2007 he had a research stay in the United States, at the University of California, Berkeley, one of the cradles of the open source software movement. He assimilated the core notions of the movement and eventually become the leader of the NYPIPE community, a group of programmers who use the Python language to develop neuroimaging software.[4]

"I'm not interested enough in code for its own sake [...]

So programming is a means, not the end.
I think programming is a way to express my work, yeah. And that is the way I think of teaching as well. Programming is just expressing the logic of what I'm doing. It's not an end in itself. It's just a mechanism that expresses the logic.

Is it an expression like language? Like verbal language?
[...] I've always written code as if I was sort of showing it to somebody else. So... Err... I think of code as kind of teaching [...] So when I'm writing code, I'm sort of working out a problem. And in working out that problem, I try to explain it to other people. So I'm always expecting that somebody is going to read what I did and try to retrace the logical steps that I was going through. And I hope that they will find it easier having read what I did. So it's kind of way of teaching, I guess [...]

So do you think that your code, in a way, represents you?
Yeah. I mean... I think whenever you read somebody's code, you get a feeling for how they think [...] So when I said I'm expressing myself, I don't mean like you have an idea of myself. It is just like: what I'd like to express is simplicity and clarity. So I'm absent from that. If I succeed, I'm absent from that. All you do is you think: 'Yes, I get that idea.' You don't think: 'I'm talking to Matthew Brett.' You are thinking: 'Yes, I get that idea' [...]."

In Cyril Pernet's (Centre for Clinical Brain Sciences, Univ of Edinburgh, UK) words: "[...] you learn a lot by taking other people's code [...] So if the code is very well-structured and transparent, it's fast to learn and you also learn the good practices because it looks good and it's well-made as well." Logic and readable code is crucial for avoiding the "design breakdowns" pointed to by Heliades and Edmonds (1999, p. 395), in which software design is interrupted due to the reader's inability to understand the rationale behind previously written code.

---

[4]Matthew Brett wrote the foreword to this book.

At this point, some consideration must be given to the fact that hackers write programs by using so-called languages, like Python or Java. In the 1950s and 1960s, hence at a very early time in the history of computing, "[...] it came to seem natural to think of programming highly complex electronic devices as a linguistic activity [...]" (Nofre et al. 2014, p. 41). Moreover, pivotal people advocated the creation of programming languages that could be more similar to natural languages and less similar to programming. For instance, in 1958 American numerical analyst John Weber Carr, one of the first champions of computing, declared: "If humans are to keep up with the voracious input capacity of the digital computers, then languages built for the humans, and not the machines, must be developed and put into operation" (apud Nofre et al. 2014, p. 55). Across the decades, these concerns have become so pressing that programming languages have actually incorporated elements that foster clarity in code writing, such as the use of commands derived from words taken from everyday language (see the Empirical Analysis at the end of this chapterThree).

As a result, today's languages do provide programmers with some resources conducive to transparency. In his interview, Dr. Guillaume Flandin (Wellcome Trust Centre for Neuroscience, Univ College London, UK) explained why the MatLab language was chosen for developing the SPM neuroimaging package. "This type of languages [...], at the end the code is quite readable. If you really want to follow out this equation we're implementing in the code, you can actually find sometimes the link between the name in the equation (the variable name) and the variable name in the code." And Dr. André Peres (Brain Institute, Federal Univ of Rio Grande do Norte, Brazil) explains that one of the reasons leading to the increasing popularity of Python is that this language has syntactic rules to enforce code clarity. In addition, he uses PEP 8,[5] an application which functions as a sort of automatic code reviewer, generating warning messages whenever code is not clearly written.

Nevertheless, much care is needed here in order not to succumb to the metaphor and conclude that programmers, by means of their

---

[5]https://www.python.org/dev/peps/pep-0008/.

code, are communicating in the same way people do when they speak English, Portuguese, or Mandarin. Everyday language, or "natural language" (Habermas 1996, p. 360), is the primary resource through which people resolve tensions produced by the existence of "a community segmented along the divisions between competing worldviews" (Habermas 2008, p. 136). For so doing, language operates at a level that includes values and beliefs. Although computer code serves to communicate some ideas and technical projects, it still lacks normative and ideological contents, thus lacking the coordinating effects possessed by natural language. In this sense, code holds some communicative traits but can, at its best, be considered as only a proto-language.

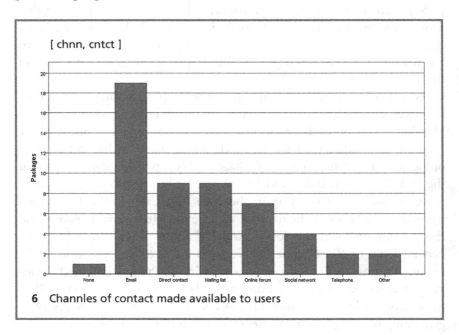

6   Channles of contact made available to users

However, it should not be considered, by any means, that software development is devoid of communicative aspects. In addition to the explanations given through code itself, clarifying tasks are performed in a direct way, as many software developers provide users with communication means to send queries. In neuroimaging, as verified in Survey 2, programming teams offer a variety of channels through which users and other developers can ask several types of questions about the package's features, its installation, its source code, and so on [chnn, cntct]. Among these channels (which include direct contacts in meetings and conferences), the email has been the preferred one. The "abstract other" of computer code gains some concreteness as soon as some means of communication is made open to potential queries. In addition, a series of interpersonal contacts are in place, allowing programmers to negotiate meanings and purposes. Because these contacts were analysed in chapter-Two, let us now return to the creative contents of software development and show how the idea of community can emerge.

9.3. The Community of Programmers

At this point, a new kind of cognitive shift can be detected. If the abstract identity (the "myself" who reads the code in the future) has become the abstract alterity (the "somebody" who inherits the code), then this imagined person can easily be framed as not just one reader but a large group of readers. The production of this abstract crowd is an expected consequence of the virtual dimension of social life, which follows the consideration of present and concrete needs and relations. We are dealing with the passage from the "immediate experience of the Other" to the "mediate experience of the social world" (Schutz 1974, p. 73).

> *"You are developing a software package that is open. The code is open.*
> Yes.
>
> *Do you think that the development process would be any different if it was not open?*
> [...] in this project we have this ideology, we defend the open science ideology [...] we would like our software to be used outside Neuromat [the research network within which the software package is developed]*

In this trajectory of this language-game, only a final step needs to be taken in order for the idea of community to appear: the assumption that the programmer and all those virtual code readers form a group bounded

> too [...] we don't want it to be a Neuromat software only. We want it to be useful for the neuroscience community in general [...] Whenever we include a new functionality, we say: 'Is this something that makes sense only within Neuromat?' And whenever we think it is so, we change it and say: 'No, let's broaden it, let's do it in a way that it will be used by more people' [...]."
>
> Dr. Kelly Braghetto (Dept Computer Science, Univ of Sao Paulo, Brazil)

by common goals and norms. In software development, and especially in open source, this idea has become very potent, hence the presence of many programmers and analysts pointing to the formation of a "developer community" (Atal and Shankar 2015, p. 1403), a "virtual community" (Torvalds and Diamond 2001, p. 166) or an "Open Source community" (Wagstrom 2009, p. 181).

From a sociological point of view, this community is characterized by three main features. First, it constitutes a space for potential collaboration. The idea of a group of developers hanging together at a distance functions as a promise that some concrete journals, meetings, conferences, and other initiatives will be generated through which people will come in touch and engage in actional collaborative work.[6] Second, as indicated by previous studies (Lindenberg 2001; Lakhani and Wolf 2005), programmers are more motivated to participate in development projects when they feel to belong to a "hacker community" (Lakhani and Wolf 2005, p. 14). Finally, the community may turn into a political reference, canalizing collective sentiments towards an increasing sense of belongingness, and against players seen as threats to the community's integrity. In this vein, Wagstrom (2009, pp. 184–185) formulated the following recommendation: "*When building an Open Source community, give the community control over the community.*" Companies are sometimes framed as potential foes, because their proprietary mechanisms (such as copyrights and patents) are at odds with the open source philosophy. Therefore, software development, in spite of frequently being realized by people working in isolation, have

---

[6]The distinction between the "actional realm" and the "virtual realm" is presented in chapterFive.

strong collective roots. One of programmers' central cognitive and intellectual references is the existence of a community from which they receive either imagined or concrete support, incentive, and inspiration.

However, it would be imprecise to portray software developers as fully other-centred, altruistic, and politicized players. As claimed in the beginning of this chapterThree, they are, to a large degree, moved by pragmatic, personal, and technical reasons. A survey by Ghosh (2005), with the participation of over 2700 open software developers, identified that 55.7% of them declared to receive from the community more than they give, whereas only 9% perceive that they give more than they take. "Thus, the responses are consistent with self-interested participation [...]" (Ghosh 2005, pp. 33–34).

So once again, it is possible to identify the presence, in neuroimaging software development, of what I previously called mediational actions (Bicudo 2012). On the one hand, researchers comply with instrumental requisites and seek practical targets, including career enhancement, acquisition of scientific prestige and research funding, technical achievements, and so on. On the other hand, they are frequently inspired by the image of a large, and even international, community from which they get coding support and to which they may recur to initiate concrete coding collaborations.

> *"For research, wouldn't it be interesting to use closed software?*
> No. In my opinion, it wouldn't. Because I wouldn't have the acceptability I have with Slicer [the software package called 3D Slicer, used in this research group], with a huge community [...] Another interesting aspect of Slicer is that it is as attractive for the open source community as it is for the industry. Why? Because it uses a license [...] that allows you to appropriate the software, make modifications, and you can even commercialize it. This has attracted the interest of the software industry as well."*
>
> Professor Luiz Murta Junior (Computational Group for Medical Signals and Images, Univ of Sao Paulo, Brazil)

This confirms the conclusions of a study by Lakhani and Wolf (2005), which identified an interplay between extrinsic motivations (related to practical and monetary aspects) and intrinsic motivations (related to the joy of participating in creative projects). This is also in accordance with the hybrid nature gained by open source software, which since the late 1990s, has been increasingly supported by private companies (Schwarz and Takhteyev 2010; Wagstrom 2009). Acting in a domain configured by a whole set of technical standards, but at the same time permeated by normative elements, software developers need to oscillate between the instrumental and the communicative rationalities, extracting from both the resources and tools needed for the proper conduct of their tasks.

section10 Empirical Example: FORTRAN, ALGOL, and COBOL

The social diffusion of computer programming (with an increasingly large number of people interested in this craft) was greatly reinforced when programming languages began to assimilate some features of everyday language (Skog 2003). In the 1940s and 1950s, coding was always performed in machine language, also called binary language: tedious and counter-intuitive arrays of 0s and 1s that programmers had to master. Not surprisingly, then, "[…] programming was relatively esoteric and poorly understood and was considered by many to be something of a 'black art'" (Friedman 1992, p. 9). However, computer engineers sought, from very early times, ways to render their work more intuitive. For example, Hungarian-American computer scientist John von Neumann, who helped to establish the bases of computing in the 1940s, proposed keyboard shortcuts that could be used instead of binary commands (Skog 2003).

In the early 1950s computer scientists created the so-called pseudocodes, which contained some characters other than just 0s and 1s. However, when programmers wanted to transport pseudocodes written in a computer to a different type of computer, the code had to be translated into machine code. In 1951 American computer scientist Grace Murray Hopper proposed the notion of compiling routine; she was referring to an automated method enabling to transform pseudocode into machine code. In the mid-1950s some coding systems were released containing some commands taken from the English language. For the first time, computer coding borrowed some elements from natural language. What needs to be stressed is that, in this slow process, code was becoming somewhat more transparent, insofar as the logic of the code writer could be transmitted to the code reader more speedily.

In this metamorphosis underwent by computer languages, from a purely technical code to the state of proto-language, a remarkable event took place in the mid-1950s, when John Backus and his team, working for IBM, released FORTRAN. The greatest novelty of this programming language was that it contained a considerable range of commands derived from everyday English words, thus inaugurating the era of so-called interpretive languages. Significantly, FORTRAN, which can in fact be considered as the first actual programming language, took its acronym from the expression FORmula TRANSlator. It became possible to *translate,* in an innovatively efficient way, machine code into something readable and understandable by people, and vice-versa. This was possible because the FORTRAN package included a so-called compiler, a cutting-edge application made responsible for the translation. Accurately, Skog (2003, p. 300) remarked that FORTRAN was "the first programming language in which the human or problem domain was the prime focus." However, FORTRAN was initially designed for a specific machine, the IBM 704 computer. To write FORTRAN code in other computers, programmers had to first develop a new compiler for the targeted machine. "[…] by 1964 there were more than 40 different FORTRAN compilers on the market" (Friedman 1992, p. 7).

It was seen that a new language was in need, one that could reproduce the advantages brought about by FORTRAN, but one that could get rid of its technical limitations. Definitely, a major challenge was the production of a language to be used in different computers. At those times, computers of different types and brands were proliferating, causing problems of portability from machine to machine. In the United States, those problems were particularly tricky for the aircraft industry and the military, which were becoming more and more dependent on computers (Nofre et al. 2014; Ensmenger 2010). In 1958, computing and mathematics organizations from Europe and the United States had meetings in Zurich. In the same year, a new language was released, named ALGOL (for Algorithmic Language). Born into the family of interpretive languages, ALGOL, like FORTRAN, made considerable use of words taken from English. Furthermore, its syntactic rules were decisively shaped by the rules of logic (Priestley 2011).

In 1959 the American Department of Defence organized a different series of meetings, with participation of the software and hardware industry, academics, and representatives of government agencies. The goal was to elaborate yet another language that would be particularly suitable for data analysis purposes. COBOL (COmmon Business-Oriented Language) came into existence in 1960. One of its greatest achievements was its portability, as it could be easily transported from one type of computer to another, with no technical hardships. The impacts of COBOL were rapidly felt among programmers. In addition to its portability, COBOL provided programmers with an attractive content of transparency, as its source code was open. As explained by Schwarz and Takhteyev (2010), IBM, surprised by the quick popularity gained by COBOL, was forced to also open the source code of FORTRAN.

With FORTRAN, ALGOL, COBOL, and subsequent languages, all developed under the same intuitive principles, computer programming became somewhat more popular while assimilating a structure closer to people's everyday reasonings. Let us imagine, for example, that a certain program contains a variable called "var" whose initial value is 1. The programmer wants to increase this value by 1 and continue to add 1 as long as the variable is less than 101. In the FORTRAN syntax, this instruction would be passed to the computer in the following manner:

```
var = 1
do while (var < 101)
     var = var + 1
end do
```

In this extremely simple example, one sees three commands formed by English words: "end," "do," and "while." With only four lines of code, it is perhaps not possible to note that these natural words become a sort of visual reference for the code reader. For example, when the "while" word is seen, it is rapidly understood that something is going on which will stop when a certain condition has been met, which is precisely the idea conveyed by the word "while." In this way, the expressive, externalizing act described in this chapterThree, through which the programmer's logic is *communicated* to all those who inherit the code, is facilitated.

These menial miracles of interpretive languages may look too menial and not sufficiently miraculous. However, they have been one of the technical pillars of the whole set of collaborative schemes described in chapterTwo. They underpin the externalization of thought, the coordinating tricks of programming, the possibility of self-display through code, and the idea of developer communities. Through these processes, the collaborative relations and code sharing practices to which programmers have become so accustomed gain momentum. Thus the creative and expressive dimensions analysed in this chapterThree have not to do with any transcendent realm but are the basis of concentre communicative deeds. However, all these communicative phenomena have not completely deprived code writing from its dark sides and scientific imbalances, hence the persistence, in software development, of theoretical inequalities, political hierarchies, and automatic behaviours, as analysed in the 2ndPart of this book.

References

Aguirre, Geofrey K. 2012. "FIASCO, VoxBo, and MEDx: Behind the code." *NeuroImage* 62:765–767.

Amabile, T. M. 1996. *Creativity in context*. Boulder: Westview.

Arendt, Hannah. 1978. *The life of the mind: One—Thinking*. London: Secker & Warburg.

Ashburner, John. 2012. "SPM: A history." *NeuroImage* 62:791–800.

Atal, Vidya, and Kameshwari Shankar. 2015. "Developers' incentives and open-source software licensing: GPL vs BSD." *B.E. Journal of Economic Analysis and Policy* 15 (3):1381–1416.

Beltrame, Francesco, and Stephen H. Koslow. 1999. "Neuroinformatics as a megascience issue." *IEEE Transactions on Information Technology in Biomedicine* 3 (3):239–240.

Beniger, James R. 1986. *The control revolution: Technological and economic origins of the information society*. Cambridge: Harvard University Press.

Berezsky, Oleh, Grigoriy Melnyk, and Yuriy Batko. 2008. "Modern trends in biomedical image analysis system design." In *Biomedical engineering: Trends in electronics, communications and software*, edited by Anthony N. Laskovski, 461–480. Rijeka: InTech.

Bicudo, Edison. 2012. "Globalization and ideology: Ethics committees and global clinical trials in South Africa and Brazil." PhD thesis, Department of Political Economy, King's College London. Available at https://kclpure. kcl.ac.uk/portal/en/theses/globalization-and-ideology-ethics-committees-and-global-clinical-trials-in-south-africa-and-brazil(ab94b2ce-b023-4ab 3-9dd4-93616db5c455).html.

Bicudo, Edison. 2014. *Pharmaceutical research, democracy and conspiracy: International clinical trials in local medical institutions.* London: Gower/ Routledge.

Brooks, Frederick P. 1995. *The mythical man-month.* Reading: Addison-Wesley.

Deci, E. L., and R. M. Ryan. 1985. *Intrinsic motivation and self-determination in human behavior.* New York: Plenum.

Dummet, Michael. 1993. *The seas of language.* Oxford: Oxford University Press.

Durkheim, Émile. 1925. *Les formes élémentaires de la vie religieuse.* Paris: Félix Alcan.

Durkheim, Émile. 1932. *De la division du travail social.* Paris: Félix Alcan.

Ensmenger, Nathan. 2010. *The computer boys take over: Computers, programmers, and the politics of technical expertise.* Cambridge and London: MIT Press.

Fazi, Beatrice. 2018. *Contingent computation: abstraction, experience, and indeterminacy in computational aesthetics.* Lanham: Rowman & Littlefield.

Frey, B. 1997. *Not just for the money: An economic theory of personal motivation.* Brookfield: Edward Elgar.

Friedman, Linda Weiser. 1992. "From Babbage to Babel and beyond: A brief history of programming languages." *Computer Languages* 17 (1):1–17.

Galloway, Alexander R., and Eugene Thacker. 2006. "Language, life, code." *Architectural Design* 76 (5):26–29.

Galloway, Patricia. 2012. "Playpens for mind children: Continuities in the practice of programming." *Information & Culture* 47 (1):38–78.

Geertz, Clifford. 1973. *The interpretation of cultures: Selected essays.* New York: Basic Books.

Ghosh, Rishab Aiyer. 2005. "Understanding free software developers: Findings from the FLOSS study." In *Perspectives on free and open source software*, edited by Joseph Feller, Brian Fitzgerald, Scott A. Hissam, and Karim R. Lakhani, 23–46. Cambridge: MIT Press.

Habermas, Jürgen. 1984. *The theory of communicative action, vol. 1: Reason and the rationalization of society.* Boston: Beacon Press.

Habermas, Jürgen. 1987. *The theory of communicative action, vol. 2: Lifeworld and system.* Cambridge: Polity Press.

Habermas, Jürgen. 1993. *Justification and application: Remarks on discourse ethics.* Cambridge: Polity Press.

Habermas, Jürgen. 1996. *Between facts and norms: Contributions to a discourse theory of law and democracy,* Studies in contemporary German social thought. Cambridge, MA: MIT Press.

Habermas, Jürgen. 2008. *Between naturalism and religion*. Cambridge: Polity Press.

Heliades, G. P., and E. A. Edmonds. 1999. "On facilitating knowledge transfer in software design." *Knowledge-Based Systems* 12:391–395.

Himanen, Pekka. 2001. *The hacker ethic: And the spirit of the information age*. New York: Randon House.

Hood, Leroy. 1990. No: And anyway, the HGP isn't 'big science'. *The Scientist*. Available at http://www.the-scientist.com/?articles.view/articleNo/11452/title/No–And-Anyway–The-HGP-Isn-t–Big-Science.

Kitchin, Rob. 2017. "Thinking critically about and researching algorithms." *Information, Communication & Society* 20 (1):14–29.

Kohanski, Daniel. 1998. *Moth in the machine: The power and perils of programming*. New York: St. Martin's Griffin.

LaToza, Thomas D., Gina Venolia, and Robert DeLine. 2006. "Maintaining mental models: A study of developer work habits." Proceedings of the 28th International Conference on Software Engineering, New York, USA.

Lewis, Jamie, and Andrew Bartlett. 2013. "Inscribing a discipline: Tensions in the field of bioinformatics." *New Genetics and Society* 32 (3):243–263.

Lindenberg, S. 2001. "Intrinsic motivation in a new light." *Kyklos* 54 (2/3):317–342.

Mackenzie, Adrian. 2003. "The problem of computer code: Leviathan or common power?" Available at http://www.lancaster.ac.uk/staff/mackenza/papers/code-leviathan.pdf.

Mackenzie, Adrian. 2006. *Cutting code: Software and sociality*. New York: Peter Lang.

Merton, Robert K. 1968. "The Matthew effect in science." *Science* 159 (3810):56–63.

Moran, T. P., and J. M. Carroll. 1997. "Overview of design rationale." In *Design rationale: Concepts, techniques and use*, edited by T. P. Moran and J. M. Carroll. Hillsdale: Lawrence Erlbaum Associates.

Naur, Peter. 2001. "Programming as theory building." In *Agile software development*, edited by Alistair Cockburn, 227–239. Boston: Addison-Wesley.

Nofre, David, Mark Priestley, and Gerard Alberts. 2014. "When technology became language: The origins of the linguistic conception of computer programming, 1950–1960." *Technology and Culture* 55 (1):40–75.

Parnas, David Lorge, and Paul C. Clements. 1986. "A rational design process: How and why to fake it." *IEEE Transactions on Software Engineering* SE-12 (2):251–257.

Priestley, Mark. 2011. *A science of operations: Machines, logic and the invention of programming*. London: Springer.

Raymond, Eric S. 2001. *The cathedral & the bazaar: Musings on Linux and open source by an accidental revolutionary*. Sebastopol: O'Reilly.

Ryan, R. M., and E. L. Deci. 2000. "Instrinsic and extrinsic motivations: Classic definitions and new directions." *Contemporary Educational Psychology* 25:54–67.

Schutz, Alfred. 1974. *The structures of the life-world*. London: Heinemann.

Schwarz, Michael, and Yuri Takhteyev. 2010. "Half a century of public software institutions: Open source as a solution to hold-up problem." *Journal of Public Economic Theory* 12 (4):609–639.

Seaver, Nick. 2017. "Algorithms as culture: Some tactics for the ethnography of algorithmic systems." *Big Data & Society* 4 (2):1–12.

Sennett, Richard. 2008. *The craftsman*. New Haven: Yale University Press.

Shapin, Steven, and Simon Schaffer. 1985. *Leviatahn and the air-pump: Hobbes, Boyle, and the experimental life*. Princeton, NJ: Princeton University Press.

Simmel, Georg. 1950 [1903]. "The metropolis and mental life." In *The sociology of Georg Simmel*, edited by Kurt A. Wolff, 409–426. Glencoe: Free Press.

Simmel, Georg. 1997 [1900]. *The philosophy of money*. 2nd ed. London: Routledge.

Skog, Knut. 2003. "From binary strings to visual programming." In *History of Nordic computing*, edited by Janis Bubenko Jr., John Impagliazzo, and Arne Solvberg, 297–310. Boston: Springer.

Stallman, Richard M. 2002a. "The danger of software patents." In *Free software, free society: Selected essays of Richard M. Stallman*, edited by Joshua Gay, 95–112. Boston: GNU Press.

Stallman, Richard M. 2002b. "Releasing free software if you work at a university." In *Free software, free society: Selected essays of Richard M. Stallman*, edited by Joshua Gay, 61–62. Boston: GNU Press.

Stallman, Richard M. 2002c. "Why software should not have owners." In *Free software, free society: Selected essays of Richard M. Stallman*, edited by Joshua Gay, 45–50. Boston: GNU Press.

Stein, Lincoln D. 2010. "The case for cloud computing in genome informatics." *Genome Biology* 11 (207):1–7.

Tönnies, Ferdinand. 1955. *Community and association*. London: Routledge & Kegan Paul.

Torvalds, Linus, and David Diamond. 2001. *Just for fun: The story of an accidental revolutionary*. New York: HarperCollins.

von Hippel, Eric, and Georg von Krogh. 2003. "Open source software and the 'private-collective' innovation model: Issues for organization science." *Organization Science* 14 (2):209–223.

Wagstrom, Patrick Adam. 2009. "Vertical interaction in open software engineering communities." PhD, Carnegie Insitute of Technology/School of Computer Science, Carnegie Mellon University.

Weber, Steven. 2004. *The success of open source*. Cambridge: Harvard University Press.

Wittgenstein, Ludwig. 1922. *Tractatus logico-philosophicus*, International Library of Psychology, Philosophy and Scientific Method. London: Kegan Paul.

2ndPart

Codifying Society

This part aims at exploring the emergence and consolidation of hierarchies in software development. These hierarchies can manifest themselves at the international level (chapterFour Owning Code: Institutional Aspects of Software Development) and at a personal level (chapterFive Using Code: The Social Diffusion of Programming Tasks).

```
% social code
% source code
```

chapterFour (Owning Code: Institutional Aspects of Software Development) {

```
1  function out = spm_run_coreg_estwrite_MIBCA(varargin)
2  %SPM job execution function
3  %takes a harvested job data structure and call SPM functions
     to perform
4  %computations on the data.
5  %Input:
6  %job - harvested job data structure (see matlabbatch help)
7  %Output:
8  %out - computation results, usually a struct variable.
9  %_____
10 %Copyright (C) 2008 Wellcome Trust Centre for Neuroimaging
11
12 %$Id: spm_run_coreg_estwrite.m 4380 2011-07-05 11:27:12Z volkmar$
13
14 job = varargin{1};
15 if isempty(job.other{1})
16          job.other = {};
17 end
18
19 x=spm_coreg_MIBCA(char(job.ref), char(job.source),job.eoptions);
20
21 M=spm_matrix(x);
```

© The Author(s) 2019   **175**
E. Bicudo, *Neuroimaging, Software, and Communication*,
https://doi.org/10.1007/978-981-13-7060-1_4

```
22  PO = [job.source(:);job.other(:)];
23  MM = zeros(4,4,numel(PO));
24  for j=1:numel(PO),
25          MM(:,:,j) = spm_get_space(PO{j});
26  end
27  for j=1:numel(PO),
28          spm_get_space(PO{j}, M\MM(:,:,j));
29  end
30
31  P = char(job.ref{:],job.source{:],job.other{:]);
32  flags.mask = job.roptions.mask;
33  flags.mean = 0;
34  flags.interp = job.roptions.interp;
35  flags.which = 1;
36  flags.wrap = job.roptions.wrap;
37  flags.prefix = job.roptions.prefix;
38
39  spm_reslice(P,flags);

40  out.cfiles = PO;
41  out.M = M;
42  out.rfiles=cell(size(out.cfiles));
43  for i=1:numel(out.cfiles),
44          [pth,nam,ext,num]=spm_fileparts(out.cfiles{i});
45          out.rfiles{i} = fullfile(pth,[job.roptions.prefix, nam,
                ext, num]);
46  end;
47  return;
```

The piece of computer code presented above belongs to a software package called MIBCA (for Multimodal Imaging Brain Connectivity Analysis).[1] Its main producers are Dr. André Ribeiro and Dr. Luís Lacerda, who started the project as students of the Institute of Biophysics and Biomedical Engineering, University of Lisbon, Portugal. I would like to draw attention to seven lines in this code written in the MATLAB language: 12, 19, 21, 25, 28, 39, and 44. The reader will note that in all those lines, André and Luís used some functions whose names begin with "spm_". SPM is the world's most widely used neuroimaging software package. To be precise, MIBCA is not truly a software package; it is an application that allows researchers to combine several functionalities present in other open source packages such as FSL, FreeSurfer, and SPM. So in the aforementioned code lines, the authors wrote code to use some functionalities available in the SPM package. In the MatLab language, comments are introduced by the percentage symbol (%). In lines 3 and 4, it is explained that MIBCA takes some neuroimaging data and process them by recurring to functions available in SPM. Interestingly, in line 10 another comment acknowledges the authorship of the Wellcome Trust Centre for Neuroimaging, the institution responsible for the SPM project.

Two aspects of software development need to be carefully pointed to. On the one hand, computer code, even if it is published as open source code, never ceases to be somehow attached to some people or institutions. On the other hand, issues of ownership and code control are always present, as illustrated by the history of MIBCA [hist, mibca].

[hist, mibca]

Dr. Luís Lacerda, formed a biomedical engineer, did the Master's Degree at the Institute of Biophysics and Biomedical Engineering, Univ of Lisbon, Portugal, under the supervision of Professor Hugo Ferreira. He developed a toolbox for the analysis of neuroimaging data. He then met André Ribeiro, another Master's Degree student who was being supervised by professor Hugo and who developed another neuroimaging

---

[1]http://www.mibca.com/.

toolbox. The students had the idea of joining the two applications, an initiative that was promptly supported by their supervisor. Satisfied with the outcome, they decided to do some further work on the application, including the use of functionalities present in some popular neuroimaging packages. Luís worked in close collaboration with André, also relying on some occasional coding inputs from professor Hugo Ferreira. At a certain point, Luís left the Institute and went to London for a PhD program. Shortly after, André also went to London for the same reason. The two students decided to carry on working on MIBCA, having frequent meetings and continuing to have some occasional support from their former supervisor. When both students completed their PhDs and obtained their first academic positions, the work had to be stopped. Nowadays, MIBCA is still attached to the Portuguese university. In this way, professor Hugo Ferreira continues to make the code available to his new PhD students, so the application can be further improved. In his interview, Luís recalled the original insights behind the application's development. He said that some original things had to be implemented, but functionalities available in other packages could also be reused.

"[...] in the beginning, we didn't want to create an original thing but we wanted to use what had been done. But... We didn't want to replace everything, but at least some things that, according to our view, we could adjust better to our needs, and then we could have more control. So it was more a question of learning what was behind and also to have better control over the inputs we had, the outputs, and how all of that happened in the whole code.

That is, today, if I have a look at the code of MIBCA, I'll find in there its code (the original code of MIBCA) and code... and other code.
Yes. Yes.

And is it possible to say what is the percentage of original code and...
[...] Much of MIBCA's code is for liaising the different toolboxes and also to manage the platform, but all the visualization part and statistical tests, which are very important in the toolbox, is all original code. And this part is still quite big [...] I'm not sure, but at least 50%."

Like MIBCA, there are many software packages that are left behind when their initial developers move to a new institution. It is not rare that the institution takes possession of the coding work so other students and researchers add new functionalities, turning the personal project into an institutional one. In this chapterFour, it is precisely this metamorphosis through which personal code becomes institutional code, as well as its political and geographical implications, that will be stressed. For so doing, six phenomena will be analysed: commercial and non-commercial aspects of software development; personal and institutional contents of computer code; the role played by universities; the consequences of institutional code ownership; the political and geographical hierarchies of software development; and the example of the SPM package.

section11 (Institutional Barriers, Code Shields) {

The BrainVoyager neuroimaging software package constitutes an interesting case because, in a field dominated by open source, it has been a successful project managed by a company [hist, bvgr].

[hist, bvgr]

The BrainVoyager project began through the initiative of a young and most talented programmer, Rainer Goebel. In the mid-1990s, working as a post-doc researcher of the Max Plank Institute for Brain Research, Germany, he felt the need to develop an analysis software that could be, at the same time, efficient, quick, user-friendly, and visually attractive. He combined the development of this new tool with the conduct of his neuroimaging studies. In 1996 Rainer had the first version ready for release. When he presented his product at the Human Brain Mapping Conference, in the United States, much surprise and excitement followed, because of the application's beauty and speediness. Recognizing BrainVoyager's commercial potentiality, Siemens approached Rainer and offered a large sum to acquire the right the explore the software commercially. The Max Plank Institute did not allow Rainer to sell the software but supported him to launch his own commercial initiative. In 1998 the Brain Innovation company was founded. Thanks to his academic success, Rainer was subsequently invited to join the Maastricht University. In 2000, he, his wife, and the company moved to the Netherlands where he continues to be a professor, in addition to being the CEO and main developer of Brain Innovation.

"So I think that BrainVoyager is the most popular software produced by a company (in the world). Why do you think that the project has been so successful?
I think that the reason is because it makes a difficult task, namely to, you know, analyse complex data from brain scanners, in a quite, you know, easy, transparent, non-black box way [...] You can select menu items. And it has a lot of tools, but they're all integrated in the same spirit, in the same interface, right? So they [users] get good speed (quite fast) without having to learn multiple tools [...]."

The history of BrainVoyager [hist, bvgr] helps see the importance of code ownership. Initially, the code belongs to its writer, in this case Rainer Goebel. However, the programmer works in an institution, which may have some control of what happens with code. In the case of BrainVoyger, the Max Plank Institute prevented Siemens from acquiring the rights to explore the software. Because code ownership can generate financial rewards, it can underpin the creation of new companies, which happened with Brain Innovation.

Companies generally need to keep some secrets and pursue financial goals quite adamantly, so they fail to follow the collaborative schemes reviewed in chapterTwo. To be sure, a company like Brain Innovation cannot simply ignore some fundamental features of the scientific domain. For example, some of Brain Innovation's programmers are PhD students who want their coding productions to be open source. This is why part of the package's source code is made available on the internet. Professor Goebel claims to appreciate this model, "because it helps to make the stuff we do more transparent to the outside." At the same time, he is preparing a book where many of the algorithms used in the BrainVoyager package will be disclosed. However, Brain Innovation cannot help assuming, on strategic points, the protective stance typical of companies. In this way, Rainer's book will not reveal the most distinctive characteristics of his package. For instance, he will not disclose the coding strategies that make BrainVoyager have a processing speed superior to most neuroimaging packages. In his words, "I will not give away the speciality of BrainVoyager." Furthermore, even if users are allowed to program toolboxes for the package, this functionality is offered in a tightly controlled way,

> "So I expose indeed the code but there's a shield. That means, it's documented, it's open source, so to say, but it's precisely determined by me. So you cannot kind of read the real C++ code. It is C++ but it's kind of... We call it an API. So any commercial software, like Apple or Google, they do the same thing. They expose their code via a so-called API, application programming interface, which documents what you can and what you cannot do, right? So I control what people can do. But I do it very liberal."
>
> Professor Rainer Goebel (Brain Innovation, Netherlands)

so that people cannot gain the knowledge of the package's most basic coding features.

For many hackers, and especially those committed to open source, the strategies adopted by Brain Innovation and other software companies would be an example of "hoarding of technology" (Torvalds and Diamond 2001, p. 194), a practice typical of "gate keepers for code" (von Hippel and von Krogh 2003, p. 219). For people voicing this kind of concern, the social and economic benefits of software development are maximized whenever programmers share all their coding productions "without attempting to make them excludable" (Schwarz and Takhteyev 2010, p. 624). This concern has been so strong that software development became one of the few domains with instruments aimed to prevent productions from being patented or copyrighted. For example, the Berkeley Software Distribution (BSD) License determines that a certain software package must be open but does not requires that future modifications also be open, while the GNU Public License (GPL) protects the open nature of the package and also obliges future modifications to be equally released as open source (Atal and Shankar 2015; von Hippel and von Krogh 2003; Stallman 2002a, b).

The creation of such legal protections can be understood by considering that, at early times of the computing history, companies were framed as players jeopardizing the collaborative and open spirit of software development. In the mid-1950s, American numerical analyst John Weber Carr made efforts to formulate a programming language that could be ported from computer to computer. One of his main motives was to protect the autonomy of universities in a moment when, according to him, companies were beginning to control the technology and trying to realize a "commercial capture of the computer" (apud Nofre et al. 2014, p. 54). By the same time, American mathematician Derick Lehmer also denounced the advance of the computer industry, fearing that its growth would impair the circulation of information. The concerns voiced by Carr, Lehmer, and other contemporaries constituted relevant and understandable warnings, but they did not turn out to be precise predictions. On the one hand, companies did realize an increasing control over the computing field. On the other, however, their presence did not always narrow the space for collaboration in software development. In the 1950s and 1960s, as showed by Nofre and colleagues (2014), many companies actually adopted initiatives fostering code sharing and co-development.

Companies are frequent promoters of instrumental rationality, because the systems of actions in which they engage are informed by a search for profits and the need to protect commercial assets. Analysts have indeed pointed out that programmers working for companies have their tasks and operations strictly controlled, and even limited, by the steering power of managers and bosses (von Hippel and von Krogh 2003; Cusumano 1992; Sawyer and Guinan 1998). This is not to say, however, that an instrumental logic cannot be fostered by other kinds of social agents. Neuroimaging software is an ideal object for the study of this aspect, because academics and universities working in this field are many times involved in schemes alien to collaborative rationales. To be sure, some academic researchers acknowledge collective goals and have a sense of political responsibilities. For example, in his interview, Professor Carlos Garrido (Dept Physics, Univ of Sao Paulo, Brazil) criticized the behaviour of some PhD students who download open source code on the internet, implement modifications, complete their degrees, and do not upload their productions back onto the internet. From his viewpoint: "The goal can be purely academic, but you have your responsibility when you take on these things... You're also committing yourself to publish so you can join a community where people share with others. It is not enough to only pick it up. You're also assuming a commitment, aren't you?" However, this communicative concern voiced by Professor Garrido is sometimes lacking not only for students but also advanced researchers and professors.

The interests and goals that prevail in the academic domain are sometimes quite purposive, with a potential to even lead to market initiatives. For example, one of my Brazilian interviewees is a professor whose group has worked on a closed code neuroimaging software package. The group has been assessing the possibility of selling the package while keeping a simpler version available for free. Moreover, universities, as institutions engaged in a competition for funds, devices, and resources, would sometimes try to gain control over the coding productions of their researchers. In the example of the BrainVoyager software, described at the beginning of this section11, it was seen how the Max Planck Institute prevented its developer from selling the product to Siemens. Indeed, universities may be intent on interfering on the story of certain packages, deciding on issues like collaborations and code ownership. Concerned about this sort of institutional interference, Stallman (2002b, p. 62) advised academic programmers to have "[...] a good footing to stand firm when university administrators try to tempt you to make the program non-free." In this respect, the emergence of platforms such as GitHub may serve as a kind of indirect shield.[2] Researchers can develop a certain package as part of their academic tasks while publishing the source code on a collaborative internet platform. In this way, the project is automatically taken to an open space and freed from institutional control, being always available for the original programmer (and potential collaborators) even if the person leaves the institution. For example, Andreas Schuh (Biomedical Image Analysis Group, Imperial College London, UK) is aware that the MIRTK neuro-imaging package, developed in the framework of his PhD, will continue to be accessible to his programming work on completion of his degree, because the source code is being published on the GitHub platform.

One of the goals of this chapterFour is to challenge a division that many analysts might feel tempted to advance or adopt. I am referring to the division between commercial and academic software. One might consider that the development logic of companies is completely different from that of academics, because in this second case the search for profit is absent. However, this division is not always valid. The decisive factor is not only the stance towards profit-making but also, and even more importantly, the presence of an instrumental rationality, characterized by a prevailing concern with technical issues, a constant discursive reference to practical aims, and a fierce pursuit of technical goals. The instrumental rationale is surely manifested when certain social agents look for market success but tends to also prevail when academic groups develop software without making source code open; when PhD students use open source code without publishing the modifications they implement; or when administrators, for the sake of their universities'

---

[2]In chapterTwo, the history and social meaning of GitHub was analysed.

benefits, prevent researchers from realizing large and free exchange of coding productions. To further explore the occurrence of instrumental actions in academic programming, the issue of code ownership is further examined in section12.

}

section12 (Geographical Space, Namespace) {

The concept of namespace comes from the software development jargon. Let us imagine a certain program divided into many classes, one of which is called "Cortex." A certain programmer decides to enhance the code and creates a second class called "Cortex." Surely, the programmer can choose a different name but the person considers that a repeated name would enhance code clarity. This repetition is likely to create bugs because whenever the programmer tries to use the new "Cortex" class, the computer is unable to specify which of the two classes is being referred to. The solution for this problem is the use of a namespace, which is a simple prefix appended to the name of the new "Cortex" class. The programmer can use the initials of a name, the name of the development group, an acronym, or any simple everyday word, which then becomes a namespace, a sort of coding signature turning different pieces of code into unique and distinguishable elements. The existence of namespaces indicates a phenomenon of which hackers are very aware: computer code has some fundamental links with the people, groups, or institutions responsible for its initial production.

The original developers of a certain package are the people who possess, to the highest degree, the "contextual design rationale" pointed to by Heliades and Edmonds (1999, p. 394), because they know the basic decisions and intentions underpinning the programming work. The coding preferences and style of a programmer will surely remain in the codebase for many years, even after the person leaves the project. Hence the identification of "the irreducible connections between programmers and their work" (Galloway 2012, p. 70).

"I had programming courses before the Master's Degree but, in practice, I learnt to program while doing the Master's Degree, because I had to do it. So when I was developing it [a toolbox], I didn't think of making it ready for other people to use it. So I did something very specific, for our needs. So I don't even know if it would be useful for other people. I don't know if it's

The deep connections between programmer and program can be clearly detected in neuroimaging, where countless pieces of code are produced by PhD students who frequently *worth making it available or not. Today it [the toolbox] is not in a state that this would be possible. Perhaps, in the future, I can take something I did there and make it available, somehow, but nowadays...*

But do you think this is going to happen at some point?
I don't know. To be honest, I don't think so [...]."

Fabrício Simozo (Computational Group for Medical Signals and Images, Univ of Sao Paulo, Brazil)

carry out a solitary work with very specific goals. Some outcomes from Survey 2 show the centrality of universities, in neuroimaging software. Of the 23 projects surveyed, 18 had partial or complete involvement from at least one university [sftw, instit]. Therefore, there is a

[sftw, instit]

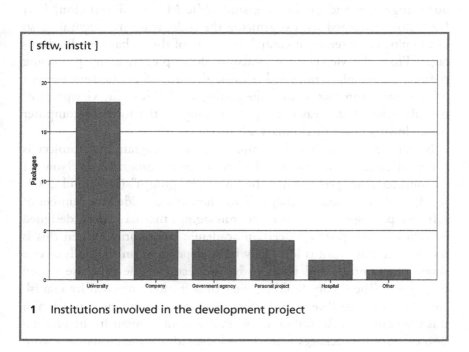

1   Institutions involved in the development project

vast population of university students writing code for neuroimaging applications.

Code produced this way can become very unclear and hardly understandable for other programmers. Furthermore, as explained in chapterThree, academics frequently frame software development as a minor scientific task, so students are encouraged to publish excellent theses but frequently fail to take good care of their coding productions. For example, Dr. Antonio Senra (Computational Group for Medical Signals and Images, Univ of Sao Paulo, Brazil) explained that during his Master's Degree, the notion of collaboration through software was unknown to him. In this way, his research group ended up not publishing the toolbox he developed at that time. "We used that tool to solve our things here and publish the study [the Master's dissertation] […] I was not concerned with organizing the code, from the beginning, so as to upload and release it easily." This state of things has two disadvantages. From the viewpoint of software development, many promising projects are simply terminated because the lack of publication prevent other people from taking over the coding work. From the viewpoint of neuroimaging, less ideas end up circulating in the form of computer code, slowing down disciplinary advances.

So far we have reviewed examples where a programming project is deeply dependent on and attached to a certain programmer. Now let us consider that "projects may be owned by groups" (Raymond 2001, p. 73). Indeed, as seen in chapterTwo, there is a considerable number of software packages, not only in neuroimaging, that have been designed by two or more people in certain academic laboratories. When this is the case, coding work is less likely to be dependent on the skills of one single programmer, and tends to be the result of the expertise shared in a group. The theory building view of software, proposed by Danish computer scientist Peter Naur, offers some tools for the interpretation of this phenomenon. According to Naur, the main element in the production of software packages is the knowledge, ideas, intentions, in a word the theory formulated by the group and crystalized as computer code. Most frequently, such knowledge fails to be fully expressed in written documents and can only be grasped in the context where the software appears and unfolds. This is why Naur (2001, p. 234) voiced the following conclusion:

> For a new programmer to come to possess an existing theory of a program it is insufficient that he or she has the opportunity to become familiar with the program text and other documentation. What is required is that the new programmer has the opportunity to work in close contact with the programmers who already possess the theory, so as to be able to become familiar with the place of the program in the wider context of the relevant real world situations and so as to acquire the knowledge of how the program works and how unusual program reactions and program modifications are handled within the program theory.

In 1968 the interpretation of software was enriched by the ideas published by computer scientist Melvin Conway. The core of his interpretation can be summarized by the following words: "Organizations which design systems are constrained to produce designs which are copies of the communications structures of these organizations" (apud Raymond 2001, p. 224, Note 10). This view came to be known as the Conway's Law. To illustrate its meaning, let us imagine a certain development group composed by five subgroups. The resulting software package will tendentially be composed by five modules, each of them developed by one of the subgroups. In this way, the Conway's Law highlights the strong bounds between the interpersonal organization guiding the coding work, on the one hand, and the technical organization of the software package, on the other. Taking Conway's thought one step further, it can be assumed that a package originally developed by a highly stratified and resourced team would pose great programming challenges if it were to be modified by another team having a simpler organization and access to less resources.

To be sure, the links between a development team and the resulting software have crucial communicative aspects. For example, it was seen in chapterTwo that many pieces of code evolve, during many years, thanks to the collaboration and interpersonal support found within development groups. Code sharing, mutual encouragement, and exchange of ideas are expressions of an event noted by LaToza and colleagues (2006, p. 496): "[...] the culture of informal communication works well and [...] the team boundaries are typically in the right places." However, the same authors suggest that communicative aspects are frequently coupled with instrumental ones, as development teams are prone to nurture a feeling of code ownership. Whenever this feeling is experienced by the members of a group, it "[...] forms a kind of *moat*, isolating them from outside perturbations" (LaToza et al. 2006, p. 497). At this point, the cognitive stimulation that fosters collaboration and communication within the group may turn into a sort of intellectual shield against outsiders. In this way, a defensive and self-centred stance can end up characterizing the group, in such a way that only ancillary ideas are allowed to flow towards other groups. This situation amounts to a distortion of the very nature of software development. As explained in chapterTwo, it becomes increasingly unviable, for both individuals and groups, to design software in isolation, because every coding effort is highly potentialized when it relies on previous achievements made by hackers who have developed libraries, algorithms, toolboxes, and packages. In this way, defensive stances are in contradiction with the mutual dependencies that, in both technical and interpersonal terms, have become traditional and meaningful in software development. In other words, defensive stances, here, lead to an "egocentric competition between individuals who are systemically chained to one another" (Habermas 2008, p. 195).

Such distortions are sometimes potentialized by the features of the academic world, which tends to be dominated by hierarchical and competitive logics. Indeed, the connection between software packages and groups is sometimes transformed into the possession of

> "Okay. Is it possible to say that the challenges in terms of data processing were sort of provoked by the increased capacity of generating data by means of these new MR scanners?
> [...] It's quite... quite clear that the... what today is the big neuroimaging packages ended up originating in centres that very early on had access to scanning equipment. So if you look at the... at the SPM world that was based on all the imaging that happened initially at the Cyclotron Centre

[Hammersmith Hospital, London] with PET and then moved into the MRI domain at the field; [...] if you look at FSL from Oxford. If you look at FreeSurfer from Boston and MGH [Massachusetts General Hospital, United States], the big packages ended up being developed (or AFNI with the NIH [National Institutes of Health, United States]) where there was very early the access to scanning equipment, simply because the presence of data generated a need for looking into ways of analysing that data."

Professor Christian Beckmann (Statistical Imaging Neuroscience group, Donders Institute for Brain, Cognition and Behaviour, Netherlands)

packages by groups. In the analysis of software development, the role, brilliance, and insights of certain programmers are sometimes put forward. By focusing on people too fiercely, one might disregard a crucial aspect: the weight of institutions and geographical contexts that enable software development. The field of neuroimaging is also appropriate for the study of this aspect. In my interviews, it became clear that people using and modifying neuroimaging packages do not frequently lose sight of the fact that certain popular projects are "owned" by such and such university and funded by such and such government agency. In this way, the FSL package is often described as a product of the University of Oxford whereas the FreeSurfer package is often seen as a product of the Harvard Medical School, to give only two examples. Therefore, computer code can at times be seen as an entity endowed with social status, institutional label, and geographical origin. In the same way that namespaces distinguish and specify some pieces of code within a codebase, the social, institutional, and geographical dimensions of code specify and distinguish the coding production. In order to further scrutinize this dimension of software development, two factors of pivotal relevance for universities will be analysed in the sequence: funding and reputation.

}

section13 ("Economic Order" & "Social Order") {

Magnetic resonance scanners are produced within an oligopolistic market (Cowley et al. 1994), being extremely costly. The most sophisticated of them, which are also the most expensive ones, are those enabling the conduct of functional magnetic resonance imaging (fMRI), in which researchers observe changes in the brain while research subjects perform some tasks. As noted by Sawle (1995, p. 136): "Although many hospital radiology departments are equipped with magnetic resonance machines, few have the hardware and software on hand for the acquisition of an fMRI signal." In the same way that such inequalities can be detected in data collection, the domain of data processing, including software development, is rife with economic disparities. This section13 aims to analyse these phenomena.

13.1. Funding for Software Development

According to Weber's (1958, p. 181) classical explanation, social stratification is determined by two interdependent "orders." On the one hand, the "social order" has to do with reputation, status or honour. On the other, the "economic order" is defined by the uneven distribution of material and economic resources. This explanation is by the way quite similar to Merton's (1968, p. 57) interpretation, which speaks of "status system" and "class system." In the academic domain, and more specifically in academic programming, these two kinds of orders or systems can also be identified. To begin to analyse them, let us first consider the financial or economic issues involved in software development.

In chapterTwo, we focused and commented on an interpretive approach that frames software development as a gift culture. In this interpretation, hackers would enjoy a state of constant abundance. For example, Raymond (2001, p. 81) claims: "[...] it is quite clear that the society of open-source hackers is in fact a gift culture. Within it, there is no serious shortage of the 'survival necessities' – disk space, network bandwidth, computing power." Because of such digital wealth, production costs would be dropping steadily, creating a "zero marginal cost society"

(Rifkin 2014). However, such affirmation can surely be tempered or even contested. By talking to programmers based in different locations, it gets clear that, in some places, countries, and institutions more than others, shortage of resources is part of the research context. This is so because in order for a software development project to be initiated and carried out satisfactorily, a series of more or less considerable resources needs to be available. Three main types of costs have to be met.

First, there is the obvious need to obtain funding to pay salaries. At first sight, this need would have no difference in relation to the everyday needs of universities, because human resources always represent an important fraction of these institutions' expenses. However, things are not so simple because, as explained in chapterThree, software development continues to be considered as a minor, and even negligible, side of researchers' tasks. Neuroimaging researchers are supposed to formulate concepts and theories pertaining to the human brain, not really to produce tools to analyse data. To be sure, conceptual goals would not be reachable without the presence of analytical goals. These latter, however, are systematically overlooked for the sake of the former. In this way, researchers will most likely fail whenever they are intent on obtaining funding specifically for tasks of code writing and code maintenance. In his interview, Professor Robert Oostenveld (MR Techniques in Brain Function group, Donders Institute for Brain, Cognition and Behaviour, the Netherlands) recalled that he once did get some funding for hiring a programmer responsible for reorganizing the codebase of the FieldTrip software package. Nevertheless, he also hastened to explain that the award of such funding is extremely rare. Generally, software development is done as an ancillary fraction of more general (and more highly regarded) research activities.

This is why most academic coding work ends up being realized by students who receive scholarships for the conduct of their (conceptual and development) activities. As claimed by Professor Alexandre Andrade (Institute of Biophysics and Biomedical Engineering, Univ of Lisbon, Portugal): "Our research ends up depending on more or less specific projects or on PhD students or on the possibility of hiring postdocs." As a consequence, software development threatens to get paralysed whenever a student finishes the program or a job contract expires.

"I think we would've struggled a lot and maybe just not achieved at all what we have achieved if we had had everybody come for three years and leave.

Because people can familiarize with the code.
Yeah. Because there's that stable sort of thing of paying people knowing the code, and having other people around that you can ask about the code and the codebase. But also because actually this cycle of developing a tool, releasing the tool, getting feedback from other people, and incorporating that feedback to make it more robust, that's also something which takes longer than three years. So I think in order to make things which work really well actually needs somebody who's going to stay around or you need people who, you know, take over things in that way. And, you know, somebody who has developed a tool, maybe they have moved on and their principal focus is doing some other work, but if they're still here, they're much more likely to also spend a bit of time, you know, thinking about the robustness of that tool and thinking about the feedback and working on it to some degree, which is crucial, whereas if they just leave to go somewhere else, that's far, far less likely to happen and it's much less likely that you'll get somebody who is going to take over the project and be invested in the same way."

Professor Mark Jenkinson (Oxford Centre for Functional Magnetic Resonance Imaging of the Brain (FMRIB), Univ of Oxford, UK)

Such interruptions can surely happen in many academic disciplines. However, they are particularly troubling in software development, because new students or researchers, those who will take over the coding work, frequently have to spend days or weeks just to grasp the rationale of the codebase they inherit. In this regard, a division appears. On the one hand, there are those institutions who have more difficult access to scholarships and funding, and whose development projects, therefore, are more likely to be interrupted or even terminated. On the other hand, there are institutions with consistent access to scholarships and funding, thus being able to improve their software constantly. One clear example of this second situation is the FSL package, which is developed at Oxford University, UK. Some key names of this programming team (including Mark Jenkinson, the team coordinator, as well as Mark Brady, Stephen Smith, Tim Behrens,

Mark Woolrich, Saad Jbabdi, and others) have been contributors of the project for decades.

Expenses with personnel are not limited to the programming team, though. In universities, it is also important to rely on support from people dealing with administrative tasks. Whenever researchers are efficiently provided with such support, their activities are streamlined. In my interviews, it was clear that in Brazil and Portugal, there is a dearth of such institutional support. As a result, Brazilian and Portuguese researchers, and especially those who have had some experience in Europe or the United States, voice many complaints about the time and amount of administrative work they have to realize. According to Professor Ana Luisa Raposo (Faculty of Psychology, Univ of Lisbon, Portugal): "People lose much time dealing with things that are not really the research they're supposed to do." This type of hurdle has also been identified for the case of Brazilian researchers working on human genetics (Ferreira 2018).

Second, software development implies more or less substantial expenses with devices. Obviously, computational infrastructure must be in place, which involves not only computers but also storage systems. Once again, the excellence of this infrastructure is proportional to the financial dynamics of each institution. After the interviews I conducted in a research group of the University of Coimbra, Portugal, I was showed the group's storage centre, which consists of two connected computers. In other universities and countries, those systems reach much higher degrees of complexity and sophistication. For example, at the Centre for Neuroimaging Sciences, King's College London, UK, neuroimaging data are processed in a virtual central system, with many software packages and toolboxes available, and with a system for automatic back-up of analyses. According to Dr. Vincent Giampietro, who is based there, "what we do, really, is mass production, here, of neuroimaging, 'cause we've got the scanners, we've got fifty or sixty research projects going on at any time…"

In addition to personnel and computational infrastructure, neuroimaging software development requires, from time to time, access to some devices for data collection. At a certain point of his interview, Dr. Brunno Campos (Laboratory of Neuroimaging, Univ of Campinas, Brazil) opened a door to a shelf close to us. He showed me a special cap

> *"Generally, when you're developing a data collection protocol... I believe it's sometimes necessary to acquire a new device. Yes [...]*
>
> **What type of device?**
> *Ahm, stimulation devices, for example. Eerr... For acquisition of biosignals sometimes. Recently we bought one, for example, to measure the breathe, the breathe itself, I mean the expansion of the chest inside the resonance [device]. I mean, if a new project arrives and we think we need to have... to continually register the breathe, then we eventually buy it. Yes.*
>
> **These things are not cheap, are they?**
> *These things are not cheap, they aren't. But we don't buy them every month, right? But sometimes, yes."*
>
> Researcher (Univ of Coimbra, Portugal)

his research group had recently acquired. It is an electroencephalogram (EEG) cap that research subjects use inside the magnetic resonance scanner, enabling researchers to collect EEG signals while neuroimaging data is collected. As there is no company producing this kind of cap in Brazil, it was necessary to pay an extremely high price for its importation. Another Brazilian interviewee, based in a private hospital, explained that it was once necessary to import from Switzerland a very expensive device that measures the levels of radiation in the research subjects' blood. In neuroimaging software development, researchers are designing tools aimed to process data. At some points, some special types of data may prove necessary to validate the accuracy of the software's algorithms. If such data cannot be accessed in public databases, then researchers can either buy the devices enabling the production of new data or give up their coding targets. This second alternative is what remained for Dr. Victor Hugo Souza (Laboratory for Biomagnetism and Neuronavigation, Univ of Sao Paulo, Brazil), the developer of a neuroimaging toolbox. The application processes data pertaining to the localization of brain processes. A special device is needed for acquisition of these data. A Canadian company called Claron and an American company called Polaris are leaders in this market. So far the toolbox developed by Victor Hugo supports only the data produced with Claron's device. Asked why the other company has been excluded from his project, he gave me a very direct answer:

"Money [...] It is expensive. It is expensive. Only the device costs around 100,000 Reais [around 26,000 US dollars] [...]."

Wagstrom (2009, p. 188) noted the financial difficulties faced by open source projects, which can nevertheless be fostered and anchored by "an ecosystem for research" whenever there is a big company or a solid funding agency backing them. Once again, things are particularly tricky in the academic context, where software development is seldom recognized as a task worth being funded. Even MELODIC, an internationally popular neuroimaging toolbox, would fail to be funded if such attempt were made. Its main developer, Professor Christian Beckmann (Statistical Imaging Neuroscience group, Donders Institute for Brain, Cognition and Behaviour, Netherlands), declared in his interview: "I don't have a grant to develop MELODIC, so a lot of the development happens as part of the research that we are interested in the group." In this way, one key element defining the fate of different software packages is the institutional savviness put in place by their academic developers, who need to wisely navigate bureaucratic seas and obtain some sort of indirect funding with which software packages continue to evolve. When this challenge seems too overwhelming, researchers may either give up their development projects or prefer to produce packages destined to be used only locally. This second option is more frequent then one might suppose, and is perfectly illustrated by the history of the XBAM package [hist, xbam].

[hist, xbam]

The XBAM software package began to be designed in the late 1990s, at an early moment of the history of neuroimaging software. Its original developers, Michael Brammer and Edward Bullmore, were doing research at King's College London. Their goal was to develop an application to be used in their own analysis, as well as those of colleagues based in the Centre for Neuroimaging Sciences. Some years after the project's beginning, Dr. Vincent Giampietro joined it while Edward Bullmore had to leave. Over the decades, the development team has never grown bigger than this. With modest financial and institutional support, the developers opted to keep their ambitions at low levels, maintaining a software package that, up to our times, continues to be used mainly at King's College. In his interview, Vincent summarized the package's story and its institutional insertion.

> "[...] it was a conscious choice. Because the Institute was not putting in any resources behind it [XBAM], we decided [...] that, you know, we were not going to publicly export it, you know, we were not going to say to people 'Hey, everybody in the world, download it,' because we didn't have a support system in place. If people went keep sending us things like 'I can't install it, it doesn't compile, what do I do?' we didn't have the resources to help them [...] At some point, there was some noise about giving us some money here to employ people. We would have gone... we would have done this. But no, because of the resources (basically, there was a two-men team) we just... we couldn't [...] Because we had the research to do as well, teaching... And we didn't have the resources to do this. So that's why we never put it out."

In contrast to XBAM, its lack of resources, and its local use, some other packages have reached high degrees of diffusion, which is due, among other things, to the financial and institutional support given to their design. It is interesting to note that some pivotal packages have even attracted funding from national agencies. For example, AFNI is the neuroimaging package of the National Institute of Mental Health, a federal research body of the United States. At the beginning the project was purely academic, being subsequently supported by federal agencies. Today, in addition to being funded by the National Institute of Mental Health, it receives some funding from the National Institute of Neurological Disorders and Stroke (Cox 2012). In Map 13.1, with data from my analysis of neuroimaging papers published in 2015, it can be seen that AFNI has been used in the most productive geographical hubs, namely those located in the United States, Europe, and Eastern Asia.

Map 13.1 Geographical hubs with papers citing the AFNI package: 2015 (*Source* Survey 1)

In spite of its considerable geographical diffusion, AFNI has not been very intensely used. Considering all the software packages used and cited in each hub (but excluding toolboxes), it generally represents 25% or less. To a large extent, this is due to AFNI's technical features, as its installation procedures are quite complicated and the package is far from being user-friendly.

The FreeSurfer neuroimaging package constitutes another interesting example. It has been developed by the Athinoula A. Martinos Center for Biomedical Imaging, a research unit of the Massachusets General Hospital, United States. In addition, the project relies on contributions from the Harvard Medical School, and the Massachusets Institute of Technology. As explained by the project's coordinator, Bruce Fischl (2012), FreeSurfer has had the financial support of the National Center for Research Resources, the NCRR BIRN Morphometric Project, the National Institute for Biomedical Imaging and Bioengineering, the National Institute on Aging, the National Center for Alternative Medicine, and the National Institute for Neurological Disorders and Stroke. With such heavy list of funders, FreeSurfer is able to display a wide geographical diffusion, as seen in Map 13.2.

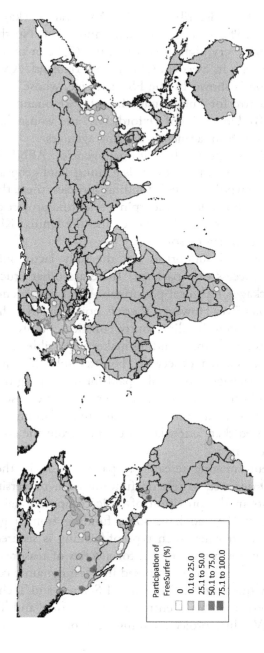

Map 13.2 Geographical hubs with papers citing the FreeSurfer package: 2015 (*Source* Survey 1)

FreeSurfer could be described as the American package, because of its intense use (25% or more) in many hubs from North to South America. However, a very high intensity of use (75% or more) can also be noticed in one hub in Sweden, China, and Thailand. A very similar pattern of diffusion is shown by the FSL package, whose team is based in the Oxford Centre for Functional Magnetic Resonance Imaging of the Brain (FMRIB), University of Oxford, UK, and whose development is funded with grants from academic funding agencies.

Being financially supported by national agencies, AFNI, FreeSurfer, and FSL receive a decisive push towards gaining solid geographical diffusion. From the viewpoint of their funding agencies, then, the financial investment makes it possible to underpin a scientific project placing the country (in these cases, the United States and the United Kingdom) at the forefront of neuroimaging software.

To be sure, financial support is not the only factor guaranteeing the international success of a package. There are other crucial factors, including the package's technical performance and user-friendly nature, as shown in chapterFive. However, there is no doubt that the presence of a strong and consistent funding source constitutes a pivotal help. Let us consider, for example, that funding makes it possible to pay salaries that could not be secured otherwise. For example, the development team of the SPM package is not really paid for developing the software. The team is based in the Wellcome Trust Centre for Neuroscience, Univ College London, UK. In his interview, Guillaume Flandin, the project's coordinator, declared that "our salary is coming from the core grant we get from Wellcome Trust."

Of the 23 neuroimaging researchers participating in the Survey 2 [prjt, fndg], 10 declared to rely on funding from universities, which frequently consists in the provision of basic computational infrastructure. Most projects (14) do have some funding from a government agency. However, even when such type of funding is secured, international success may not be guaranteed for a series of reasons; for example, the development team may be based in a less dynamic country. For instance, in 2001 the Brazilian Ministry of Science and Technology, by means of the Renato Archer Centre for Information and Technology, launched the InVesalius project, aiming at producing a software for

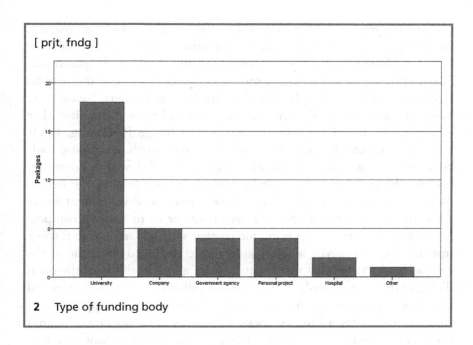

[prjt, fndg]

Packages

2   Type of funding body

visualization of technical devices. Some years later, a new and completely different toolbox started being developed, for localization of stimulated brain areas. In spite of the commitment of a national Ministry, InVesalius has displayed a very modest diffusion. In Survey 1, only one citation to the package was identified in the 2015 neuroimaging literature.[3] Moreover, the InVesalius project has had to face the reduction of funding for scientific activities that has marked Brazil after the 2016 coup d'etat.

The relatively limited diffusion of InVesalius, as well as the wide geographical reach of projects like AFNI, FreeSurfer, FSL, and SPM indicate that the international diffusion of neuroimaging software may be facilitated by the access to solid and consistent funding from agencies of dynamic countries. To be sure, it is crucial to rely on the work of

---

[3]According to the inVesalius team, in 2018 the package had been downloaded over 20,000 times, in 145 countries. However, such diffusion has not been reflected in international publications.

talented programmers able to produce efficient and innovative systems. However, it is equally important to mobilize resources and infrastructure to underpin software development. At this moment, the potentialities of open source software become unevenly distributed, as they can be fully explored in certain places and institutions, not others. Researchers based in less dynamic countries are fully aware of these inequalities. For example, Dr. Ivanei Bramati (Laboratory of Image Processing, D'Or Research Institute, Brazil) told me that it is not worth developing toolboxes similar to those of wealthy universities of dynamic countries, because the competition cannot be gained. "If you look at the investment that those universities make in personnel, in development infrastructure, it is enormous. So it is very hard for us to wish to compete unless it is in some very specific niches that we find." As a result, some institutions, and maybe whole countries, get excluded from the methodological discussions of some cutting-edge areas. In this way, the open dialogue that can flourish in open source software is eventually blocked by monetary questions. As claimed by Habermas (1996, pp. 39–40), "money" unleashes a kind of social integration that operates "'behind the backs' of participants," rather than generating clear communication.

In this regard, it is interesting to come back to the issue of software flexibility, as discussed in chapterTwo. Rigid packages are those whose source code is tightly controlled, especially by companies. Flexible packages are those which can be easily accessed and modified, with quick and effective means available for the incorporation of coding contributions into the old codebase. Semi-flexible packages are those whose source code can be easily accessed and modified but whose development team prefers not to provide contributors with quick and effective means to turn external contributions into part of the official codebase. Survey 1 showed that whenever a development project relies on strong funding from national agencies, the resulting package tends to be semi-flexible. This is the case of FreeSurfer, FSL, and AFNI. All these three packages' codebase can be easily downloaded, but coding modifications do not manage to gain publicity, as there are no quick and efficient means to incorporate them. As a result, the official toolboxes of these packages have, for the most part, been developed by the main development group, which eventually constitutes a sort of gatekeeper of code. In this way, semi-flexibility and solid financial support reinforce each other. Having focused on the financial aspect of software development, it is time to consider a second and equally relevant aspect: reputation.

}

13.2. Software Development and Scientific Reputation

By hosting the development of a largely successful neuroimaging software package, an institution can acquire scientific prestige, joining a short list of internationally renowned development centres. On this point, the history of the FieldTrip package is most illustrative [hist, fldtrp].

[hist, fldtrp]

Strictly speaking, the FieldTrip software, in spite of processing brain data, is not exactly a neuroimaging package, as it processes electroencephalogram data, without dealing with images. The project was initiated unconsciously. Professor Robert Oostenveld (MR Techniques in Brain Function group, Donders Institute for Brain, Cognition and Behaviour, the Netherlands), at the time of his PhD, wrote a relatively large piece of code to analyse data. Subsequently, he received some code contributions. As the resulting application proved most useful, he began to let people use it. The number of both users and contributors grew quite rapidly, and this is how the FieldTrip project was eventually formed. In 2014 Robert decided to manage the project by using GitHub, the online platform for collaborative software development.

"Hm. Is it possible to say that Fieldtrip enhances your scientific prestige? Oh, yes, certainly. Yes.

Hm. Hm. So it's good for you in terms of... It's a scientific achievement. Yes... I think... I think the Donders... The Donders Institute that's now working in Nijmegen is... Part of... Part of the international reputation that the Institute has is due to Fieldtrip.

> *Hm. Hm.*
> *Everyone knows that Fieldtrip is developed here in Nijmegen. Nijmegen is a small town at the edge of Netherlands. Most people don't know where the Netherlands is. And if they know where the Netherlands is, they only know about Amsterdam but not about Nijmegen. Radboud University is also something people haven't heard about. And then there's Donders Institute and there's Fieldtrip, and they, like internationally, like worldwide, they are known. And I think that's... These go hand in hand. Like, had it not been for the Donders Institute, the Donders Centre, then Fieldtrip would not have been a success but also the other way around. So I think that we've been contributing to each other's success."*

"And do you think that somehow your involvement with the FSL project enhances your prestige here in this Centre...

Not sure about prestige. I mean, it's... it's certainly... Living in an academic world where people are being quantified in terms of output metrics, obviously being involved in FSL has had a great benefit, also career-wise, in terms of papers that are heavily cited. So yes. I mean, it's been very beneficial having been involved in FSL. If you look at... at kind of the high-citation papers from all of the core group that initially started developing FSL, many of these papers are indeed related to the fact that they're being cited because people use the [computational] tools [...] But [...] there's very, very high-impact papers that are not about tools and not about techniques, they're not about methodology, they're all about neuroscience."

Professor Christian Beckmann (Statistical Imaging Neuroscience group, Donders Institute for Brain, Cognition and Behaviour, Netherlands)

Software developers are generally based in academic institutions, which have gained different status in the international scientific hierarchy. In neuroimaging, some institutions have managed to secure much academic recognition by creating calculations or analytical tools that have become almost gold standard. For example, the Montreal Neurological Institute (MNI), Canada, formulated the so-called MNI space, a 3D coordinate system used to localize structures and events in the brain. The MNI space has been used by a large number of software packages, which indirectly accrues more scientific reputation to the Canadian institution. This example, as well as other examples reviewed below, show that the neuroimaging field becomes increasingly marked by what Merton (1968, p. 57) called the "the stratification system of honor in science." Academic institutions would benefit from the acquisition of scientific prestige in at least two ways. First, the best programmers would try to be based there, thus enhancing their personal scientific standing. Second, these institutions gain more scientific visibility, which can be translated, for example, in a large number of citations for the papers they publish, a performance that has become crucial in times of consolidation of the bibliometric rationale. Slowly, the institutions from which the greatest achievements are expected do acquire the resources to fulfil those expectations. Eventually, this constant and uneven distribution of prestige "[...] violates the norm of universalism embodied in the institution of Science and curbs the advancement of knowledge" (Merton 1968, p. 62).

> *"When you're going to perform some [data] processing, what is the criterion you use to choose the software to be used?*
> It is mainly what the literature has cited. We have always to check in the literature what program has been used, if it is leading to good results. It is always based on what is in the literature, the most recent things in the literature. Because if we're going to design a software package... For example, if I'm going to design a new software package, I have to process my data with a program that is well established in the literature and with mine, and show: 'Both led to similar results.' I have to do it several times before I start citing mine and then I can use mine only [...] We take the most recognized packages, learn to do use them well, and use them."
>
> Dr. Fábio Duran (Laboratory for Neuroimaging in Psychiatry, Univ of Sao Paulo, Brazil)

Some analysts, such as Atal and Shankar (2015, p. 1395) have pointed to the "[...] race to win the reputational prize from design innovation among developers," stressing that such competition is triggered whenever programmers present their products to the community. However, analysts have failed to highlight an equally important phenomenon: prestige, rather than being the outcome of the publication of development work, frequently precedes such publication. When developers are based in certain renowned institutions, reputation is from the beginning guaranteed by their affiliation, even if that initial prestige is indeed eventually enhanced by the work they carry on. In this way, it is not completely correct to suppose that at the initial phase of the reputation race, developers have the same point of departure. Eventually, a prestigious university can enhance its prestige by means of its software. All the publications related to that software heighten the university's prestige and help make the application used at the global level. Indeed, frequent citation in the literature is one of the main criteria that researchers mobilize to use a software package. Moreover, the reputation race is made less of a fair game by the fact that different institutions can put different material resources and infrastructure at the disposal of their researchers. As pointed out by both Merton (1968) and Weber (1958), reputational and material hierarchies nourish each other.

Financial support, on the one hand, and scientific prestige, on the other, are crucial resources for software developers willing to produce analysis packages. Both types of resources are found in the universities where those programmers are generally based. However, whereas some hackers struggle to access devices, infrastructure, and high scientific standing, others find all the conditions needed for the production of internationally successful projects. In the following sections, it is precisely those projects that will be focused on.

}

section14 (Social Paralysis, Software Evolution) {

14.1. Fostering Code Evolution

In the literature analysing software development, authors seem to be aware of only two trends: open source and closed source software. They often work with the implicit assumption that when open source software is designed, the principles of collaboration and knowledge sharing are fully and automatically upheld. However, a more refined view can be promoted if the issues of flexibility and semi-flexibility are taken into account. For even if programmers are producing open source software (that is software whose source code can be largely downloaded), such production, instead of promoting wide and equal collaboration (flexibility), may be consolidating the scientific dominance of certain institutions (semi-flexibility). If it is true that the number of open source packages has been increased, it is also true that the most popular packages have not been designed through fully collaborative schemes. Therefore, it is not sufficient to consider whether the source code is open or not; it is also necessary to ask whether there are expedient channels to foster dialogue between developers while expanding the number of potential contributors. If developers can realize an effective distribution of their modifications only at a local scale, actual flexibility cannot be pointed to.

In neuroimaging, software development sometimes reinforces the principles of semi-flexibility. Table 14.1 brings some information on the most widely used neuroimaging software packages.

Table 14.1 World's most popular neuroimaging packages

Package	Classification	Release	Main development institution	Ctiy and country
SPM	Flexible	1991	Wellcome Trust Centre for Neuroimaging, Univ College London	London, UK
FSL	Semi-flexible	2000	Centre for Functional MRI of the Brain (FMRIB), Univ Oxford	Oxford, UK
FreeSurfer	Semi-flexible	1999	Athinoula A. Martinos Center for Biomedical Imaging, Massachusets General Hospital	Boston, United States
AFNI	Semi-flexible	1994	National Institute of Mental Health	Bethesda, United States

Source Literature and interviews

These four packages have been adopted by many neuroimaging researchers in different world regions. Interestingly, three of them have been designed through semi-flexible arrangements. Basically, they share two characteristics. First, they are produced by groups that greatly emphasize the academic nature of their work, which is also the case for the group responsible for the Caret software, developed at the Washington University in St. Louis, United States (van Essen 2012). Second, the vast majority of toolboxes designed for the three semi-flexible packages of Table 14.1 have been created by their core development team, even though access to the source code is largely guaranteed. For example, Bruce Fischl (2012, p. 779), the leading programmer of the FreeSurfer project, declared that "[...] we have recently adopted a liberal open source license that allows great freedom in the use of our source code [...]." Nevertheless, there have been few examples of toolboxes being developed outside the main development institution and its partner institutions: the Harvard Medical School, and the Massachusets Institute of Technology. In a similar way, the AFNI

> *"[...] I suppose that the [FSL] code is being read and analysed by many people in many places, people who want to understand what the code is doing, people who want to learn with the code...*
>
> *[...] we have an internal C++ set of classes and libraries which people use within the code and within... but it's very rare, I think, that people use that outside of... of this institution. So I think when people do stuff in FSL, they do it in the Unix way that they will combine different things, different tools, but they aren't necessarily writing within our class structure. And it wouldn't be easy for them to do that, to be honest.*
>
> *Why?*
>
> *Because it was not... Because the original development was done when the most important thing for us, as fairly new academics, was the research, was the novel methodology, was not to invest a lot of time in the software engineering aspects of writing particularly well structured and documented code. So I think the code works well; I don't think it is documented well internally, and I don't think it is necessarily structured that well."*
>
> Professor Mark Jenkinson (Oxford Centre for Functional Magnetic Resonance Imaging of the Brain (FMRIB), Univ of Oxford, UK)

package has received some features making it easier for external contributors to add new functionalities to the software, as explained by Cox (2012). "However, my dream of having significant contributions to AFNI from outside developers has only been partially realized. A few such programs have been donated and are distributed with AFNI, but most people (understandably) prefer to work independently" (Cox 2012, p. 744). Unfortunately, the author does not explain why he said "understandably," an affirmation that is intriguing when it is considered that many programmers would be most happy to have their coding contributions integrated into a globally successful package like AFNI.

Another outstanding example is the FSL package. In his interview, Professor Mark Jenkinson, who coordinates the FSL team, explained that most of the package's toolboxes have been designed inside the institution. There is little collaboration not only in terms of relations with external contributors, but also within the development team itself. Generally, FSL developers work on their own. When internal collaboration does happen, this is frequently the outcome of conversations and code sharing between two programmers only. Thanks to this organization, even programmers who have left FMRIB can carry on contributing code at the distance. This is what happens with Christian Beckmann (Statistical Imaging Neuroscience group, Donders Institute for Brain, Cognition and Behaviour) who continues to be a key FSL developer while based in the Netherlands. According to him, FSL is a "by-product" of a higher goal, namely the production of knowledge related to the brain. "I'm not being paid to be a programmer [laughter]. And I don't publish about programming [...] So FSL, in my mind, is [...] a tool developed to address my research interests, and it's then made available. But I'm not programming [...] for the sake of programming. I'm doing it to look at interesting projects that... where I want to contribute in terms of generating output."

> "**But do you sometimes have to analyse or study the code of these [internationally successful] packages?**
> Yes. Because I'm so much into the software engineering part, I definitely look into the code of some of them every once in a while. Yes.
>
> **Is it frequently easy to understand the code?**
> Absolutely not! No, no. It's terrible.
>
> **Really?**
> Every single major... Every single neuroimaging software is just a typical example of what we call evolved design, which is basically no design, it's just evolution of code through the ages. So... Well, the FSL software library, for example, is a library of separate tools that all have different interfaces and different people writing the code and therefore completely different styles of writing. Well, SPM, the same thing [...] And that is still very clear when you dive into the code, and that is a major hurdle for people that want to, well, improve and develop in that direction."
>
> Dr. Tim van Mourik (MR Physics Group, Donders Institute for Cognitive Neuroimaging, Netherlands)

In the software engineering jargon, software evolution is the technical history of a package, since its creation to its current state, involving all the modifications made to the codebase and the ways in which such modifications have been implemented. For some packages, a linear and rational evolution can be detected whereas other packages are modified in a more intuitive, organic or even messy fashion. The most popular neuroimaging packages (namely SPM, FSL, FreeSurfer, and AFNI) have followed this kind of hectic evolution, which is greatly due to the low degrees of true collaboration between their developers. As a consequence, the source code of these packages has become technical puzzles, with poor documentation and precarious organization. In this way, external developers have little incentives to solve those coding puzzles and provide their contribution.

When the researchers' emphasis is heavily put on science and knowledge production, then the relevance of code, and especially the collaborative potentials of code, is downplayed. The scientific knowledge that is produced in this way may be so potent that the software (the digital "by-product" of science) ends up incorporating an intricate scientific reasoning that may be hardly understood when one looks at the computer code itself. Scientific publications (which have only modest collaborative potentials) are treasured whereas computer code (whose evolution can potentially involve large numbers of people) is disregarded. This is so even if code works well and, therefore, the software package fulfils its analytical functions. In these conditions, effective code may be satisfying the technical (instrumental) needs of its producer but disappointing the collaborative (communicative) wishes of potential contributors.

14.2. Halting Social Change

To be sure, largely popular neuroimaging packages make room for some communicative phenomena, two of which are particularly worth mentioning. First, let us remember the words written by the leader of Linux, the largest collaborative online project: "I control the Linux kernel, the foundation of it all, because, so far, everybody connected with Linux trusts me more than they trust anyone else" (Torvalds and Diamond 2001, p. 168). When a certain software package acquires large diffusion, whether it results from large or limited collaboration, a considerable amount of trust must be in place. This is true for both flexible packages, where trust in deposited in the collaborative work of the community, and semi-flexible packages, where the renowned institutions responsible for software development is the target of trust. Thus the popularity of packages like FSL and AFNI is also produced by the scientific beliefs that people imparted to the University of Oxford and the American Institute of Mental Health, respectively.

Second, it should be considered that those popular packages constitute an indirect coding support that may underpin future development work. From an instrumental viewpoint, Raymond (2001, pp. 92–93) is right to claim that large projects may turn into "category killers": "People who might otherwise found their own distinct efforts end up, instead, adding extensions for these big, successful projects". From a communicative perspective, however, these large projects can be seen as "category creators," because they prevent developers from rewriting large, tedious, and basic pieces of code. In this way, programmers can focus on the specific task they have chosen to perform, developing the neuroimaging field, and using the previous code as technical support.

In the development of semi-flexible packages, there tends to be no large circulation of what Naur (2001) called the "theory" of the program, that is the intimate knowledge necessary to fully comprehend the codebase. When communicative rationality prevails and flexible software is produced, the theory of the program is largely shared, which also enlarges the group of designers able to implement future modifications. When instrumental rationality is predominant, as tends to be the case for semi-flexible packages, the theory of the program is controlled by small groups of designers. Because this situation secures scientific prestige to those groups, they tend to protect their program, safeguarding the theory and blocking a wide decision-making process.

The protective stance just described can be the outcome of conscious decisions, because development teams are surely aware of the scientific prestige brought about by their successful software packages. However, it is crucial to stress that, to a considerable degree, the low levels of collaboration entailed by semi-flexible packages derive from those packages' technical history, not really from strategic choices clearly taken. As previously explained, packages that are nowadays largely used began to be designed in the late 1990s, a relatively early moment in the history of both the modern open source software movement and modern neuroimaging. At that time, many of the collaborative resources available today (such as online platforms for code sharing) had not yet been created. Moreover, the programming philosophy of that moment gave priority to technical factors, as code sharing and collaboration had not yet been put forward by the subsequent open source movement. In this way, issues of code clarity and documentation were not carefully attended to by pioneering software developers, many of whom continue to work today. In his interview, Professor John Ashburner (Wellcome Trust Centre for Neuroscience, Univ College London, UK), one of those neuroimaging-programming pioneers, declared: "I'm a bit old-school in terms of the way that I write things." One of the consequences of being "old-school" is the superficial knowledge of coding strategies that have appeared in recent years, aiming at making code clarity and collaboration feasible. For example, Professor Ashburner revealed that he is not very familiar with object-orientation, a feature that enhances the organization, clarity, and coherence of code, as explained in chapterThree.

The point that needs to be highlighted is that widely used packages (mainly FSL, FreeSurfer, and AFNI), in spite of being semi-flexible packages with development teams quite closed towards full collaboration, have become almost mandatory analysis tools for neuroimaging researchers. These packages contain some basic analysis steps that must be taken in almost every neuroimaging study. For example, researchers generally have to realize normalization, so that different data, collected from different research subjects, can be displayed in a final image that becomes a sort of "average brain" of the sample studied. Normalization procedures can be easily found in packages such as AFNI. What is more, the scientific literature has elevated those packages to gold standard tools for normalization. Neuroimaging researchers, instead of spending time developing new normalization toolboxes, simple use what is already available and concentrate on more innovative tasks. As a consequence, the most popular packages have been, and will most likely continue to be, unavoidable tools.

Ironically, though, the pioneering packages, which have gained such legitimacy in the scientific literature, are not necessarily the best option in terms of technical efficiency. My interviewees, and particularly those with more profound programming knowledge, stressed two issues. First, newer packages have been launched which sometimes contain more efficient data processing approaches. However, the widespread use of old packages prevents those more efficient packages from gaining popularity. Second, largely used packages are sometimes too slow, because they result from obsolete programming techniques. Because of these two aspects, an interesting situation emerges where the majority of neuroimaging researchers help consolidate the dominance of packages that, from a technical viewpoint, are not always the best option.

"They [the oldest packages] are [...] not necessarily the most modern tools, like how they are developed, and I think there's still quite some possibility to provide better tools, like newer algorithms [...] For example... So FSL is something, I think, broadly used but, personally... Ahm, no, I don't want to talk about that [laughter].

But that would be my next question. I would ask you: do you think MRITK [the package developed by him] is better than FSL in some aspects?
[...] it really serves a very different purpose. The one thing that MIRTK is better is the particular registration algorithm that's implemented there, like for deformable registration [...] I mean, it has been compared already, in a few studies, how that [FSL] algorithm, how that tool compares to free-form deformation or others, and it just doesn't produce as good as results, for example [...]."

Dr. Andreas Schuh (Biomedical Image Analysis Group, Imperial College London, UK)

"**So, then... But is there any moment in which you have to have a look at the code of FSL to understand something?**
I looked at the code of FSL some times by curiosity, to see why it was so slow in some parts [...] These things should be very fast, but they are extremely slow. And then when you go to the code to try to understand why, you see they are doing many things manually where it would be much worthier taking libraries that are already available, you know. For example [...], there's an algorithm that's called Fourier transform, FFT, and there are libraries for doing that. They're extremely fast. And then you go to the code of FSL and you see that everything is kind of manually implemented [...] There are simplified things. You say: 'How can they do that manually!?' [...]

But do you think they do that manually because these tools are recent whereas their code is old?
[...] I think that it's maybe because those libraries were not very known at that time, because the internet should be extremely slow, right? But, so, this is now ready, and the code has never been modified [...] this is one of the reasons why the tools are extremely slow. You take hours to complete any little thing. So... [laughter]."

Researcher (Donders Institute for Cognitive Neuroimaging, the Netherlands)

On this point, another specific aspect can be pointed out. In flexible software, the swiftness with which new code can be integrated into the old codebase, as well as the access to the codebase by a potentially wide pool of contributors, make the constant update of source code easy (although not certain). In rigid software, the presence of an institution moved by practical concerns (frequently a company) guarantees a constant renewal of the codebase. This is what happens, for example, with the BrainVoyager neuroimaging package, produced by the Dutch company Brain Innovation. Since its initial release, in 1998, the package has been famous by its frequent incorporation of new features. In his interview, Professor Rainer Goebel, the company's CEO and main developer, declared: "[…] I hate nothing more than keeping legacy code and losing the edge that we had once […] Therefore, I rewrite […] old code constantly." Compared to these two types of packages, semi-flexible ones tend to be very slow in terms of code development, a characteristic found even in SPM, a software that can be classified as flexible.

In fact, whereas some packages display quick code development, other packages, which are by the way the most widely used, seem to be plodding. In order to understand the paradox, it is necessary to consider that most of those slow packages have semi-flexible development and are therefore dependent on inputs from small and busy teams. Thus the more or less pronounced sense of code

> *"So as a last question: SPM is now a quite old piece of software, a piece of code, if I can say that.*
> *Yeah.*
>
> *Do you think that this old age of the package, of the code, is an advantage or a disadvantage?*
> *[…] There is a lot of legacy stuff in there and we don't really have the resources to kind of bring up things up to date, so we just keep using the old code […] there is probably a half million lines of code, depending on how you count it… well, three hundred thousand to half a million… […]*
>
> *Changing it would be too costly.*
> *Ah, yeah, and it wouldn't really be science, and we're supposed to be here to*

> be doing science, publish papers. And, you know, bring the SPM code up to kind of new standards, it wouldn't be classified as science or... There isn't really a funding mechanism for that kind of thing. So we just kind of leave alone, fix little things that stop working [...] and, yeah, let the code being alone."
>
> Professor *John Ashburner* (Wellcome Trust Centre for Neuroscience, Univ College London, UK)

ownership nurtured by those teams ends up blocking a smooth improvement of the codebase. Furthermore, those teams are based in universities. Even when these latter can provide programmers with considerable technical and funding conditions (as seen in section 13), this cannot be comparable with the huge sums of resources invested by specialized multinational software companies. For example, one of the SPM developers, interviewed by me, explained that the development team has not had the time and resources to fully incorporate object-oriented features into the package's codebase. "There are some parts of SPM that use object-oriented code but for the most part, no [...] it could make various things easier but the cost of getting that, in terms of rewriting the entire codebase, essentially... we just... I don't think we have the manpower for that. So most of us working on it are doing research and developing the method and implementing the software and supporting the software, but we don't have dedicated software engineers [...] Basically, to start from scratch, which is what that would take, is probably too expensive." It is also important to remember that the SPM team, as well as other academic groups responsible for neuroimaging software design, are supposed to formulate theories and models about the brain, which turns software development into an ancillary aspect of their work.

In this way, there emerges a social dilemma that is present not only in the neuroimaging domain. On the one hand, some prestigious packages have their dominance consistently reinforced by the reputation of their institutions, as well as their frequent citation in the scientific literature. They have become a kind of coding support, as they provide researchers with basic functionalities that must not be redesigned over and over again. On the other hand, however, this very scientific success prevents other packages from becoming popular. What is more, these obscured packages sometimes display more robust analysis approaches and faster analysis performances. According to Atal and Shankar (2015, p. 1385), the open source development model, contrary to the proprietary model, generates the software even when "the cost of effort provision" is very high, because in open source the provision of a "socially beneficial" product is a major goal. It is then possible to further refine these authors' conclusion by pointing to an important difference between the two kinds of open source software: flexible and semi-flexible packages. In the design of flexible packages, the assimilation of new ideas and approaches is more efficient, which makes it easier to streamline the code constantly. In their turn, semi-flexible packages are to a large degree dependent on efforts and ideas by small groups of developers, which slows down the codebase's improvement. Therefore, flexibility, in addition to favouring the communicative process and fostering social collaboration, ends up being beneficial from the viewpoint of technical efficiency. Moreover, when the field is not dominated by companies, as is the case of neuroimaging, flexibility reinforces the possibility of a consistent renewal of available software packages, undermining the social dominance and technical inefficiency generally produced by semi-flexible schemes.

}

section15 (Leading Institutions, Main Classes) {

15.1. Basic Hierarchies in Software Development

The findings and ideas presented in section14 may have given the impression that the predominance of semi-flexible packages has technical implications only. In order to counter such false impression, it is crucial to scrutinize the close connection between technical and social phenomena. Most frequently, hierarchies found in computer code are products, as well as generators, of social hierarchies. Let us consider, for example, the issue of main classes. In programming languages that allow the division of computer code into different classes, there is always one of such classes that is considered as the main class. If one compares software with orchestral music, the main class can be said to be the music score. Even though no sound is produced by the written music, it is the music score that determines what each instrument has to play, when to play, how to play, when to stop, and so on. Equally, the main class of a software package usually performs minor functions, but it sends crucial messages to other classes, urging them to begin and stop their job. When the production of software is analysed from a sociological viewpoint, hierarchies can also be detected. For example, the main class is always designed by the group of original or leading programmers. In this way, those programmers, by controlling the package's most basic features, have a decisive influence on its future development, as well as on future modifications to the source code.

From a technical viewpoint, these considerations have been stressed by many analysts. While designing a software package, hackers have to define the conditions in which it will be used and modified in the future (Naur 2001). Programmers, and mainly those working on previously written code, depend on other people who have previously set objectives, provided resources, or controlled information (Brooks 1995). Therefore, whenever technology is tightly controlled, its future use gets considerably limited (Torvalds and Diamond 2001). However, the author whose analysis focuses on these issues in the most detailed way is Canadian software engineer David Parnas. He identified a "hierarchical structure" whereby software development consists of some basic decisions which lead to the production of some basic modules and which subsequently limit the scope of future decisions and future modules (Parnas 1972, p. 1057). In this way, the technical coherence of software is underpinned by a logical coherence.

> The last piece of code inserted may be changed easily, but a piece of code inserted several months earlier may have 'wormed' itself into the program and become difficult to extract. These considerations suggest that the early decisions should be those which are the least likely to change; i.e., those based on 'universal' truths or reasoning which takes into account little about a particular environment. (Parnas 1971, p. 5)

This explanation can be associated with the philosophical reasoning proposed by young Wittgenstein, for whom human reality is constituted by a chain of propositions. "Elementary propositions" are the ones which assert basic truths or facts whereas subsequent propositions must be stated in accordance with those basic facts. Hence the conclusion that "[…] *all* propositions are generalizations of the elementary propositions" (Wittgenstein 1922, p. 55). However, not all programmers engaged in collaborative design have been responsible for the assertion of "elementary propositions." In other words, basic modules and libraries, which establish the pillars of software development, are generally created by a small number of programmers and groups. In this way, the ability to steer software development in the future tends to be unevenly distributed. As a result, technical hierarchies, which are technically necessary in code writing, end up being coupled with series of social and geographical hierarchies that could be avoided and got rid of.

The first manifestation of such hierarchies lies in the relations between different researchers. As explained in chapterTwo, most neuroimaging software projects have mailing lists through which programmers and users can share information. Clearly, the mailing list of the biggest packages has a strong hierarchical organization, because users play the role of apprentices looking for knowledge whereas the members of the main development group appear as knowledge holders, frequently having the final say in technical discussions.

This is in strong contrast to the example of more collaborative projects where a more balanced discussion is allowed and even encouraged. For example, Professor Robert Oostenveld, the initiator of the FieldTrip software package, is willing to foster wide collaboration and discussion by means of the project's mailing list. "It's not a moderated email discussion list [...] So before you're allowed to ask a question, you at least should put yourself in a position such that you can also answer questions. It is something that I've seen failing in other open source environments [...] the way that I've set up the mailing list is that [...] every question that gets asked is received by 1500 people, and those people are the community and they jointly can hopefully go answer the question [...]." This simple example shows that, in some projects more than others, open and balanced discussion is promoted. If such discussion is absent, strategic behaviour, rather than communicative action, tends to be in place. The occurrence of such strategic behaviour, which generates hierarchies, can be verified not only at the local level but also internationally.

15.2. International Hierarchies in Software Development

Obviously, the conditions in which software development occurs, in different countries, mirrors the scientific inequalities between countries. In brain studies, this has to do, for example, with the fact that, in less dynamic countries, researchers of the human brain are many times neurologists with a vast population to serve clinically, thus having relatively less time to devote to research (Khadilkar and Wagh 2007), not to speak of software development. In section13, it was explained why funding is a pivotal dimension of software development. Some countries have bigger difficulties in terms of providing researchers with continued funding lines. For example, according to Professor Ana Luisa Raposo (Faculty of

Psychology, Univ of Lisbon, Portugal), of the main difficulties faced by Portuguese neuroimaging researchers, "the biggest difficulty is to obtain funding." Even when funding is secured, there is a limited number of research groups that manage to conduct cutting-edge studies, because of the knowledge, technical and scientific hardships found in the country. This is why Professor Alexandre Andrade (Institute of Biophysics and Biomedical Engineering, Univ Lisbon, Portugal) claimed that "in Portugal, honestly, research in the field of neuroimaging is too scattered." This lack of a "critical mass" was also pointed out by two of my Brazilian interviewees, namely Professor Li Li Min (Laboratory of Neuroimaging, Univ Campinas) and Professor Alexandre Franco (Laboratory for Images-Labima, Pontifical Catholic Univ of Rio Grande do Sul).

However, these infrastructural difficulties, so to say, are not really the focus of this section. Here, some other difficulties will be stressed, which stem from the convergence between the technical requisites of code writing and the institutional arrangements prevailing at most universities. To begin this analysis, one has to consider a distinction in code writing that has been overlooked by most analysts. On the one hand, source code is the basic text that animates the operations of any software package. On the other hand, there is another modality of code called script. Let us imagine the existence of two toolboxes, one responsible for measuring the concentration of blood in different brain regions, and the other one representing this uneven distribution of blood by means of dots on the computer screen. Now let us imagine that a neuroimaging researcher wants to use those toolboxes but, instead of focusing on the whole brain, the study will concentrate on the frontal cortex only. This researcher can then write a couple of lines of code sending instructions to both toolboxes, so they can perform their tasks as they would generally do, but excluding all the brain areas outside the frontal cortex. These few lines of code is what data analysts call script.

Therefore, a script is a quick and simple modality of code writing that, instead of creating something new, simply uses code that has been already created, implementing a more focused or specialized use of software. Whereas actual code writing requires advanced skills and knowledge in software development, scripting can be done with only basic knowledge and skills. Over the last years, most academic researchers have given up huge ambitions in terms of code writing and been satisfied with the acquisition of scripting notions, which is due to three main factors: the complexification of software packages; the rapid and constant progress in the field of software engineering; and the over-specialization of academic studies, making it necessary, for most students, to use only fractions of each software package. As a result, there has been a proliferation of script writers, described as "light programmers" by Skog (2003, p. 308). A knowledge hierarchy is then created, separating those "light programmers" from other people who could perhaps be described as "heavy programmers."

What needs to be stressed is that, many times, this new hierarchy manifests itself in the geographical space. By using data collected in Survey 1, I verified the number of "applications" (software packages plus toolboxes) developed in each geographical hub. It was then possible to identify six geographical levels, represented in Map 15.1.

Map 15.1 The six geographical levels in terms of number of applications developed: 2015 (*Source* Survey 1)

Map 15.1 shows the clear geographical concentration of the most productive hubs. Composing the first level, the least productive hubs (which developed from 1 to 10 of the applications cited in the 2015 neuroimaging literature) are located in some areas of the United States and Europe but are mainly found in less scientifically dynamic countries such as Brazil, Mexico, South Africa, and India. The second level is composed by emerging hubs (which produced from 11 to 50 applications) located in countries such as the United States, Japan, and China. The hubs of the third level constitute a rare intermediate situation, being located in only three countries: the United States, South Korea, and Japan. The fourth level is composed by only two hubs: the one comprising the Northwest of the United States and a small fraction of Canada, and the one formed by the UK and the north of France. The fifth level is formed, exclusively, by the big hub of the American Atlantic Coast. Finally, the sixth level, representing the most intense production (526 applications), is composed by the huge Central European hub, which encompasses nine countries. (Considering the countries included in my fieldwork, Brazil is at the 1st level, Portugal at the 2nd level, the UK at the 4th level, and the Netherlands at the 6th level. This variety of situations was one of the reasons why these countries were selected for fieldwork.)

One of the questions asked in Survey 1 was: "Why did you and your colleagues decide to use" the software package cited in your 2015 publication? A total of 14 options were given to respondents, who could select one or more options. During the analysis process, I classified these options in seven groups of reasons: academic (the package is largely cited in the scientific literature); technical (the favourable package's performance); practical (the package is user-friendly); access (the package is available in the institution); institutional (the package is largely used in the institution); support (the existence of online support for users); and programming. This last reason goes: "The source code is available, so it is possible to make modifications and add plug-ins." This is the reason that would be selected by the "heavy programmers" referred to above, because this motivation clearly implies the possession of sophisticated programming skills. According to the country where the respondent was based, I divided all the questionnaires into the geographical levels of Map 15.1. And according to the answers given by the respondents, I gave scores to each group of reason, and prepared statistical boxplots. The outcomes were different for the different hub levels presented in Map 15.1. The following boxplots bring the results pertaining to two geographical levels:

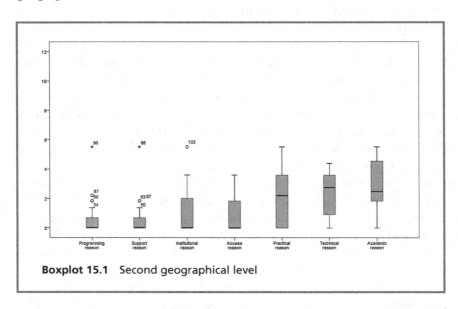

Boxplot 15.1   Second geographical level

For the second level (Boxplot 15.1), practical, technical, and academic reasons seemed to receive relatively higher scores. A statistical test showed that technical and academic reasons have indeed statistically significantly superior scores.[4] A similar pattern was verified for the first, third and sixth levels.[5] The fifth level (Boxplot 15.2) presents a different pattern. It can be seen that differences are attenuated here. The statistical test showed no statistically significant difference, enabling us to conclude that in this very dynamic hub, the programming reason comes to equal the relevance of other reasons. In this way, when the hub manages

---

[4]Friedman test, with $p = 0.000$.

[5]A small number of questionnaires were obtained from the fourth level, which prevents me from drawing conclusions about it.

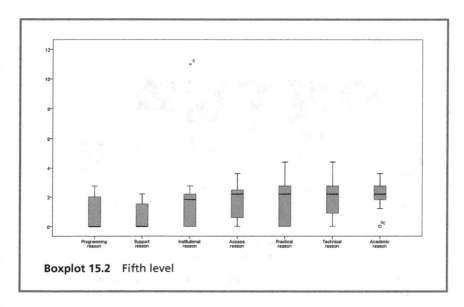

Boxplot 15.2   Fifth level

to become very productive, and when such production is coupled with the prospective and cutting-edge nature of the studies conducted locally, the programming logic and the role played by "heavy programmers" tend to be accentuated.

This is not the case for the 6th geographical level, though. This is so because in this level, code writing has become so widespread that it reaches levels of massification, which also increases the weight of "light programmers" writing simple scripts. This phenomenon is clearly expressed in the number of programming languages mastered in this level [lang, levels]. Generally, people with deep programming knowledge have had experience using different languages and would therefore declare that they know more than one language. In the sixth geographical level, even though the proportion of survey respondents declaring that they know three languages or more is high (almost 20% of 41 respondents), the highest proportion of people with no programming knowledge (almost 50%) is found. On this point, the Donders Institute for Cognitive Neuroimaging, the Netherlands, visited in my fieldwork, is very representative. Most of its researchers have background in psychology, lack deep programming training, and can therefore use neuroimaging software very basically, without being even able to write simple scripts.

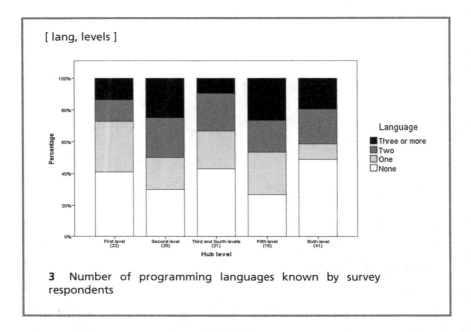

3 Number of programming languages known by survey respondents

"[...] I did all the [PhD] analysis by mainly pressing the little buttons [of the SPM package]. However, as I had to make small changes... There were several people in the laboratories where I have been, including in the PhD, there were several colleagues with programming experience and therefore they also helped us to create Batch scripts based on the analyses we wanted to perform. So, for example, to run the analysis in a loop for several participants, one after the other. Also, when the analyses were more complex and took longer to press the buttons, they helped us to prepare those...

To automatize it.
... those automatic processes."

Professor Ana Luisa Raposo (Faculty of Psychology, Univ of Lisbon, Portugal)

The situation of the 2nd level deserves to be highlighted. There, the proportion of people knowing three or more languages (more than 20% of 20 respondents) almost equals that of the dynamic 5th level (where by the way the lowest proportion of researchers with no programming knowledge is found). However, the 2nd level is modestly productive in terms of applications developed. This expresses a sort of "wasted talent" at this level, because programming knowledge cannot become actual coding work, which is related to the funding and infrastructural limitations mentioned in section15. However, the 2nd level also holds around one-third of researchers with no programming skills. For those people, two alternatives are available. First, they can run their studies by using the basic, user-friendly features provided by software packages. For example, a researcher I interviewed in the University of Coimbra, Portugal, has used the BrainVoyager and FreeSurfer packages. "So far I have managed to do it with the tools available on the software [...] But in terms of changing algorithms, no, I don't do it [...]." Second, those people can ask for help, so friends can write scripts or design basic applications for them.

This second type of help is much more frequent than one could imagine, because in addition to happening via personal contacts, it can happen through digital means. By using the online fora or mailing lists of popular packages, researchers can send queries pertaining to analysis difficulties they are facing. In this way, they are provided with help by the developers of the package, which sometimes include the writing of some "tailor-made scripts" which, in spite of aiming to solve one very particular problem, are usually kept on the website, so other people can also use them. One of the SPM developers I interviewed (University College London, UK) is very much used to providing people with this kind of support. "I've got so many... dozens, hundreds, probably, of emails where I've written bits of code for people [...] People can now just search that and copy and paste my code, often to do what they need to do [...]." This state of things, although it constitutes one example of indirect personal help, also consolidates the international hierarchies of software development. In addition to diffusing code through their main software package, those institutions that produce popular packages can also diffuse some simple and small pieces of code (scripts) by means of their mailing lists. In this way, those development groups become, at the same time, sources of lines of code and sources of ideas that solve technical problems experienced by other groups. Thus a state of scientific and coding hegemony is built up, drawing on the scientific and coding dependency of many groups and countries. In software development, as in other areas of social life, collaboration is not truly collaboration when it does not take the form of mutual collaboration.

15.3. Knowledge Hierarchies in Software Development

In chapterTwo, much attention was given to the issue of modularity (the division of software packages into more or less independent parts called modules, toolboxes, or plug-ins). It was then shown how this approach helps bring about collaboration and communication to software development. However, in a historic period when mediational actions tend to gain space, modularity could not help also having its instrumental implications, fostering purposive actions and even information hoarding.

Once again, it was David Parnas that analysed the technical dimensions of modularity in the most detailed way. He concluded that programmers implementing new modules to a certain software package should not know how previous modules work. Only really necessary information should then be shared. "We should not expect a programmer to decide not to use a piece of information, rather he should not possess information that he should not use. The decision is part of the design, not the programming" (Parnas 1971). Therefore, Parnas makes the crucial distinction between design (pivotal decisions about the program's structure) and programming (addition of modules). This distinction has not only a personal dimension (different programmers being responsible for either design or programming) but also a geographical manifestation (design decisions being taken mainly in some countries, namely those where the most popular packages are developed).

Parnas (1972) goes on to explain that software packages are composed by modules that hide information from each other. Indeed, there are technical features enabling this kind of secrecy. For example, when a program is written in the C++ language, even if it is an open source program, a list of all the classes is published but the actual code of some classes may not be published. This is so because the original developers may consider some classes as perfect or crucial, thus hiding their code in order to safeguard the program's technical integrity. This approach has been recommended by celebrated programmers. For example, Brooks (1995, p. 272) claimed: "I am now convinced that information hiding, today often embodied in object-oriented programming, is the only way of raising the level of software design."

My interviews showed that programmers, even if they create new modules for certain software packages, generally have a rather superficial knowledge of those package's source code. The irony of modularization is that it creates expert-ignorant programmers, those having in-depth knowledge of the modules they design but little information about preceding modules that underpin their work. To a large extent, modularization follows the logic of data sharing (data flowing from one module to others), not the logic of dialogue (previous modules being adjusted to new ones). In this way, collective development is speeded up, as seen in chapterTwo, but scientific and technical hierarchies are likely to emerge. This is one of the reasons why collective development, and even open source software, has been explored by large corporations. For example, in 2001 IBM decided to explore open source software and created the Eclipse project, a pool of companies devoted to the collective production of some software packages. According to Wagstrom (2009, p. 184), the project is characterized by collaboration but also by a hierarchical structure that he defines as "vertical interaction": "Modular architectures [...] reduce the amount of communication and coordination necessary as developers can treat substantial components as black boxes." The logic of information hiding, black boxes, and hierarchical relations can be perceived not only in rigid packages (those produced by companies) but also in many semi-flexible packages (those whose source code is subjected to institutional control). Eventually, the logic of dialogue is manifested within either the original developing group or the group that implements new modules, being rarely present in the conversations between different groups, especially when they are based in different countries.

The revolutionary and collectivist implications of open source software and the internet have been stressed (Barbrook 2003). However, such enthusiasm, which was very much widespread until the beginning of the twenty-first century, can now be analysed with caution. Even in open source software, there is a great deal of information that remains hidden, being fully acknowledged and understood only by core software designers willing to protect their product's stability, guarantee their knowledge hegemony, or increase their scientific prestige. It was noted that open source developers are sometimes not free from "[…] the development of strong leadership structures in a highly modularized environment" (Ghosh 2005, p. 36). In this way, Raymond (2001, p. 102) argues, development projects are usually organized in "two tiers of contributors": that formed by "co-developers" who share the most fundamental information about the project, and that formed by "ordinary contributors" who produce ancillary code and have superficial knowledge of the project. Indeed, some few developers are usually responsible for the production of the vast portions of a package's source code (Ghosh 2005). In this way, it is not sufficient to ask whether a certain package is closed or open source. When open source software is produced, actual collaboration may still be compromised if semi-flexible (instead of flexible) development arrangements prevail.

The hierarchies and inequalities analysed in this chapterFour have negative consequences for both the technical evolution of software packages and the social organization of hacker communities. As for technical aspects, it was seen that the prevalence of some packages ends up creating a state of slow renovation of source code, in such a way that the most widely used packages are sometimes the ones whose approaches are the most backward and inefficient. From a social viewpoint, hierarchies slow down or block the chain of communicative actions, thus halting the growth of knowledge in a field (neuroimaging software) whose main purpose is the production of knowledge of the human brain. As claimed by Harvey and McMeekin (2010, p. 487), the growth of knowledge depends "[…] on an absence of divisions between producers and users of that knowledge for the production of further knowledge." Limited knowledge advances derive from a situation where a state of feeble collaboration is frequently noted, and where a few countries, groups, and people are supposed to be the main producers of innovative knowledge whereas the vast majority of countries, groups, and people are supposed to produce only complimentary or imitative knowledge. In order to further explore these issues, it is important to delve into the features of one particular software.

}

section16 (R&D, SPM) {

In this chapterFour, the hierarchies of software development have been highlighted. This is not to say, however, that inequalities always prevail. Most of times, we are dealing with a set of mediational actions incorporating, at the same time, the instrumental and the communicative rationalities. In order to unravel this blend, it is interesting to focus on the example of a package that can be described as the world's most successful neuroimaging software: SPM. Financially underpinned by the Wellcome Trust, a biomedical research charity, as well as grants from funding agencies, SPM is vastly used by researchers engaged in neuroimaging data analysis. Map 16.1 shows the package's geographical diffusion.

Map 16.1 Geographical hubs with papers citing the SPM package: 2015 (*Source* Survey 1)

The strong presence of SPM in different world regions can be noticed in Map 16.1. Interestingly, SPM's diffusion in the hubs of the United States is only modest, indicating that it cannot surpass the dominance of FreeSurfer in this country, as shown in Map 13.2. Even with this weakness, the geographical diffusion of SPM is not equalled by any other neuroimaging package. A very high use intensity (75% or more) is noted in countries as different as Brazil, Mexico, the United States, Norway, Finland, India, and China. Why has the SPM project reached such high degrees of success? To answer this question, it is necessary to acknowledge the mediational features of SPM, as the package derives its force from both communicative and instrumental phenomena.

As for communicative aspects, the SPM development team, based in the Wellcome Trust Centre for Neuroscience, University College London, UK, formulated a very particular way of incorporating external contributions. Outsiders can design new toolboxes, which are publicized on the package's website, with links for the websites of the institutions where each toolbox has been developed [spm, tlbx]. By following those links, interested people can have further information

---

[spm, tlbx]

Quick Links

Toolboxes:

AAL | AAL2 | ACID | AICHA | ALI | ALVIN | AMAT | AnalyzeMovie | Anatomy | AQuA | ArtRepair | aslm | ASLtbx | at4fmri | aws4SPM | BFAST3D | BrainNetViewer | Brainnetome | BRANT | BredeQuery | Bruker2nifti | bspmview | CAT | CCAfMRI | CLASS | Clinical | Complexity | conn | ConnExT | CPCA | DAiSS | DICOMCD_Import | Diffusion_II | DPABI | DPARSF | DRIFTER | EMS | ExtractVals | FASL | FAST | fECM | FDR | FieldMap | fieldmap_undistort | FieldTrip | fMRIPower | fOSA | gPPI | GraphVar | GridCAT | Grocer | HV | IBASPM | iBrainAT | iBrainLT | IBZM_tool | ImaGIN | INRIAlign | ISAS | lead-dbs | lesion_gnb | LI | LogTransform | Mantis | MARINA | MARS | MarsBar | MASCOI | mfBox | Masking | Masks | MIP-C | MM | multifocal | MRTOOL | MRM | NIRS-SPM | NPBayes | NS | PETPVE12 | Ortho | PhysIO | PSPM | QModeling | REST | rfxplot | RobustWLS | SAfE | SAMIT | SCRalyze | SDM | SimpleROIBuilder | SnPM | SpikeDet | spm_wavelet | SPMd | SPMMouse | SSM | SUIT | SurfRend | SwE | TDT | TOM | UF2C | Unwarp2 | VarTbx | VDB | Volumes | WBM | WSPM | WFU_PickAtlas | xjView | XMLTools | ASLtbx | BENtbx | SVRLSMtbx | GIFT

4 List of SPM toolboxes developed by external collaborations, in a screenshot of the package's website (https://www.fil.ion.ucl.ac.uk/spm/software/)

on each toolbox, including documentation and access to source code. The publication of this list of toolboxes constitutes a great incentive for people willing to produce new SPM toolboxes, because their production will be announced on an online space visited by large numbers of neuroimaging researchers.

One of such toolboxes, named UF^2C, has been developed by Dr. Brunno Campos (Laboratory of Neuroimaging, Univ of Campinas, Brazil). From the beginning of his work to the moment when the toolbox became publishable, the development process took four years. He then contacted the SPM team via email. In three days, his toolbox was included into the list of SPM's website. Every time UF^2C is downloaded (on the website of the Laboratory of Neuroimaging), the system tries to identify the place from where the download is realized. Identification has failed for 35% of cases. For the cases where the download place could be identified, it was seen that most users (17%) have downloaded the toolbox from Brazil itself. However, there is also a considerable proportion of downloads in the United States (14%), China (6%), Australia (5%), and Germany (4%).[6] Therefore, the toolbox has obtained a considerable diffusion for a specialized application of its kind.

More detailed information could be obtained for two other SPM toolboxes (called VBM and REST) which, contrary to UF^2C, got citations in the 2015 neuroimaging literature. VBM has been developed by Christian Gaser at the University of Jena, Germany. It was released in 1995, four years after the release of SPM. Map 16.2 shows the geographical diffusion of this toolbox.

---

[6]https://www.lniunicamp.com/uf2c.

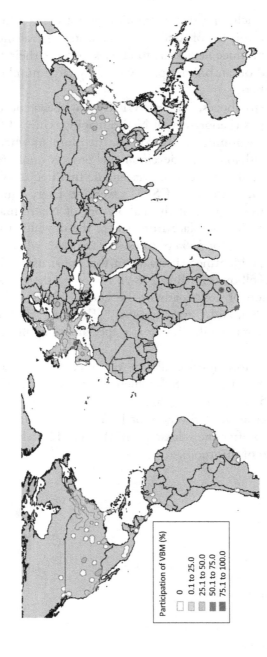

Map 16.2 Geographical hubs with papers citing the VBM toolbox: 2015 (*Source* Survey 1)

It can be seen that VBM has reached considerable geographical diffusion, including high intensity use in hubs of South Africa, Canada, and France.

The SPM toolbox called REST, released in 2011, has been developed by Yu-feng Zang at the Hongzhou Normal University, China. Its geographical diffusion is depicted in Map 16.3.

Map 16.3 Geographical hubs with papers citing the REST toolbox: 2015 (*Source* Survey 1)

Even though the REST toolbox has been mostly used in China itself, some use was also verified in Europe, as well as countries such as Pakistan, Saudi Arabia, and the United States. Its intensity use is smaller than those seen for the VBM toolbox, which is explained, among other factors, by its more recent release.

UF^2C, VBM, and REST are only three examples of the effectivity with which coding work based on SPM gets to be publicized. In this way, SPM has become a sort of programming pillar supporting work undertaken by several programmers in different parts of the world. Without such support, those programmers would face great difficulties to make their applications and coding ideas known. The proliferation of SPM toolboxes shows that the package's success is also due to an arrangement that strongly suggests the presence of flexibility in software development. In this way, all the mediations described in chapterTwo (personal, historical, and geographical mediations) are facilitated by SPM's popularity.

However, in a quite reserved manner, the SPM team does not really analyse the toolboxes' code or gauge their quality. In this way, the list of toolboxes functions as a kind of reference so that people are informed about coding work that has been based on the package, without representing any kind of technical endorsement from the SPM team. Thus there is no dynamic collaboration between this team and external producers of toolboxes, not to speak of version control. In this way, considering the features of software flexibility, SPM constitutes a rather intriguing case. While it is surely not a rigid package, it is hard to classify it as either flexible or semi-flexible. One might even say that it lies at the midpoint between flexibility and semi-flexibility, because even though some dialogue with external contributors is allowed to happen, a truly dynamic and open dialogue is lacking.[7]

SPM's development, in addition to being underpinned by some proto-collaborative work, is marked by three instrumental phenomena. First, the package, initially released in 1991, was actually the first one of its type. In the beginning, SPM was aimed to analyse data produced by positron emission technology (PET) and was used by a small group

---

[7]In my quantitative analyses, SPM was classified as a flexible package.

of researchers based in the Hammersmith Hospital, London, UK. The group decided to distribute SPM, which rapidly became "[…] the community standard for analyzing PET activation studies […]" (Bandettini, n.d, p. 3). When magnetic resonance technology was invented, the package was adjusted to operate with this new kind of data. According to Ashburner (2012, p. 792), SPM's development followed the principles of open science, so the package "[…] was then distributed to collaborators and other interested units around the world. Within a few years, SPM had become the most popular way to analyse studies of rCBF [regional cerebral blood flow] changes". In his interview, Ashburner recalled the initial moments of the package's history [jhn, ash].

[jhn, ash]

Professor John Ashburner has background in biochemistry. After doing the Master's Degree in information technology, he worked as computer officer at the Hammersmith Hospital in London, in the late 1990s. At that time, the hospital was one of the pioneering research centres in neuroimaging, by means of positron emission technology. Professor Ashburner met Karl Friston, who had recently begun to develop SPM. Inspired by Karl's ideas, John published, in 2001, a paper that launched the voxel-based morphometry (VBM) approach. This method, which consists in the study of the brain by considering small volumetric units called voxels, is nowadays widely used. In terms of software development, one of John's greatest achievements was the design of DARTEL, an SPM toolbox that enables the normalization of brain scans taken from several subjects. In the neuroimaging literature, DARTEL is today one of the most frequently cited applications. Nowadays, John is professor at Univ College London, being based, as a researcher, in the Wellcome Trust Centre for Neuroscience.

"Is it possible to say that SPM changed the history of neuroimaging?
I think it had a big impact in the 90s. [Pause] Yeah, a lot of people using SPM and people doing... You know, if we suggested people doing things in one way, then 'Yes, we'll do it that way' and people just... people kind of followed. But I think there is a lot more methodologically savvy people in the neuroimaging field now. So there is probably a lot more different methods people can use. So, now, I think the influence it has on the way people do things is... it isn't as great as it used to be [...]."

Because of this early appearance of the package, as well as the pioneering coding work undertaken by its developers, SPM gained an indisputable scientific prestige. It soon became a sort of yardstick for those involved with neuroimaging, a trend reinforced by its association with the MatLab programming language (see the Empirical example in the end of this chapterFour). In his interview, a researcher based in the University College London, UK, recalled the force gained by SPM in the 1990s. He claimed that "everyone trained in neuroimaging, at least at that time, you were trained to do something better than SPM." Even though the reputation of the SPM team has been softened by the emergence of other packages, it has not been completely destroyed. The sheer geographical diffusion of the package, as shown in Map 16.1, serves to underpin such prestige. In addition, the frequent citation of SPM in the scientific literature helps consolidate its image of scientifically sound software. This is why, in universities and research centres of many countries, people who learn to conduct neuroimaging studies are initially introduced to the SPM package. For example, Professor Alexandre Franco (Laboratory for Images-Labima, Pontifical Catholic Univ of Rio Grande do Sul, Brazil) explained to me that every year, in a short neuroimaging course offered by the Brain Institute of Rio Grande do Sul, students learn to process neuroimaging data with SPM.

Second, it was already explained that the collaboration between the SPM team and external hackers is not absent but is limited when the package is compared with truly collaborative projects. Currently, nineteen programmers are working in the latest version of SPM. Of those people, twelve are not considered as part of the core development team. In this way, one might have the impression of considerable levels of external interaction. However, let us consider that those twelve external contributors are also based in the University College London, where the core team works. In addition, the seven main developers (who form the so-called Methods Group) are those who possess the deepest knowledge of the codebase. According to an SPM developer interviewed in my fieldwork, the knowledge of more than 80% of SPM's codebase is held by the four most experienced programmers of the Methods Group. By the way, collaboration is rather uncommon even within the core development team. According to Professor John Ashburner, he and other members of the Methods Group seldom discuss the coding strategies they use while developing SPM. This is so because computer code written by one member has little interference, if at all, in code produced by other members, as "things are organized in a fairly modular way."

> "There are extremely clever algorithms in SPM for warping brains into the same space. So it's pushing and squashing them, so they all have the same form, so they comply to a standard, a stereotactic space [...] Those [...] algorithms, I'm just accepting that I'm never going to understand them [...] So I appreciate that some parts of this code, you have to be kind of... err... at least mentally, kind of compartmentalized [...] So you do get a bit of, I guess, ghettoization, possibly, within the code, and you kind of get people whose expertise is in one or more part of it. And that is by necessity, because it's... because the theory is different for each different area."
>
> One of the main SPM developers (Wellcome Trust Centre for Neuroscience, Univ College London, UK)

There is a third instrumental phenomenon marking SPM's development. As a result of the modest collaboration just described, the SPM source code is far from having the syntactic clarity which is typical to collaborative code writing. On the contrary, as noted by many of my interviewees, its code is hard to understand and lacks even internal coherence, as SPM developers are let free to follow their own rules in terms of code structure, documentation, and comments. Eventually, there emerges a phenomenon of "ghettoization" of the SPM code, to use an interesting expression used, in an interview, by one member of the Methods Group. Thus even though the SPM code has constituted a working platform for developers willing to implement new toolboxes, such complementary work does not go without much effort necessary to decipher the secrets of SPM's code lines.

The SPM development team has already begun to use GitHub, the online platform for code sharing. Nevertheless, only past versions of the code will be published. In other words, the most recent and innovative coding work being conducted on SPM will continue to be accessed by the main developers only, a strategy in line with the defensive stance of teams designing semi-flexible packages. This state of things has triggered much criticism from other programmers, and particularly those committed to open source.

"Is there any ideology behind the NIPY [open source neuroimaging software] community, something like Richard Stallman's Free Software movement?
Yeah, absolutely [...] It's definitely about ownership, about who owns the software; in my case, who owns the software to do neuroimaging. Is it BrainVoyager people? Is it, you know, is it the Oxford lab? Is it the London lab that makes SPM? Is it even the NIH lab which has AFNI? [...] So, for example, you see in our licensing that it's completely open [...] We do not ask you to sort of recognize us when you use the software [...] So it's... it's... it's the result of a very strong idea of what science should be and how it should be performed, yeah. And a resistance, I think, to the way that software's being used in imaging. It's being used for political purposes, I think.*

For political purposes?
Yeah. I mean, I can't find it now but when SPM released its code [...], the main developer, who had written most of the code at that point, wrote a little paper explaining why it was in the lab's interest to do it. He called it enlightened self-interest [...] He said: 'It's better for science and it's better for us, because people will use our stuff and then they'll know our stuff and we will be, you know, more influential in the field.' That's been true, I think [...] [The SPM development team] achieved its dominance not just because of the scientific papers but also because that's where the software is written [...]*

Even though it's open source.
Yeah... The thing is, the open source doesn't cut it [...] You can get the development repository for SPM, but the stuff they're working now is secret. They'll release it when they're ready, but you can't see what they're working on now. I don't know whether the same is true with FSL but it wouldn't really matter because their code is not something that it's easy for other people to work on. So they haven't... There is a natural barrier for both FSL and SPM in the sense that the code is very hard to read and edit, so...*

Do they do it intentionally?
No. No, no, no. I think, in both cases, the people who wrote the code came from a much older... well, what we would now recognize as being old-school engineering. You know, you write the C code and you put the

> *C code in your website. That's how things worked [...] I mean, our stuff is much less successful, obviously, than their stuff, but when we're writing stuff, we do everything so that you can see everything of what I'm doing at every moment, like things I'm just playing with would appear on GitHub, as soon as I'm finishing that thought, and you can read it. And we put a big emphasis on code review. We try... you know, we care a lot about people being able to read our stuff and being able to edit our stuff [...]."*
>
> Dr. Matthew Brett (School of Biosciences, Univ of Birmingham, UK)

In this way, the example of SPM reveals the increasing weight of mediational actions in software development. On the one hand, the communicational rationality is fostered when people are indirectly encouraged to write new toolboxes which will be published on the package's website. On the other hand, the instrumental rationality is consolidated by means of a protective stance whereby the SPM team prefers to have exclusive access to the most cutting-edge ideas being elaborated by its core developers. This paradoxical process is not exclusive to SPM, though. Considering the whole neuroimaging domain, many examples can be found of collaboration and hierarchy occurring simultaneously. The tension is likely to be consolidated in the years to come, as instrumental and communicative actions, instead of preying on each other, emerge in a complementary state of mutual preservation.

section17 Empirical Example: The MatLab Programming Language

For the sake of simplicity, I will describe MatLab as a programming language. However, it is more than just this. It is actually a computing environment allowing scientists to develop algorithms and perform several types of complex quantitative analyses. In its origin, MatLab carried out only operations with mathematical matrices, hence its name, which derives from "Matrix Laboratory." American mathematician and computer programmer Cleve Moler wrote MatLab's original pieces of code. Before this work, in the 1960s, he used the available programming languages, mainly Algol and Fortran.

At that early time, British mathematician James Hardy Wilkinson, working with collaborators, developed two computer libraries (called EISPACK and LINPACK) containing many mathematical operations. The subsequent development of MatLab by Moler was decisively influenced by these two libraries. In addition, Moler relied on findings made possible by his collaborations with George Forsythe and John Todd, two researchers in computing. In the late 1970s, Moler mobilized all this knowledge to write the first version of MatLab, using the Fortran language. Although he was very satisfied with the outcomes of his work, he was not completely happy with the use of Fortran, which he saw as a technical limitation to his application. The main issue with Fortran was its difficult portability from computer to computer, a hardship that Moler was able to circumvent, so the first version of MatLab was fully portable. Initially, MatLab was not framed, by many scientists, as an outstanding application. However, engineers, who frequently have to deal with matrices, rapidly identified MatLab's potentialities. Gradually, the application was adopted in engineering and related fields (Moler 2004). Even today, MatLab is largely used in the so-called hard sciences.

When the application's scientific worth was being recognized by more and more people, Jack Little, an American control engineer, gave a commercial spur to this story. In the early 1980s, he used MatLab to develop a commercial product. Subsequently, helped by his colleague Steve Bangert, he translated the MatLab's code into the C language, implementing many optimizations. The two colleagues sought then the collaboration of Coler, MatLab's original developer, and, in 1984, the three collaborators decided to create MathWorks, a company that continues to be responsible for the package's development. With headquarters in the city of Natick, United States, MathWorks launched a large number of parallel products but MatLab has always been its main commercial achievement. In 2004, twenty years after the creation of the company, MatLab had around 1 million users worldwide (Goering 2004).

Due to its early appearance, as well as the robustness of its algorithms and graphical interface, MatLab was gradually diffused among scientists. Repeating what has happened to many software packages, MatLab has received much endorsement in the scientific literature. However, we are not dealing with a simple software package; MatLab is also a computing environment that can be used to build up software packages. This technical power provides scientists with many possibilities, turning the application into a kind of gold standard in many academic disciplines. Nowadays, in several universities across the world, students have joined programming courses where MatLab is the language taught. In my fieldwork in Brazil, the UK, the Netherlands, and Portugal, I did come across many examples of researchers who learnt to program by joining university courses on MatLab.

One might consider that MatLab, as a proprietary language, would have its geographical diffusion hindered by the financial limitations of institutions unable to pay its licence. Indeed, the prices determined by MathWorks are very high. In addition, the application's toolboxes are sold separately. According to Dr. Fábio Duran (Laboratory for Neuroimaging in Psychiatry, Univ of Sao Paulo, Brazil), the general licence costs around 5000 US dollars whereas additional 2000 dollars must be paid for each toolbox. Therefore, only wealthy institutions are able to access the whole range of MatLab's functionalities. As a consequence, some academic departments, or even whole universities, have been unable to provide their researchers with access to the application. For example, the University of Lisbon, Portugal, does not have a licence for MatLab, according to Professor Ana Luisa Raposo. She is nevertheless able to use the application because the laboratory where she is based managed to secure funds for purchasing the licence.

Therefore, the commercial nature of MatLab helps reinforce some institutional and scientific inequalities. By the way, the hierarchical logic is embedded in the package's technical structure. Object-orientation had not yet been created when MatLab was developed. In this way, code written in this language could not

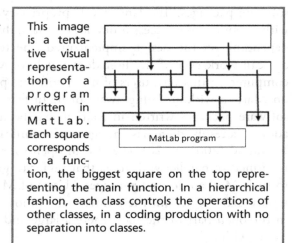

This image is a tentative visual representation of a program written in MatLab. Each square corresponds to a function, the biggest square on the top representing the main function. In a hierarchical fashion, each class controls the operations of other classes, in a coding production with no separation into classes.

be separated in classes, being rather rendered in a single file formed by large numbers of lines. As a result, MatLab code, compared with code written in object-oriented languages, assumes a particularly heavy hierarchical structure, with some functions playing strongly hypertrophic roles. Recently, MatLab gained object-oriented features. However, because most MatLab programmers have learnt to work without such features, most programs written in this language continue to be organized in the old, hierarchical way.

In spite of its price barriers, MatLab is nowadays largely used in many countries. One could say that it has been used both directly and indirectly. In terms of direct use, MatLab has been accessed because universities strive to purchase the licence of a product with increasing scientific legitimacy. For example, neither Dr. Brunno Campos (Laboratory of Neuroimaging, Univ of Campinas, Brazil) nor Dr. Fábio Duran (Laboratory for Neuroimaging in Psychiatry, Univ of Sao Paulo, Brazil) have met any difficulty to use MatLab, as their institutions guarantee this access. In terms of indirect use, it is interesting to note that MatLab has been the main tool for the design of many software packages. The language is proprietary but packages built up with it can be open access if developers make this choice. Furthermore, depending on the design implemented, the resulting software package can work even if the user does not have MatLab installed in the computer. In this way, users of many software packages may be relying on MatLab without being aware of it. In fact, a pool of open source software packages written in MatLab has been formed over the decades. As a result, there has emerged the technical divide pointed out by Raymond (2001, p. 28) whereby an open source MatLab community is formed in parallel with users that prefer to give a commercial shape to their MatLab-derived packages. "Users of MATLAB [...] invariably report that the action, the ferment, the innovation mostly takes place in the open part of the tool where a large and varied community can tinker with it."

The field of neuroimaging holds many examples of open source use of the commercial MatLab application. The language is not only very popular, it also seems to be the preferred tool for the design of neuroimaging packages. MatLab has been used, either exclusively or in combination with other languages, in the design of most of the 23 packages involved in Survey 2 [lang, pckg]. In my fieldwork, I found many examples of researchers writing code and developing toolboxes with MatLab. This is so because the language has been an important part of their training as data analysts. For example, Dr. Marcel Swiers (MR Physics group, Donders Institute for Brain, Cognition and Behaviour, the Netherlands) has taken MatLab as his main working language, in spite of his growing interest in Python. "[...] for my data analysis? So I tend to... Still, I have a lot of it still in MatLab because I'm more fluent, I'm more experienced in MatLab. And also I have developed, over the years, more code in there [...] Like I said, I'm moving away from it, but I still have a legacy codebase [...], so I can do it more quickly in MatLab."

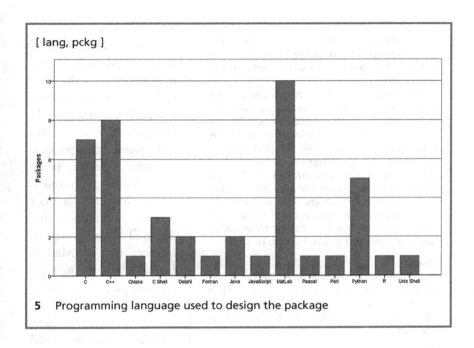

5   Programming language used to design the package

The popularity of the MatLab language in neuroimaging research can be explained by its aforementioned scientific prestige and pioneering role but also by two concurrent factors. On the one hand, the application contains a powerful graphical interface to represent data visually, a feature that can only be highly appreciated by researchers willing to produce brain images. On the other hand, MatLab offers a wide range of libraries, which are digital repositories of mathematical and statistical operations. Once again, Dr. Marcel Swiers explains that "in MatLab you also have toolboxes that you sometimes use but... Yeah, a lot of build-in functions." Much coding work is therefore spared, as researchers can simply use the operations embedded in the application.

To give just a final idea about the relevance of MatLab for the neuroimaging domain, let us consider that SPM, the most widely used neuroimaging package, is developed in this language. It is surely possible to use SPM without having MatLab installed in the computer. However, two serious

> **"Is it possible to say that MatLab is the most appropriate language (or at least one of the most appropriate languages) for scientific analysis?**
> *Well, scientific analysis is too broad a term, but if we talk about neuroscience analysis specifically, I would say that I think it's the most widely used tool currently. And I say this because I coordinate the undergraduate research stay program, which implies contacting many laboratories in Europe and sometimes in America as well trying to position our students. And many times, either directly or indirectly, I know that they... One of the first things they ask is: 'Does the student know MatLab or not?'*
>
> **Really?**
> *Yes. I would easily say that 50 per cent, roughly, of potential [foreign] supervisors of research stays or Master's Degree thesis are very much willing that students know MatLab [...]*
>
> **Okay. That is, it's almost turning into a second language...**
> *Yes, but I think this is not recent, I think this is not recent [...] My view is that, in this specific field of bran signal processing, neurosciences, and so on, MatLab is currently dominant. There are many applications (most of which I use myself) that are MatLab-based [...]."*
>
> Professor Alexandre Andrade (Institute of Biophysics and Biomedical Engineering, Univ of Lisbon, Portugal)

limitations must be faced in this case. First, without access to MatLab, it gets impossible to write new code in order to automatize and customize SPM's operations. Second, and even worse, the package becomes unstable, being even prone to crashing, when MatLab is not present in the computer. Thus SPM promotes an indirect use of MatLab in addition to strongly encouraging its direct use.

Therefore, some similarities between SPM and MatLab can be pointed to. Both applications were released at a relatively early moment of modern neuroimaging studies. Both have obtained wide geographical diffusion greatly underpinned by their frequent citation in the scientific literature.

Both have been part of the academic formation of many scientists. Many times, students learn these two applications in tandem. For example, Professor Alexandre Andrade's students are taught basic notions of programming with MatLab, and neuroimaging data processing with SPM. Because of this close relation, some researchers I interviewed curiously describe SPM as a toolbox of MatLab.

From a sociological viewpoint, the main difference between the two applications is the fact that SPM can be considered as a flexible package whereas MatLab, because of its commercial nature, is surely a rigid application. Anyway, the connected history of SPM and MatLab has much to reveal about a historical moment when divergent rationalities (in this case, the flexible and the rigid logics) can be mingled to form thriving manifestations of mediational actions.

References

Ashburner, John. 2012. "SPM: A history." *NeuroImage* 62:791–800.

Atal, Vidya, and Kameshwari Shankar. 2015. "Developers' incentives and open-source software licensing: GPL vs BSD." *B.E. Journal of Economic Analysis and Policy* 15 (3):1381–1416.

Bandettini, Peter. n.d. A short history of Statistical Parametric Mapping in functional neuroimaging.

Barbrook, Richard. 2003. "Giving is receiving." *Digital Creativity* 14 (2):91–94.

Brooks, Frederick P. 1995. *The mythical man-month*. Reading: Addison-Wesley.

Cowley, L. Tad, Hope L. Isaac, Stuart W. Young, and Thomas A. Raffin. 1994. "Magnetic resonance imaging marketing and investment: Tensions between the forces and the practice of medicine." *Chest* 105 (3):920–928.

Cox, Robert W. 2012. "AFNI: What a long strange trip it's been." *NeuroImage* 62:743–747.

Cusumano, Michael A. 1992. "Shifting economies: From craft production to flexible systems and software factories." *Research Policy* 21 (5):453–480.

Ferreira, Mariana Toledo. 2018. *Centro(s) e periferia(s) na produção do conhecimento em genética humana e médica: um olhar a partir do Brasil*. PhD thesis, Department of Sociology, University of Sao Paulo, Brazil.

Fischl, Bruce. 2012. "FreeSurfer." *NeuroImage* 62:774–781.

Galloway, Patricia. 2012. "Playpens for mind children: Continuities in the practice of programming." *Information & Culture* 47 (1):38–78.

Ghosh, Rishab Aiyer. 2005. "Understanding free software developers: Findings from the FLOSS study." In *Perspectives on free and open source software*, edited by Joseph Feller, Brian Fitzgerald, Scott A. Hissam, and Karim R. Lakhani, 23–46. Cambridge: MIT Press.

Goering, Richard. 2004. Matlab edges closer to electronic design automation world. *EETimes*. Available at https://www.eetimes.com/document. asp?doc_id=1151422.

Habermas, Jürgen. 1996. *Between facts and norms: Contributions to a discourse theory of law and democracy*, Studies in contemporary German social thought. Cambridge, MA: MIT Press.

Habermas, Jürgen. 2008. *Between naturalism and religion*. Cambridge: Polity Press.

Harvey, Mark, and Andrew McMeekin. 2010. "Public or private economies of knowledge: The economics of diffusion and appropriation of bioinformatics tools." *International Journal of the Commons* 4 (1):481–506.

Heliades, G. P., and E. A. Edmonds. 1999. "On facilitating knowledge transfer in software design." *Knowledge-Based Systems* 12:391–395.

Khadilkar, S. V., and S. Wagh. 2007. "Practice patterns of neurology in India: Fewer hands, more work." *Neurology India* 55:27–30.

LaToza, Thomas D., Gina Venolia, and Robert DeLine. 2006. "Maintaining mental models: A study of developer work habits." Proceedings of the 28th International Conference on Software Engineering, New York, USA.

Merton, Robert K. 1968. "The Matthew effect in science." *Science* 159 (3810):56–63.

Moler, Cleve. 2004. The origins of MATLAB. MathWorks. Available at https://www.mathworks.com/company/newsletters/articles/the-origins-of-matlab.html.

Naur, Peter. 2001. "Programming as theory building." In *Agile software development*, edited by Alistair Cockburn, 227–239. Boston: Addison-Wesley.

Nofre, David, Mark Priestley, and Gerard Alberts. 2014. "When technology became language: The origins of the linguistic conception of computer programming, 1950–1960." *Technology and Culture* 55 (1):40–75.

Parnas, David Lorge. 1971. *Information distribution aspects of design methodology*. Pittsburgh: Computer Science Department, Carnegie-Mellon University. Available at http://repository.cmu.edu/cgi/viewcontent.cgi?article=2828&context=compsci.

Parnas, David Lorge. 1972. "On the criteria to be used in decomposing systems into modules." *Communications of the ACM* 15 (12):1053–1058.

Raymond, Eric S. 2001. *The cathedral & the bazaar: Musings on Linux and open source by an accidental revolutionary*. Sebastopol: O'Reilly.

Rifkin, Jeremy. 2014. *The zero marginal cost society: The internet of things, the collaborative, and the eclipse of capitalism*. New York: Palgrave Macmillan.

Sawle, Guy V. 1995. "Imaging the head: Functional imaging." *Journal of Neurology, Neurosurgery and Psychiatry* 58 (2):132–144.

Sawyer, S., and P. J. Guinan. 1998. "Software development: Processes and performance." *IBM Systems Journal* 37 (4):552–569.

Schwarz, Michael, and Yuri Takhteyev. 2010. "Half a century of public software institutions: Open source as a solution to hold-up problem." *Journal of Public Economic Theory* 12 (4):609–639.

Skog, Knut. 2003. "From binary strings to visual programming." In *History of Nordic computing*, edited by Janis Bubenko Jr., John Impagliazzo, and Arne Solvberg, 297–310. Boston: Springer.

Stallman, Richard M. 2002a. "Free software: Freedom and cooperation." In *Free software, free society: Selected essays of Richard M. Stallman*, edited by Joshua Gay, 155–186. Boston: GNU Press.

Stallman, Richard M. 2002b. "Releasing free software if you work at a university." In *Free software, free society: Selected essays of Richard M. Stallman*, edited by Joshua Gay, 61–62. Boston: GNU Press.

Torvalds, Linus, and David Diamond. 2001. *Just for fun: The story of an accidental revolutionary*. New York: HarperCollins.

van Essen, David C. 2012. "Cortical cartography and Caret software." *NeuroImage* 62:757–764.

von Hippel, Eric, and Georg von Krogh. 2003. "Open source software and the 'private-collective' innovation model: Issues for organization science." *Organization Science* 14 (2):209–223.

Wagstrom, Patrick Adam. 2009. "Vertical interaction in open software engineering communities." PhD, Carnegie Insitute of Technology/School of Computer Science, Carnegie Mellon University.

Weber, Marx. 1958. "Class, status, party." In *From Marx Weber: Essays in sociology*, edited by H. H. Gerth and C. Wright Mills, 180–195. New York: Oxford University Press.

Wittgenstein, Ludwig. 1922. *Tractatus logico-philosophicus*, International Library of Psychology, Philosophy and Scientific Method. London: Kegan Paul.

```
# social code
# source code
```

chapterFive (Using Code: The Social Diffusion of Programming Tasks) {

```
583     void MainWindow::createActions()
584     {
585         m_newAct = new QAction(tr("New"), this);
586         m_newAct->setShortcuts(QKeySequence::New);
587         m_newAct->setStatusTip(tr("Create a new file"));
588         connect(m_newAct, SIGNAL(triggered()), this,
                SLOT(newFile()));
589
590         m_openAct = new QAction(tr("Open..."), this);
591         m_openAct->setShortcuts(QKeySequence::Open);
592         m_openAct->setStatusTip(tr("Open an existing
                file"));
593         connect(m_openAct, SIGNAL(triggered()), this,
                SLOT(openFile()));
594
595         m_saveToJsonAct = new QAction(tr("Save..."), this);
596         m_saveToJsonAct->setShortcuts(QKeySequence::Save);
597         m_saveToJsonAct->setStatusTip(tr("Save the document
                as XML-file to disk"));
598         connect(m_saveToJsonAct, SIGNAL(triggered()), this,
                SLOT(saveFileToJson()));
599
```

© The Author(s) 2019
E. Bicudo, *Neuroimaging, Software, and Communication,*
https://doi.org/10.1007/978-981-13-7060-1_5

```
600        m_exportAct = new QAction(tr("Export as…"), this);
601        m_exportAct->setShortcuts(QKeySequence::Print);
602        m_exportAct->setStatusTip(tr("Export worflow"));
603        connect(m_exportAct, SIGNAL(triggered()), this,
             SLOT(exportFile()));
604
605        m_loadNodesAct = new QAction(tr("Load Dictionary…"),
             this);
606        // m_loadNodesAct->setShortcuts(QKeySequence::);
607        m_loadNodesAct->setStatusTip(tr("Load new nodes into
             the library"));
608        connect(m_loadNodesAct, SIGNAL(triggered()), this,
             SLOT(loadNewNodes()));
```

The piece of computer code presented above belongs to a software package called Porcupine,[1] which has been developed by Dr. Tim van Mourik (MR Physics Group, Donders Institute for Cognitive Neuroimaging, Netherlands). These lines were extracted from a class called "createActions" (line 583). Initially, I would like to draw attention to lines 585 and 586 where two functions are used whose name start with Q: "QAction" and "QKeySequence." These two functions were not written by Tim: they belong to a package called Qt (pronounced "cute"). The C++ language, used to build up the Porcupine package, enables a particularly rapid data analysis but it does not provide hackers with a graphical interface. Thus everything happens in a console with black background on which inputs are given by means of written commands and outputs also assume the form of written letters and numbers. In this way, by using the C++ platform only, it is impossible to create graphics, draw figures, apply colours, and so on. For neuroimaging researchers, who aim to produce brain images, this constitutes an obvious limitation. To circumvent it, software developers have recourse to graphical platforms with which they transform raw and dry data into cute images. Qt is one of such platforms. By using them, developers of neuroimaging software packages can attend not only to the need of producing visual representations of the brain, they can also make their packages become user-friendly: instead of having to type in written commands, users are provided with buttons, tabs and other intuitive references.

This is what Tim van Mourik realized for the production of his Porcupine package. In the piece of code presented above, he is creating widgets (icons on the computer screen) for making the navigation of his package easier. The functions of the Qt platform appear in lines 585, 586, 590, 591, 595, 596, 600, 601, 605, and 606. In addition, some written instructions are made to appear on the computer screen, such as "Create a new file" (line 587) and "Open an existing file" (line 592). By analysing this source code, two intentions show through. First, the code was intended to be very clear, organized, and

---

[1] https://github.com/TimVanMourik/Porcupine

readable, which is achieved through outstanding care in terms of organization and comments. Second, the final package should be easy to use, as its development was decisively motivated by some needs felt by users [hist, prcpn].

[hist, prcpn]

Formed a computer scientist, Dr. Tim van Mourik has solid programming skills, knowing six languages: C++, Java, MatLab, Python, Ruby, and JavaScript. In his Master's Degree, he focused on the games industry and film industry. For a short period, he worked as a research assistant at the Donders Institute for Cognitive Neuroimaging. In 2013 he changed the focus of his attention and began a PhD program on data analysis and brain studies. Initially, he was mostly interested in the technical aspects of data collection and the operation of magnetic resonance scanners. Gradually, he became interested in the anatomical and functional aspects of the brain itself. The beginning of the Porcupine project was greatly motivated by a gap he identified in the range of available neuroimaging packages.

"[...] I was just in general annoyed that there was no clear way of linking up one function to the next. And I was very used to that from my previous experience. For example, in the movie industry most programs have some type of graphical interface of all the pipelining things that you do in data processing in general. And I was quite amazed actually that neuroimaging did not have anything like that. So I just started out to program that as a kind of hobby project but that grew up [...] and slowly started becoming something real and basically now is a very useful tool [...]."

The example of Porcupine illustrates a frequent concern of software developers: their packages are mainly devoted to solving research needs. As most developers of neuroimaging software are, at the same time, users of neuroimaging software, the creative drives of code writing are deeply connected with the analytical needs felt during the conduct of particular studies. However, not all software users are also software developers. A growing number of researchers do lack programming skills, being able to only use ready-made packages, especially those which are intuitive and user-friendly. Slowly, scientific divisions appear whereby some researchers fail to comprehend all the dimensions of software while becoming more and more dependent on the technical support offered by it. To analyse these issues, this chapterFive explores seven phenomena: the growing production of user-friendly packages; the central role played by software users; the difference between the virtual realm and the actional realm; the spontaneous teaching initiatives of the neuroimaging domain; the technical mysteries of software before many users' eyes; standards (and lack thereof) in neuroimaging; and the cognitive inequalities in the universe of software, as well as their political implications.

section18 (Intuition, Automation) {

In chapterFour, it was argued that over the last decades, an important difference has emerged between developers able to write computer code and create new software packages, on the one hand, and researchers who can write simple scripts to automatize some data analysis tasks, on the other. In this chapterFive, it will be necessary to further analyse this difference between "code writers" and "script writes," so as to explore some social, geographical, and political aspects of software development. It is then important to have a deeper understanding of the distinction between code and script. Let us then imagine a numeric group formed by ten numbers, from 0 to 9. Somebody wants to add all these numbers together. In the C++ programming language, this is how such an operation would be realized:

Example 1

```
1    int sum = 0;
2    int number = 0;
3    while (number < 10){
4            sum += number;
5            number++; }
6    cout << sum << endl;
```

By writing these six simple lines (92 characters), and putting them to run, the user would see the correct outcome (45) printed on the computer screen. Now let us see how the very same operation would be performed with the R programming language:

Example 2

```
1    sum (0:9)
```

This time, with only one line (9 characters), the user obtains exactly the same outcome. What is the difference between C++ and R?

Every programming language is a tool used by hackers to "speak with the computer," transmitting commands and instructions to the central processing unit (CPU). However, each language realizes this conversation from a certain distance. If a so-called low-level language is used, the programmer writes code by using 0s and 1s, which is the language spoken by the CPU, so the conversation happens from very close and no translation is needed. For the so-called high-level languages like R, there is a distant conversation, making it necessary to usher the instruction into different layers of technical translation. C++ might be classified as a middle-level language. It is actually one of the languages used to build up R. In this way, when the line presented in Example 2 is read by the machine, it is automatically transformed into the operation represented in Example 1. From there, the instruction is turned into machine language to be finally realized by the computer. In this way, Example 2 does not really contain a piece of computer code: it is actually a script. Therefore, a script can be defined as a simplified piece of code whose operation activates some underlying and more complexes pieces of code. When scripts are used, a chain of technical translation is triggered, with increasing complexity, leading to the final realization of the desired functionality [lang, lvls].

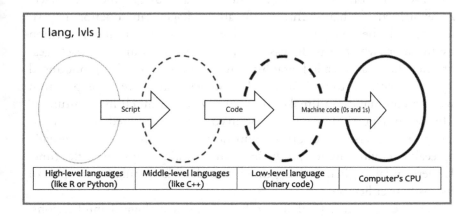

[lang, lvls]			
→ Script →	→ Code →	→ Machine code (0s and 1s) →	
High-level languages (like R or Python)	Middle-level languages (like C++)	Low-level language (binary code)	Computer's CPU

Because of the presence of so many layers, one of the main concerns of people who use high-level languages is computing time. When simple tasks are performed, such as the one presented in Example 2, speediness is guaranteed. However, if complex calculations and visual operations have to be called for, much time may be spent in passing from one layer to the next one. This is why high-level languages are commonly used for less demanding operations such as data processing whereas middle-level languages are optimal for the construction of whole software packages. For example, Dr. Danilo Dias (Centre for Images and In Vivo Spectroscopy through Magnetic Resonance, Univ of Sao Paulo, Brazil) has been developing a neuroimaging package in Python. He appreciates the several resources and libraries provided by the language but is aware that his package may at times work slowly because a high-level language is being employed.

In 1996 at the Human Brain Mapping conference, Rainer Goebel communicated the release of the BrainVoyager software package. Some participants got very intrigued with the communication whereas others were simply excited (Goebel 2012) because of two main reasons. First, the analyses made possible by the software proved unusually quick. This is so because Rainer Goebel combined his programming talents with the speediness made possible by a middle-level language, C++. Second, BrainVoager worked with a particularly beautiful graphical interface. As explained in chapterThree, software packages can function in two manners. Either they receive inputs by means of commands written by the user (which was the case of the earliest neuroimaging packages) or they have a so-called graphical user interface (GUI, pronounced "gooey") with buttons, icons, and windows that prevent people from writing commands and make software navigation much more intuitive. This second possibility is what BrainVoyager has offered to neuroimaging researchers. It performs sophisticated data analysis in a visually agreeable environment, producing fancy brain maps in little time. Understandably, then, it has gained considerable popularity in spite of its costly user license.

The presence of a GUI has technical and sociological implications. As for technical aspects, it is as though programmers were adding yet another layer to the construction represented above. In this way, much programming savviness is needed so the application can work rapidly while being user-friendly. In terms of sociological implications, a GUI makes the package be more inclusive, so to say, as people with no programming knowledge can use it too (Gold et al. 1998). This social dimension is indirectly recognized by software developers. For example, one of my Brazilian interviewees, a participant in the development of a neuroimaging package, claims that it makes no sense to conduct purely theoretical work, because the concrete application of knowledge is also pivotal. For this researcher, application includes the development of software packages with the potential to be used by many people, thanks to the presence of a GUI.

Many programmers seem to agree with this view, as neuroimaging packages with GUIs have multiplied. This is made possible by enhancements in graphics hardware, enabling the design of increasingly sophisticated graphical interfaces that work even in mobile devices (Saad and Reynolds 2012). Packages have been designed with the willingness to help users automatize operations in ways that must be as intuitive as possible. Automation and intuition have to go hand in hand. In the AFNI package, this is achieved by means of a design that is user-friendly in terms of data organization, definition of statistical thresholds, choice of colours, and access to intermediate data (Cox 2012). In the SPM package, it is guaranteed by a code writing where much attention was given to visual interfaces and the provision of resources for the production of scripts (Ashburner 2012). Similar goals were pursued in the design of the SUMA package. "The SUMA GUI is fully scriptable, making it uniquely suited to navigate and summarize results from large numbers of datasets with minimal effort" (Saad and Reynolds 2012, p. 773).

In chapterThree, it was seen that code writing sometimes involve a thought process in which a concrete figure (the code writer) becomes an abstract figure (the future code reader). A similar process can now be identified. Here, the abstract figure that comes into scene is the potential user, frequently thought of as "somebody with little programming knowledge." Ideally, the software package would be automatic and intuitive to the point of being transparent for as many researchers as possible. Even if people have only neuroscience knowledge and the willingness to process neuroimaging data, but no programming skills at all, they should still be able to use at least some basic features of the package. This possibility involves two main things.

On the one hand, developers know that the hurdles of less knowledgeable users begin even before the software package is looked at, because software installation has always been a major challenge posed by scientific applications. In recent decades, programmers have strived to circumvent these difficulties. The development team of the FSL package, for example, have made efforts "[...] to provide pre-compiled self-contained downloads for the most common operating systems [...]" (Jenkinson et al. 2012, p. 784). By the same token, one of the main goals pursued by Dr. Victor Hugo Souza (Laboratory for Biomagnetism and Neuronavigation, Univ of Sao Paulo, Brazil) is to turn his neuronavigator into a full-fledged package, including an automatic process for its installation.

On the other hand, the package should be user-friendly once it has been installed and put to work. Here the goal is to take packages from the state of "crude, command-line driven tools" to the condition of "sophisticated packages that even scientists with limited computer literacy can use easily" (Fuchs 2000, p. 491). On this point, the SPM package represents an interesting case. In chapterFour, it was argued

> *"I once read a text about these packages and there is a... AFNI is famous for being difficult to use. Is that true?*
> [Laughter] Look, it... How can I say... It is scaring at first sight [...] SPM is surely much easier than AFNI in the beginning, but AFNI becomes easier as you learn it. Because you can write a script and run it. But SPM, in the end... You go there and use it. You can process data on SPM even if you don't know almost anything, even if you're a complete novice. You go pressing the buttons, following their tutorials, and you do it.*
>
> *It is intuitive, isn't it?*
> *Yes, there are buttons, so it's very easy."*
>
> Luiz Fernando Dresch (Laboratory for Images-Labima, Pontifical Catholic Univ of Rio Grande do Sul, Brazil)

that its diffusion is guaranteed by the consistent financial support enjoyed by its development team. This is not to say, however, that other

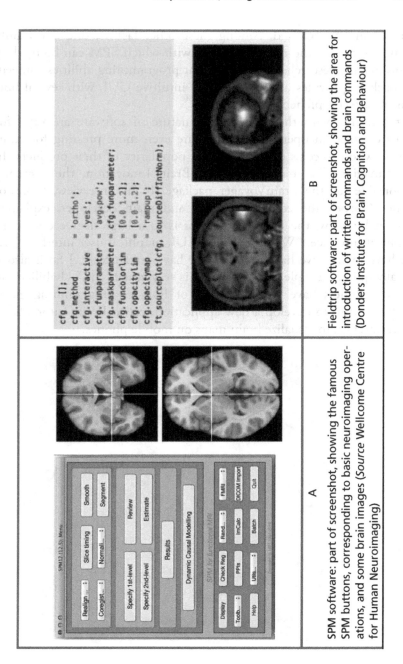

A

SPM software: part of screenshot, showing the famous SPM buttons, corresponding to basic neuroimaging operations, and some brain images (*Source* Wellcome Centre for Human Neuroimaging)

B

Fieldtrip software: part of screenshot, showing the area for introduction of written commands and brain commands (Donders Institute for Brain, Cognition and Behaviour) factors can be neglected. Among these other factors, one is certainly bound to include the relative easiness with which SPM can be used by neuroimaging researchers with different programming abilities. Indeed, the package provides users with a very intuitive GUI, with several buttons of easy comprehension.

If concerns with the user-friendly nature of software are valid for teams engaged in open source, they are even more pressing for companies, whose success depends on the popularity of their products. In neuroimaging, the main example is Brain Innovation, the company responsible for the BrainVoyager package. In his interview, Professor Rainer Goebel, the company's CEO and main developer, explained that, in his view, the package's success is to a large degree due to its graphical interface. "We have a nice GUI, graphical user interface [...] We have menus, we have graphics, we have guidance [...] So it also is visually quite... People like it also that is very visual [...]." Usability and user-friendliness have become tenets of Rainer's programming goals, which led him to develop a new application, called Brain Tutor, allowing researchers to visualize brain maps on mobile phones.

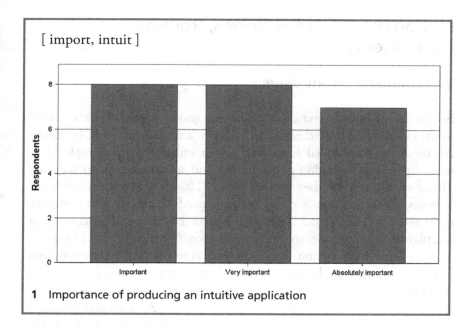

[import, intuit]

1 Importance of producing an intuitive application

Whether they are moved by commercial purposes or willing to share coding achievements, software developers are surely concerned with producing user-friendly applications, a conclusion that is confirmed by Survey 2 [import, intuit]. Asked how important is it to produce an intuitive application, none of the respondents selected "unimportant" and "not important at all." This is another indication that old times when programmers were happy with the production of command-line-based packages are really gone. In this way, the field of neuroimaging software becomes increasingly inclusive, in the sense that growing numbers of researchers, even those with basic computing knowledge, can now use most software packages. At first sight, this phenomenon would signalize nothing but the generous stance of hackers. However, as it will be seen throughout this chapterFive, the popularization of neuroimaging packages, made possible by their intuitive features, has manifold consequences in terms of social relations, knowledge hierarchies, and scientific evolution.

}

section19 (Everyday Norms, Routine Parameters) {

19.1. Requests from Users

By means of source code, programmers endow software with several functionalities. Each of these functionalities are made possible by different blocks of code called functions. Let us imagine a very simple function responsible for taking two numbers and multiplying them together. Therefore, if it receives the numbers 5 and 7, it will return the number 35. Whenever this function is called upon, it needs to receive two numbers, otherwise it will not work. The elements needed by a certain function, so it can play its technical role, are called parameters. We might think of a development team as a function whose specialty is to produce a certain output (a certain software package). Like functions, teams can also receive some parameters, which may assume two forms: feedbacks and requests.

At a certain point of many software package's history, the user base begins to grow, and this is when the development team starts receiving external views pertaining to the package's performance, usefulness, and problems. Because users apply the package's features by mobilizing several types of data, they frequently discover frailties and inconsistencies that the development team frequently could

> *"So I suppose that in order to validate SPM, you need much data.*
> *[Pause] It depends on what you mean by validating [...] If you're talking about validating some influences, you can do it with data [...] But then you might want to have a large number, you know... to run your code on a large number of datasets. But, at the same time, that's something you can do by... You put your code out there and many people will try it out and will tell you: 'Oh, I've got this MRI and your segmentation code failed.' And then you get the data, you look at it and then you need to refine it [the software] in some way and then you improve your model, you make it more robust [...] Somehow you are helped by your users that will help to make... to... to tell when things are working fine or not [...]."*
>
> Researcher (Univ College London, UK)

not have anticipated. This kind of user feedback has been crucial for the enhancement of the FSL package since its first release, as acknowledged by some of its developers (Jenkinson et al. 2012, p. 783): "We certainly feel that software written without this immediate feedback about when it does and does not work, is highly likely to fail on many datasets when released to the community at large [...]." As explained in chapterThree, most development groups keep an open channel of contact with users, in order to receive questions and feedback.

Development teams also receive inputs from external users whenever these latter request features to be implemented in the package. On this point, an apparent paradox presents itself. For, on the one hand, it is understandable that users, as they subject the package to new analysis contexts, conceive of new features that could be added to the application. On the other hand, it is known that many such users have programming skills and could develop the desired features themselves, because the source code is available. This situation can be quickly understood by considering the point made by Naur (2001): the original development team, simply because it possesses deep knowledge of the codebase, can formulate coding strategies to implement new functionalities more quickly and efficiently than any other programmers. As a result of this aspect, as well as other important factors (such as the possible lack of material resources faced by external users or the sheer lack of programming skills), requests of new functionalities are very common in neuroimaging software, as confirmed by the outcomes of Survey 2 [ftr, rqsts].

[ftr, rqsts]

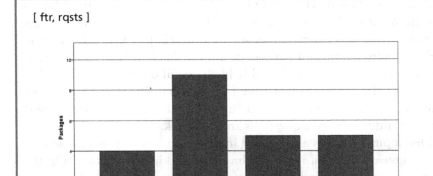

2 Frequency of feedback from users

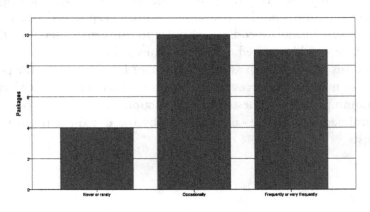

3 Frequency of modifications provoked by external requests

It can be seen that feedback from users is a considerably frequent event and that packages do receive frequent modifications triggered by external feature requests.

> "The way XBAM [package] was developed initially was really organic, so it was not 'Oh, let's implement this great statistical method' and spend a year doing it. It was more psychiatrists coming to see: 'I want to do this. I've got the data. How do I do it?' And it was like 'Well, yeah, we can't do that. Okay, give us two weeks' or whatever. And then we do... we add this bit, we add this extra function and another extra function [...] So I would say, every couple of years, we were going through... we were changing... every version, we were changing all the menu systems. Because we kept adding new options [...]."
>
> Dr. Vincent Giampietro (Centre for Neuroimaging Sciences, King's College London, UK)

In chapterThree it was seen that software developers use different channels of contact to receive feedback and queries from users. This is also how some users ask for new functionalities. For the most widely used neuroimaging package, SPM, this has been frequent. "Over the years, many changes to SPM have been a direct result of commonly recurring queries on the SPM mailing list" (Ashburner 2012, p. 795). For smaller projects, these requests are also recurrent. For example, IF^{2}C, the SPM toolbox designed by Dr. Brunno Campos (Laboratory of Neuroimaging, Univ of Campinas, Brazil), has received many functionalities requested by Brunno's colleagues.

Neuroimaging packages can sometimes be used in clinical activities. In this way, requests of new functionalities are quite common for developers in constant contact with physicians willing to use the application while having no programming knowledge. In the research group led by Professor Carlos Garrido (Depart Physics, Univ of Sao Paulo, Brazil), new functionalities and even whole packages are usually designed as a result of ideas introduced by physicians based in the university hospital. Equally, the PhD project carried out by Dr. Antonio Senra (Computational Group for Medical Signals and Images, Univ of Sao Paulo, Brazil) derives from the demand voiced by a physician who pointed out the limitations of techniques used to identify brain lesions in multiple sclerosis.

In spite of the solitary nature of software development, there are crucial factors boosting its collective dimension: the relations between code writers (chapterTwo), the institutional interests behind it (chapterFour), and, as we can now see, the key relations between software developers and users. In this way, such relations will be sought and cultivated by hackers desirous of enhancing their packages consistently. "Given a bit of encouragement, your users will diagnose problems, suggest fixes, and help improve the code far more quickly than you could unaided" (Raymond 2001, p. 27). This kind of help has been frequent for the AFNI neuroimaging package whose developers have benefited from "close interaction with users," mainly those based in the American National Institutes of Health (Cox 2012, p. 746). According to Professor Kelly Braghetto (Dept Computer Science, Univ of Sao Paulo, Brazil), the NES package, whose design she coordinates, has undergone considerable advances thanks to the relations between the development team and key users based in the Deolindo Couto Neurology Institute, in Rio de Janeiro. "If there was not involvement of a team that is so engaged with the functioning of the software, it would be much more difficult, we wouldn't have managed to develop NES to the point it is today [...]."

"And generally, how do you have the idea for a new development project?
Personally, the need. From a computer perspective, the need. So I needed to do this type of analysis [...] and then thought a new piece of code I was writing for this can transform into a whole project [...] You have ideas by reading or going to conferences and say: 'I would be interested in solving that problem.' And the same ideas for computers, which means we've got a problem, we've got software and we wouldn't be able to do that and: 'Let's make a project and develop some software

These relations invite to think about the relevance of needs in software development. In a sense, software is a computational expression of certain needs. "Software is what makes a computer useful. Software transforms the latent power of the theoretically general-purpose machine into a *to do it.' In the same way, while I do some experiment to investigate that problem. You identify a problem. That's an idea. That's it."*

Dr. Cyril Pernet (Centre for Clinical Brain Sciences, Univ of Edinburgh, UK)

specific tool for solving real-world problems" (Ensmenger 2010, p. 5). In his frequently direct and bold vein, Stallman (2002a, p. 176) claimed that "the user is king in the world of free software." In a way, this affirmation overestimates the weight of users, especially when the package can be copied and modified but is not produced under flexible schemes. However, Stallman points to an important issue. When there is no fierce commercial or institutional control over the codebase, there may emerge an efficient system of communicative actions whereby the needs felt by some people are quickly communicated, understood, and attended to. It has been noted that user needs may trigger efforts of software development, especially in open source (Lakhani and Wolf 2005), and that such circumstance has potentially inclusive effects because a certain specific need (or problem) sometimes "turns out to be typical for a large class of users" (Raymond 2001, p. 49). In software development, the possibility of large and clear communication, and consequently democratic participation, is deeply connected with the possibility of normatively chaining dispersed needs to collective fora where they resonate and become a sort of collective problem deserving solution.

In this way, a context is created where intercomprehension becomes possible. As claimed by Aguirre (2012), the open attitude of open source development teams creates a tradition that can live on when some development projects have to be terminated. Nevertheless, this situation also creates problems. In addition to increasing the institutional power of some development teams (as shown in chapterFour), it overburdens those very same teams, which are expected to respond to demands coming from countless places. Another crucial aspect, focused on in the sequence, comes to alleviate such tensions.

19.2. Solutions from Users

According to Schwarz and Takhteyev (2010), the modifications requested by software users always entail some costs, which can never be precisely foreseen. To be sure, there are financial costs involved, because, for example, groups that accumulate many programming tasks may feel obliged to hire new programmers and pay their salaries. However, other kinds of costs can also be thought of. One example would be the political costs generated by the knowledge power accumulated in certain groups. Such a situation is not taken to its last consequences, though, because of a crucial feature of scientific programming: software users are frequently also developers.

It was already seen that most of today's neuroimaging packages have graphical interfaces making them user-friendly. However, the conduct of cutting-edge studies becomes unviable if analysis procedures are not accurately automatized, which is true even if researchers use packages that are very intuitive and laden with buttons. As claimed by Luiz Fernando Dresch (Laboratory for Images-Labima, Pontifical Catholic Univ of Rio Grande do Sul, Brazil) "[...] if you're working in the area, like doing research, you can't go pressing buttons [...] If it is a small study, you can do it, but if it is a bigger study, like for example 30 patients or 60 patients, imagine you pressing buttons several times." Therefore, powerful studies with large samples do require that researchers automatize operations at the code level, or at least at the script level. Hence the conclusion to which neuroimaging researchers get sooner or later: research tasks become extremely tedious and inefficient if programming knowledge is lacking. For example, this lesson was soon learnt by Dr. Gustavo Pamplona [gstv, pmpln].

[gstv, pmpln]

Dr. Gustavo Pamplona did his Bachelor's Degree in medicine, receiving no training in programming. At the Master's Degree level, he began to work in the field of neuroimaging. This degree involved a two-month stay in the United States. To solve his data analysis needs, he initially used BrainVoyager precisely because of its intuitive nature. However, this strategy rapidly showed its limitations.

"I've always had difficulties with programming, during the whole Bachelor's Degree, but when I began the Master's Degree, I tried to flee from it by using that program [BrainVoyager] that doesn't require any programming language, but, as I told you, things are not so simple. I realized that if I couldn't program, I would have to do much manual work. That's when I started feeling the need for learning programming [...] And I have learnt only MatLab. I haven't learnt other languages."

In the same way that software packages have been designed to be user-friendly, they have received many features allowing people to implement new functionalities themselves. This is actually a pre-condition for software success because researchers are always looking for innovative approaches and would not have recourse to packages unable to accommodate new kinds of data analysis. Furthermore, interested users can design whole modules with very specialized functionalities. Such possibilities are guaranteed, for example, by the AFNI software: "Users with extensive programming skills may appreciate the flexibility of a modular organization, allowing a package to be tailored to individual needs by interchanging routines from other packages or creating and adding routines" (Gold et al. 1998, p. 82). In this way, software developers are considerate not only towards the situation of passive users, so to say, who are able to only follow predetermined analysis protocols and rely on the support ensured by graphical interfaces. They are also concerned with the needs of more advanced users who will be willing to modify the package and make it perform more robust tasks.

In the 1960s, according to the explanation given by Weber (2004), the increasing use of computers provoked a separation between the functions of code writer and computer operator. However, as Weber himself recognizes, this shift took place mainly in companies. In universities, where scientific coding is realized, the figures of software user and software developer continue to be considerably mingled. This is definitely true for neuroimaging software where "SPM and other related software are developed by academics" (Ashburner 2012, p. 797). In his interview, Professor Robert Oostenveld, asked why his FieldTrip software package has become so popular, claimed: "[...] it has become so successful because it addresses the needs of experimental neuroscientists, because it has been developed largely by experimental neuroscientists [...] We developed Fieldtrip because we needed to address our own research needs. And then our own research needs happened to be aligned with the research needs of a lot of people outside our Centre." Such alignment, which happens for various neuroimaging packages, reinforces the utility of a large community of users who instead of being passive receptors of innovations, are likely to play active roles in a certain package's design. Another crucial aspect of neuroimaging software is that the vast majority of packages have been developed as open source products, enhancing the role played by users. Indeed, "[...] software users rather than software manufacturers are the typical innovators in open source" (von Hippel and von Krogh 2003, p. 213).

Frequently, developers of neuroimaging packages are most aware of the potential role that users can play. In order to foster such participation, some tactics may be mobilized, such as the production of easily readable code, and the use of more intuitive, high-level programming languages

> "The characteristic of MatLab that made Fiedtrip possible is that it's a language that very easily allows people to mix and match code. It's not so structured as C or C++, so you can... so for people who don't have a computing science background, it's relatively easy; it's more accessible than C or C++or other languages, so it's not so restrictive. And I think that's why... One of the motivations that we continue to use MatLab in the development of Fieldtrip is that MatLab makes the boundary between the developers of the tools and the users of the tools fuzzier. So I can clearly put myself, as a physicist,

> on the side of the developers, but a lot of people that have contributed to Fieldtrip would not qualify themselves as a developer, they would see themselves as a user [...]."
>
> Professor Robert Oostenveld (MR Techniques in Brain Function group, Donders Institute for Brain, Cognition and Behaviour, Netherlands)

like MatLab and R. The successful application of such tactics has created a field where roles become increasingly mixed up. A clear example is the career of Professor Rainer Goebel (Brain Innovation, Netherlands), an academic who developed a commercial software application and launched a company to manage his production. Nowadays, in addition to being the CEO of Brain Innovation, he is an academic researcher, a Professor at the Maastricht University, a member of the Netherlands Institute for Neuroscience, Royal Netherlands Academy of Arts and Sciences, and a supervisor of both Master's Degree and PhD studies.

The diffusion of software packages cannot go without a parallel diffusion of some tenets of the programming logic. For example, as part of my fieldwork, I was invited by Dr. Matthew Brett (School of Biosciences, Univ of Birmingham, UK) to observe a workshop on neuroimaging data processing at the School of Bioscences, University of Birmingham. The teachers were Dr. Cyril Pernet (Centre for Clinical Brain Sciences, Univ of Edinburgh, UK) and Dr. Chris Gorgolewski (Stanford Center for Reproducible Neuroscience, Stanford Univ, United States). One of the tasks of the workshop consisted in running a small data analysis protocol by using an online platform called OpenNeuro.[2] Whenever the attendees faced problems, the teachers were available for helping them. One interesting event occurred when a student decided to perform a test. The person was trying to compare two variables by writing: "if x = y." As the operation was not working, the person was given the explanation that the correct instruction is actually: "if x == y." Puzzled, the person was then explained that this is the way computers compare variables. Indeed, in many programming languages, variables

---

[2]https://openneuro.org/

are compared with two equal signs. If only one sign is used, the computer might understand that the user is equating the variable, that is, making the value of "y" equal that of "x." This technicality, which is typical to low-level languages, goes up to high-level languages and comes to be integrated into some user-friendly software packages. Therefore, the small event I observed in that neuroimaging workshop reveals an interesting phenomenon: even in domains populated by intuitive software packages, some specks of deep programming knowledge need to be assimilated by the end user.

This is so because the user, instead of representing a passive pole, becomes responsible for the final layer of code writing, even if this layer is aimed to perform technically simple tasks such as the processing of a particular dataset. This phenomenon has become frequent not only in scientific data analysis but even in the use of everyday digital applications. As shrewdly noted by Skog (2003, p. 304): "The use of modem devices like mobile telephones, television sets, kitchen stoves, etc. require people without training to perform simple programming in operating these computerized gadgets [...] In its simplest form, millions of people unconsciously do non-professional programming simply by operating menus and setting options." It is then possible to point to a socialization of programming whereby some specks of programming become widespread routines in social life.

In chapterFour, the expression "light programmers" (Skog 2003, p. 308), referring to people able to write simple scripts, was used. Based on it, another expression was proposed: "heavy programmers," describing hackers with more advanced knowledge and able to design toolboxes or even whole software packages. The proliferation of users who may turn into collaborators, as described in this section, does not blur the distinction between light and heavy programmers, which is many times reinforced. However, it also produces strong bonds between these two types of programmers. Increasingly, complex software packages, deriving from intricate lines of code, are produced in order to be complemented by simple scripts written by users willing to meet specific analysis needs. As a result, software packages cease to be, like in the past, stringent productions to which users could only adjust, and turn into more malleable tools that not only accommodate changes but also make invitations for creative modifications.

Once again, it is possible to consider the relevance of needs in software development. According to Stallman (2002b, pp. 47–48), software packages that are truly needed by society are the ones which "people can read, fix, adapt, and improve, not just operate," which is in contrast with proprietary software delivered as "a black box that we can't study or change." Indeed, the design of modifiable software is likely to solve some social needs but, again, it is crucial to consider that such possibility can only be fully explored when flexible development schemes, instead of rigid and semi-flexible ones, are put in place. If this condition is met, then all the potentialities brought about by users can be explored. Speaking of software users, Raymond (2001, p. 26) affirmed: "Properly cultivated, they can become co-developers." This may hold true because in addition to being co-developers, users are co-users: the package originally produced by a development group may be taken over, used, and modified by another group, in such a way that the modified version may be subsequently used by the original development team. This is what happens in the development projects in which Dr. Matthew Brett (School of Biosciences, Univ of Birmingham, UK) is involved, as well as in those joined by Professor Luiz Murta Junior (Computational Group for Medical Signals and Images, Univ of Sao Paulo, Brazil), to give only two examples. In this way, flexible software has the potential to offer a space for a distant communication whereby every round of modification constitutes a sort of technical utterance provoking new utterances, in a permanent dialogue. Such systems of communicative actions are only possible because each utterance is meaningful, because, among other things, they carry the marks of the local context where it is voiced, as seen in the sequence.

}

section20 ("Actional Realm", "Virtual Realm") {

Throughout this book, the existence of several neuroimaging packages, which may perform very similar tasks, have been pointed out. To be sure, some packages have acquired some scientific dominance and widespread diffusion, as explained in chapterFour. However, their presence does not come to completely overshadow smaller development projects, which have also managed to gain some popularity. Therefore, the neuroimaging domain has witnessed the formation of some groups that stick to certain software packages. This is a typical feature of free software in which, according to Ghosh (2005, p. 36), "[...] the community is organized into not a single core with a vast periphery, but a collection of cores each with their own smaller, overlapping peripheries."

How can these user communities be formed in the first place? In his interview, Professor Carlos Garrido (Dept Physics, Univ of Sao Paulo, Brazil) mentioned three factors leading people to learn to use a certain software package: the application's performance; the frequency with which it is cited in the scientific literature; and the researcher's personal trajectory. This last factor has to do with the laboratories where the researcher has worked and the colleagues the person has met. Depending on these places and people, a different set of packages come to be known and learnt. In my fieldwork, I heard many stories of people getting a new academic position, or having a research

> "I am a big AFNI user [...] Why am I an AFNI user? Because my boss, the psychologists with whom I learnt neuroimaging, from whom I learnt everything [in the United States], did his PhD in a university where the creator of AFNI was then working [...] So I am an AFNI user because my boss is an AFNI user and I learnt this way [...]
>
> **It's more of a habit.**
> Exactly. It's a habit. And I also know well the AFNI developers. I have a good relation with them.
>
> **Personally.**
> Personally. I often see them in conferences. I have a very good relation with them [...]."
>
> Professor Alexandre Franco (Laboratory for Images-Labima, Pontifical Catholic Univ of Rio Grande do Sul, Brazil)

stay abroad, and coming into touch with a new software package. When a researcher moves from a laboratory to another one, the adaptation will be faster and smoother if the same set of software packages are used in the new location. This is why Galloway (2012, p. 58) claimed that programs can serve as a sort of "literary intertexts," facilitating communication between users. Furthermore, some packages get more popular among people with a particular academic background. For example, Professor Rainer Goebel (Brain Innovation, Netherlands), main developer of the BrainVoyager package, declared: "We have not many physics people using BrainVoyager. They are more neuroscientists, cognitive scientists, psychologists... [...] We have medical people... So people who are not so much engineering people, right?"

User communities, which are surely international, can have considerable stability because some effort is always needed to learn new packages, so researchers would keep using the applications they have become accustomed to if they can do so. In Professor John Ashburner's (Wellcome Trust Centre for Neuroscience, Univ College London, UK) words: "Researchers don't like to change what they've been doing." The findings of Survey 1 confirm this idea. Participants were asked how many publications they had had citing the same software application used in their 2015 publication. Whereas 22 people declared that they had had none, 52 declared they had had one to five publications, and 45 said they had published six or more papers citing the same application. In this way, people seem to really use the same software packages in successive studies. Because of this tendency to become technically attached to some software packages, researchers end up forming virtual user communities. This attachment seems to be even stronger in less dynamic geographic areas. Considering the six levels described in chapterFour (section15), it is in the less productive hubs that researchers stick to the same application for the longest time. By looking at the group of people declaring they had cited the same application in six or more publications, a (statistically significant) reduction is verified from the 1st to the 6th level [nmbr, public]. This may be due to the greater dynamism of the most productive levels (in terms of both research infrastructure and personal contacts), making researchers more likely to get in touch with new applications.

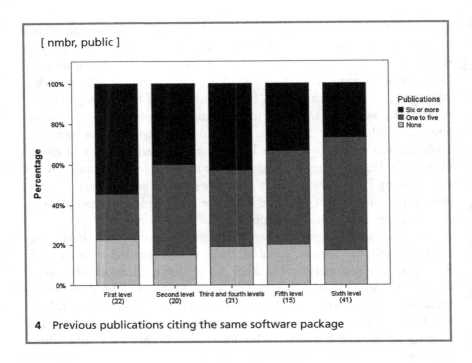

4   Previous publications citing the same software package

Now let us consider this expression (virtual communities) more carefully and ask: what expression would convey the opposite idea of virtual community? Arguably, many people would hasten to say that "virtual communities" are the opposite of "real communities." The division between the virtual and the real has been recalled, revisited, and consolidated in many academic studies (Loginova 2009; Malecki 2017; Swade 2003). In this way, the social reality is now commonly described as having two different layers, a virtual and a real one. For some analysts, an increasing mixture between these layers can be pointed to (Briggs 2002; Jones 2006). However, this is not essentially different from saying that the virtual and the real constitute distinct realities, in the same way that by saying that water results from a mix of hydrogen and oxygen, one is still acknowledging the difference between these two compounds.

At the core of this division between the virtual and the real, lies a materialistic view according to which reality is constituted by things that can be sensed by our five senses whereas other kinds of things, such as thoughts, dreams, software, and informational connections would be out of our reality's scope. This assumption leads us to the strange notion that with the diffusion of digital technologies and the internet, many human beings are now partially immersed in reality and partially living an imaginary, dream-like life. From a communicative point of view, this notion of social life does not make sense. When looked at from a communicative perspective, reality is not what results from people's sensory experiences (which can be referred to as "world"); it is rather constituted by all the issues that come to considered, discussed, and negotiated by means of language-games. Therefore, the communicative perspective does not exclude or negate materialism; it claims that the reality of the material world goes beyond the pure phenomenal nature of objects, including all the debates and normative statements generated by social actors who sense those objects.

In this section20, it will be claimed that, in the sociological terminology, the pair virtual-real is less meaningful than the pair virtual-actional. In order to grasp this point, it is necessary to understand the main features characterizing virtual user communities.

20.1. The Virtual Realm

In 2005, Van Horn (p. 57) stated: "[...] databasing and data sharing remain relatively unfamiliar ideas to many in the field of functional neuroimaging." Nine years after this publication, Dinov and colleagues (2014) identified solid initiatives for the constitution of global networks through which neuroimaging researchers can share analysis tools, algorithms, and data. Therefore, the first two decades of the twenty-first century have been marked by efforts towards making data more conducive resources for sharing within some user communities. Such initiatives resulted from the conclusion that separate research groups can generate a limited amount of data whereas large data sharing initiatives would allow researchers to access more data, reach greater statistical power, and ask more ambitious questions (Das et al. 2016). One of such initiatives has been the 3D Slicer software development project, which is coordinated by American institutions but has been joined by researchers based in many countries. Professor Luiz Murta Junior (Computational Group for Medical Signals and Images, Univ of Sao Paulo, Brazil), along with his supervisees, have been one of these contributors. According to him: "A very interesting aspect of [3D] Slicer, is that, in addition to the software repository, you can also upload your dataset that you used in your study." So this is one of the main characteristics of virtual communities in neuroimaging software: by joining one of them, people have the potentiality to access large and comprehensive datasets.

In chapterTwo, it was seen how software developers have created schemes to provide each other with online coding support. This kind of help is extended to people who instead of writing new code, need only to use software packages in a basic way. This is vital in times when neuroimaging applications reach higher and higher levels of complexity, as explained by Professor Ana Luisa Raposo [anls, rps].

[anls, rps]

Professor Ana Luisa Raposo (Faculty of Psychology, Univ of Lisbon, Portugal) is a psychologist interested in language, semantic memory, and cognition. Her studies involve the use of some neuroimaging tools. In the last year of her Master's Degree, she had a research stay at the University of Cambridge, UK, where she also did the PhD and learnt to use the SPM neuroimaging package. Before becoming a Professor, in 2013, she assumed a post-doctoral position in the United States. Nowadays, she continues to have a rather basic knowledge of SPM, which makes her appreciate the online support provided by the SPM development group.

"[...] Over the last... over the last decades, these neuroimaging software packages have become very popular, especially SPM. They have been used by psychologists, even social scientists. Is it possible to see this popularization of neuroimaging software as a good tendency?

> *Ahm... I would say yes [...] And I think it has good support in terms of mailing list and discussion forum. I think this is the most valuable thing when somebody has any doubt or difficulty: being able to access a big community where, for sure, the probability of somebody having already faced a similar difficulty is big. And this, for me, is the biggest advantage of SPM, it's the fact that there is this large community where we can... to which we can have recourse with doubts or difficulties that appear [...] So these doubts and difficulties are well received, and people respond, help, and try to contribute to advance [...]."*

This is the second main feature of user communities: by joining them (not in any official way but by only becoming a frequent user of a certain package), people become acquainted with the channels of contact through which they can have technical support.

The potentiality to access data and the potentiality to be helped are central characteristics of user communities. This is why people, when speaking of the good side of communities, generally voice expressions like "you can," "we can," or "it's possible." This helps us see a crucial dimension of the virtual realm: the presence of potentialities, capacities, expectations. From a communicative perspective, the denotational meaning of the word "virtual" can then be taken seriously: a virtual thing is the one which is "almost complete," "impending," or more precisely, "potential." Therefore, the virtual realm is not composed of unreal or imaginary things; it is rather part of reality, because in human life, all things real have to initially assume an impending, potential state. To put it simply, a person would not send an email without having the capacity to use a computer and without relying on internet infrastructure installed locally.

According to the trailblazing explanation given by physicist Max Planck, energy is not something that exists as a block; it is rather composed of small packages of energy called quanta. Communicative actions are not rigid blocks either: they are formed by different stages of action (or specks of action), one of which is precisely the stage where all the necessary contextual factors come together and make actions possible, impending or potential.[3] The development and diffusion of digital technologies and the internet have not created something completely new, but they have transformed the potential state of actions (which used to be transitory) into something permanent. In the case of neuroimaging software user communities, the potential for data sharing and technical help, which used to depend on personal meetings and practical arrangements, can now become constant because of the easiness and frequency with which researchers access the online fora used by their communities. If this characteristic of the virtual realm is well understood, then it gets easy to also understand the characteristics of the actional realm.

20.2. The Actional Realm

The expression "actional realm" comes from Bashford (1999, p. 133) and makes reference to situations where actions are performed. It can be adjusted to the present analysis and become the complementary pair of "virtual realm." This latter is left behind, and the actional realm is reached, whenever potentialities produce their expected effects. Thus whenever data comes to be shared, and whenever technical support comes to be given, potentiality turns into action. It must be clear that, from this point of view, the virtual realm cannot be seen as unreal or imaginary, because concrete actions could never become concrete without being potential in the first place. Therefore, we are not really dealing with a state of opposition but with one of complementarity.

As previously mentioned, data sharing has been sought by some initiatives. In neuroimaging, some international projects have made progress, such as the International Consortium for Brain Mapping

---

[3]This is probably the source of a confusion marking some recent theories such as non-representational theory, new materialisms (including actor–network theory), and other frameworks in which humans are conceptually put on a par with objects, and these latter are said to also act. Systems that create a potential for action are mistaken for the action itself.

(ICBM), the Alzheimer's Disease Neuroimaging Initiative (ADNI), and other projects described by Dinov and colleagues (2014). In development and user communities where the open source spirit is particularly strong, such as the 3D Slicer project, data sharing has also been a major concern. In terms of technical support, results have been even more evident. Provision of support for users with superficial programming skills is an important part of the work realized by the developers of key neuroimaging packages such as SPM and FSL. In this way, the virtual expectations introduced by the formation of user communities do not simply constitute promises; they have frequently generated actional results.

What is important to consider is that, in the actional realm, phenomena are always specific because they manifest themselves in particular contexts. Languages cannot be performed outside a context, in the same way that words cannot really signify without appearing in sentences or meaningful contexts. "There are no strictly context-independent references to something in the world" (Habermas 2008, p. 35). If it is true that computer scientists and software developers have long searched for portability, general programming languages, and technical standards, it is also true that this search has often been realized by small groups that value their specific working organization and technical traditions. When it comes to programming languages, for example, many of them have been created to be used locally, leading to a quick proliferation of languages. In the early 1990s, it was already possible to identify the existence of "literally hundreds of programming languages" (Friedman 1992, p. 16).

In neuroimaging software development, some groups have formulated particular ways to write and organize code. For example, in the SPM development team, an internationally famous neuroimaging package, a certain style of code writing was established when Professor Karl Friston took the MatLab programming language and wrote the package's first code lines in the late 1990s. One of the SPM developers I interviewed spoke of Karl's coding style: "[...] he has particular ways of naming the variables and of laying things out [...] It was quite hard to begin with [...] in terms of me learning it, yeah, it took some time to be able to know the dialect, if you like, of MatLab that Karl speaks."

It is as though many development groups, in spite of producing highly standardized and internationalized applications, have done so by using their own local programming dialect.

Interestingly, the creation of so-called open source communities, which were destined to become international, happened in small and geographically restricted groups such as the Bell Labs, the MIT Artificial Intelligence Lab, and the UC Berkeley (Weber 2004). In neuroimaging the same phenomenon was verified whereby some pioneering packages, which would subsequently become products of global diffusion, were designed by small groups decisively moved by their own specific analysis needs. Therefore, the production of old and new packages that enable the formation of virtual communities would not be possible, in the first place, without a moment when a series of local and specific features came to be engraved in the body of those technical productions. Thus it is indeed correct to claim that "algorithms perform in context" (Kitchin 2017, p. 25).

At this point, one key difference between the virtual and actional realms can be pointed to. The virtual realm creates unspecific potentialities that present themselves in the same way to everybody. Because, for example, it is impossible to precisely know, at an initial moment, which software user will need simple or complex technical support. Thus the development group has to be equally open to queries sent from different places and institutions. In its turn, the actional realm makes salient the specific needs faced in specific contexts. Because the analysis needs voiced by a certain researcher, who is based in a certain institution, has certain programming skills, works on a certain dataset, and uses a certain set of software packages, will be different from the needs voiced by a different researcher based elsewhere, with different programming skills, processing different data, and using different packages. In the actional realm, the potentialities that linger in the virtual realm come to be specified, turning the generic into specific. This is why neuroimaging researchers, who speak an international scientific language and have to comply with the analysis tenets of their disciplines, still have the possibility to make choices and work in technical contexts that are far from being standardized.

According to Aguirre (2012, p. 765), the domain of neuroimaging software, in the late 1990s, could be portrayed as "the perfect arena for a Darwinian code struggle," because many people expected many struggles between all the pioneering software packages, which threatened to sweep one another out of the scientific arena. However, this crude war between packages has not been witnessed. In this regard, the account given by Cox (2012, p. 746) is very telling: "When I started at the NIH in 2001, the [...] Scientific Director at the time [...] was enthusiastic about the idea of a central software platform to which all developers would 'plug in'. A decade later, this vision has not been realized, or even seriously attempted." Indeed, both software packages and online analysis platforms have appeared in considerable numbers (van Essen 2012), making it now possible to point to the existence of "many popular neuroimaging software packages" (Nieuwenhuys et al. 2015, p. 2570).

Why has there been such proliferation of neuroimaging packages? Software produced in universities is the outcome of projects undertaken by developers or groups specialized in certain areas. It is not rare that they perform a quite limited number of analyses. According to Gold and colleagues (1998), "no one package is comprehensive" (p. 74), explaining the fact that "more than one package may be necessary to complete all processing and analysis steps" (p. 83). One example is the neuroimaging application called Caret, which was originally designed to focus on the specific analysis of

> *"And do you know if there are similar packages that do similar things [to the Draw-EM package]?*
> FSL is the biggest one that does similar things. So FSL and also SPM, they have a segmentation software, but these packages are not typically very successful in the neonatal brain. So they are very good in analysing the other brain but when you move to the neonatal and fetal brain, because it is very different from the other brain, these software packages do not usually work. So that is why we developed the [Draw-EM] software.
>
> *Ah, okay. So you have specific algorithms working in your software.*
> Yeah. Yeah. So they are specialized for the, yeah, neonatal and fetal brain."
>
> Dr Antonios Makropoulos (Biomedical Image Analysis Group-BiomedIA, Imperial College London, UK)

post-mortem brain sections, thus usually requiring the use of complementary applications for other kinds of analyses (van Essen 2012). Over
the last years, neuroimaging software developers have taken this phenomenon to higher and higher degrees, designing packages that explore
niches of brain studies. To be sure, some projects have broader scope,
aiming at different neuroscience analyses. However, the packages produced this way frequently have weak points that are dodged by users.
"Of course each software has its weaknesses and there are analysis sce-
narios where other software packages provide more appropriate tools"
(Goebel 2012, p. 755).

This situation has technical and social consequences. From a technical point of view, there emerges what academic researchers call analysis
pipelines. In this strategy, the researcher combines two or more packages, taking advantage of the best features of each of them. In the group
led by Professor Alexandre Franco (Laboratory for Images-Labima,
Pontifical Catholic Univ of Rio Grande do Sul, Brazil), there is a preferred package for every kind of analysis need: to study the brain's white
matter, there is FSL, especially the toolbox called TBSS; for brain segmentation, there is FreeSurfer, as well as a toolbox called Siena; for
images generated by positron emission technology, there is SPM; and
so on. In the same vein, a researcher based in King's College London,
UK, told me: "now I do things in a complete mix." It is relatively easy
to write some scripts by means of which the analyst can fish in the available pool of packages for the right portion of different packages, thus
building up automatized pipelines that can be reused for future data
collected in future studies.

The social consequence of this situation is the slow construction
of a state of things where competitions between packages, instead of
being fostered, are constantly softened. In effect, developers of recently
released packages are aware that it is no longer necessary to redesign
functionalities which have been successfully implemented in traditional
packages. They can focus on some neglected areas, so to say, which have
not been addressed by previous coding work. This rationale has influenced even the developers of some old packages who now prefer to get
rid of some functionalities and streamline the strong sides of their applications. Specialization is the main tendency in neuroimaging software,

forcing users to make extensive use of analysis pipelines. In this way, many institutions, and not only those located in the most dynamic countries, are potentially able to join this international enterprise, for there are various analysis gaps to be filled by the design of new function-alities or the addition of new modules.

> "What people do is, there's a lot of pipelining, you know, they take the best of one software and the best of another software and just arrange them together [...] [the] XBAM [package, originally launched in 1995] is really good for the statistics part. Its best thing is its non-parametric statistics core [...] No other software is as flexible, in terms of being able to do all types of non-parametric analysis when you're comparing groups, all the different types of contrast with all the nuances of stats, you know... All the others are quite limited in what they can do in terms of non-parametric statistics, but... So we know that this part, the high level of that, when we're comparing groups, we've nailed this one down really, testing it deeply [...] All the pre-processing, this is now getting old, 'cause we haven't put any work on that. So the stats are top-notch and the rest is getting old. So what people are doing now, even here, they've been using SPM or FSL for all the pre-processing thing [...], but then, they do their final steps, their final analyses, in XBAM [...]
>
> **But don't you want to update the parts that are becoming old?**
> No, because it's... [...] The plan [...] is actually to drop this, to drop these parts completely and just to have a toolbox that just does the interest-ing... the stats at the end, basically [...] We are on the statistics end of it. So let's just disconnect the tool [...]
>
> **So the tendency is to have a very specialized software package.**
> Yeah, because some people... [...] Some packages are really good at one thing [...] I don't think any package is brilliant at everything [...]."
>
> Dr. Vincent Giampietro (Centre for Neuroimaging Sciences, King's College London, UK)

Even though, as noted in chapterFour, some packages come to dom-inate the neuroimaging domain, there is much space for software diver-sity, as packages continue to exist in growing numbers. Studies gauging the performance and usefulness of various packages were published in the late 1990s (Gold et al. 1998) and have also been published recently (Mueller et al. 2005; Ribeiro et al. 2015).

Another interesting effect of such software diversity is the appearance of so-called wrappers. They can be described as applications that instead of containing their own analysis algorithms, allow users to quickly combine algorithms found in other applications. Thus a certain wrapper could, for example, grab pieces of code from three different software packages and "wrap them together" in order for an analysis pipeline to be quickly produced. Let us briefly look at only two examples. First, the VoxBo package, designed at the University of Pennsylvania, United States, and released in 1998, was a traditional package that slowly gained features of software wrapper. In the beginning of the twenty-first century, those features were very pronounced, allowing users to "[...] combine, for example, motion correction from FSL, anatomical co-registration from SPM, and data display with FreeSurfer, all within a VoxBo script. Thus, as the years passed, VoxBo became less a neuroimaging software package, and more a framework in which software tools produced elsewhere could be mixed and matched" (Aguirre 2012, p. 767). Second, the MIBCA package, designed at the University of Lisbon, Portugal, and released in 2010, evolved the other way around. Born as wrapper, MIBCA started gaining original code and features, a process that had to be slowed down when its creators left the university. The existence and multiplication of wrappers testify to the possible coexistence of different packages, as well as to the massification of programming tasks through script writing.

In the virtual realm, the homogenization of analysis tasks is lurking, because all software users share the same status and could, in theory, adopt the same analysis tools and procedures. However, the actional realm introduces the possibility of plurality. In this way, different researchers and research groups can have their local traditions. For example, academic groups still have what one of my Dutch interviewees called "the software of the lab," defined as "what people have been using [in that lab] in the last ten years." At the Institute for Nuclear Sciences Applied to Health-ICNAS, Univ of Coimbra, Portugal, the software of the lab is BrainVoyager, whose license was acquired by ICNAS many years ago. One of my ICNAS' interviewees learnt to use that package precisely by joining the Institute. Depending on the research preferences and priorities of a certain academic team, certain packages may be overused while others may be almost neglected. Once again, the issue of needs comes to the foreground. "[...] since there are many available options for data analysis, a new fMRI research group will need to consider, 'Which available software package will best meet our needs?'" (Gold et al. 1998, pp. 73–74). Moreover, the same package may be used in different ways depending on the research group that mobilizes it. Thus the technical species of which the package is a representative may be subjected to specific research rationales, confirming Simondon's (1969, p. 23) conclusion that "[...] technical species exist in a very narrower quantity than the usages to which technical objects are subjected."

This coexistence of software packages reflects the coexistence of different technical cultures, which can propose their own ways to work with neuroimaging data. Cox (2012, p. 746) is right to claim: "Each software package for FMRI [functional magnetic resonance imaging] is not just a set of tools, but is also a social ecosystem, with its own style of support and types of users." Interestingly, some development projects may lose momentum and eventually fade away. This is so because some types of users and needs disappear with time. Inversely, new development projects never cease to be launched because certain emerging types of users and needs have to be taken care of. From a political viewpoint, this means that the neuroimaging domain has not been captured by the projects of large corporations (yet?). Should that be the case, less diversity would surely be verified, because the slow destruction of diversity is a frequent result of strong market rationales. For the time being, however, the domain of scientific software development is still, and to a large extent, protected from the erosion faced by other more applied scientific areas, such as pharmaceutical research, where an association has been promoted between the political and economic dominance of some institutions and companies, on the one hand, and the technical and methodological unification advocated by the champions of the hard science approach, on the other hand (Bicudo 2014). In this way, scientific homogenization, which is frequently presented as the only way to advance "global" and "scientific" knowledge (Bicudo 2012), and which eventually serves a certain hegemonic political project, has not produced its effects in neuroimaging software (yet?). The relation between the local and global scales have long been explored by philosophers and social scientists. Braudel (1979) argued that the evolution of local forms happens in tandem with that of global forms. Subsequently, the mixtures between global and local phenomena were carefully analysed, leading to the formulation of the concept of "glocalization" (Robertson 1992). More recently, other analysts have shown that the more globalized some technical and economic schemes become, the more profoundly they are mixed up with local contexts (Castells 2004; Santos 2000). Geographers are the ones who explored this pair of concepts the most deeply, reaping important explicative benefits from this. From the viewpoint of sociological explanation, another important (and to a considerable degree analogous) division can be identified: that between the virtual and the actional. According to Santos (2000), the global scale contains the possibilities that may or may not be realized in some places whereas the local scale represents the space where possible events become effective. Equally, the virtual realm holds potentialities while concrete events and effective actions are typical of the actional realm. In other words, the actional realm is the one in which the hopes and promises of the virtual realm are tested. For the concrete contexts where language-games are played are also contexts where different rationales contrast, illuminate, and reshape one another. "The ideal tension breaking into social reality stems from the fact that the acceptance of validity claims [...] rests on the context-dependent acceptability of reasons that are constantly exposed to the risk of being invalidated by better reasons and context-altering learning processes" (Habermas 1996, p. 36).

Sociology's understanding of digital technologies can therefore be enriched. Instead of admitting that the virtual opposes the real, as proposed by a rapidly expanding usage, one can recognize that the virtual is actually part of the real. This is so not because the virtual realm would have some features of the material world sensed by human senses, but because it is part of human concerns, discussions, and negotiations. When software developers claim to belong to a certain virtual community, or when they act in relation to norms and expectations held in a certain community, they are not fooling themselves or following shadows; they are in fact hewing norms and assimilating jargons that guide the concrete actions of many other developers, even if these latter can happen to be too distant to be concretely heard and seen.

}

section21 (Helping Hands, User Manuals) {

By reading about software development, as well by talking to hackers, one is confronted by a recurrent claim: we are dealing with a stimulating activity, in which people are always assimilating new knowledge. For Brooks (1995, p. 7), developers have "[...] the joy of always learning, which springs from the nonrepeating nature of the task." In the survey carried out by Ghosh (2005, p. 34), with participation of over 2500 open source developers, respondents claimed that one of their main motivations to join the community and give their contribution is the possibility to "learn and develop new skills."

Nevertheless, the acquisition of knowledge and skills is not only the outcome of personal motivations; it can also be an obligation for researchers whose training story may have had strategic gaps. In neuroimaging, three kinds of gaps can be identified. First, it was seen in chapterThree that many researchers doing neuroimaging lack deep programming knowledge. They often have to catch up by learning some scripting or coding strategies, which is even more likely to occur when new research projects are initiated. As the mastery of a certain language constitutes a knowledge basis for the assimilation of new languages, there is the very frequent case of researchers who learn one or more programming languages on their own. Such phenomenon occurs not only in neuroimaging. In a survey conducted by Lakhani and Wolf (2005), with over 600 hackers involved in open source, forty per cent of respondents declared to be self-taught in code writing.

Second, for the analysis of neuroimaging data, it is important to have notions of a broader research area: that of processing of biomedical signals. In some university programs, like medicine or medical physics, there are mandatory courses on the analysis of signals taken from the human body, whether these signals correspond to bones, lungs, brains, or any other bodily part. However, this kind of training is seldom present in the degrees of many researchers who may direct their careers towards neuroimaging, such as engineers, computer scientists, or psychologists. In this way, it is only at an advanced stage of their careers (the PhD or the post-doctoral phase) that some neuroimaging researchers acquire basic knowledge of biomedical signals processing.

Third, across the world the number of universities with courses and degrees specifically devoted to neuroimaging is still rather small. Most of the times, neuroimaging researchers are based in a variety of academic departments, having, at the best, some laboratories dedicated to its practitioners. In spite of the development displayed by this research domain, as well as the wide publicity gained by some brain studies, training in neuroimaging software and data processing continues to be heavily dependent on personal and institutional adjustments and improvisations.

In front of these three gaps, three solutions have been formulated, guaranteeing the quick diffusion of neuroimaging software packages witnessed over the last decades. The first solution was conceived of by the development groups of the most popular packages. Twice a year, for example, the SPM development team organizes a course where people learn to use the application. Each time, around fifty to sixty people join the course, which involves the payment of a fee. Equally, the development team of the FSL package offers a five-day course every year. Around 150 people from various countries and with different academic background participate. A fee must also be paid. The FSL team tries to run the course in the place where the Human Brain Mapping conference happens, either a week before or after this event. In the past, the FSL team used to organize courses in partnership with the FreeSurfer development team (Fischl 2012; Jenkinson et al. 2012), due to the similarities and complementarities between both packages. Because this course required a long period, it has not been offered over the last years. Conferences represent an opportunity also for Brain Innovation, the company responsible for the BrainVoyager package. Once a year, it tries to offer a paid course on its application. The company has also organized courses in some European countries, focusing on research groups having BrainVoyager as their preferred package. Furthermore, there are some free courses in the company's headquarters in Netherlands.

The second solution has a purely individual character: the effort made by researchers who just sit at their computer and explore the features of a targeted software package. Nowadays, even for neuroimaging applications, much information can be found on the internet, in the form of user manuals, tutorials, and online courses. For example, Dr. Justin OBrien, who is based in the Brunel University London, organized a whole course on the SPM software, with 20 tutorials that can be watched on YouTube.[4] By means of online courses and information, one of my Brazilian interviewees learnt to use four neuroimaging applications: FSL, FreeSurfer, BIC, and Neuro Lens. In addition, this researcher also analysed some scripts written by colleagues, which is another way of learning to use an application.

The third solution has a collective nature. Within research groups, informal teaching activities are very frequent, as senior researchers introduce software packages to students or less experienced colleagues. Before the interview with Dr. Brunno Campos (Laboratory of Neuroimaging, Univ of Campinas, Brazil), I could see eight of his colleagues working at their computers. He told me that he taught all of them to use the SPM package. One of my British interviewees, a PhD candidate at King's College London, has taught the FSL package to some Master's Degree students working in the same laboratory. The opportunity to learn from colleagues seems particularly interesting when one of such colleagues is a figure of prominent role in the design of a certain package. For example, a researcher based in the Donders Institute for Cognitive Neuroimaging, Netherlands, has deepened the knowledge of the FSL package due to the presence, in that same institute, of the group led by Professor Christian Beckmann, one of the main FSL developers. Equally, Professor Alexandre Andrade (Institute of Biophysics and Biomedical Engineering, Univ of Lisbon, Portugal) enhanced his knowledge of the SPM package during his PhD, when he was supervised by Jean-Baptiste Poline, one of the SPM initial developers. These accounts are interestingly underpinned by findings from Survey 1 [tchng, sftw].

---

[4]http://www.brunel.ac.uk/people/justin-obrien/teaching.

[tchng, sftw]

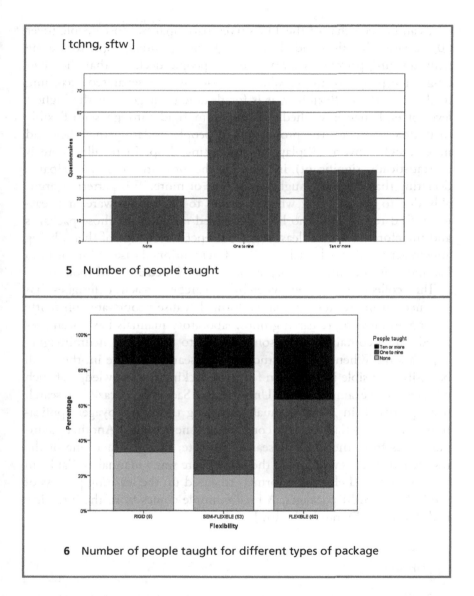

5   Number of people taught

6   Number of people taught for different types of package

It can be seen that of the 119 survey participants, most people (over 60) declared that they had already taught one to nine people to use the neuroimaging package whereas over 30 people declared that they had taught ten people or more. When the packages are separated according to the criterion of flexibility, it is for flexible packages that the highest level of exchange is verified. Passing from rigid through semi-flexible to flexible packages, the proportion of people declaring that they had never taught anyone displays an interesting drop. In parallel there is a (statistically significant) increase in the proportion of participants declaring that they had taught ten people or more. This pattern is arguably due to the fact that when it comes to flexible software, it is easier to find researchers who have developed modules for those packages and therefore have an at least slightly deeper knowledge of them, being able to act as informal teachers. The aforementioned case of Dr. Brunno Campos falls into this category, as he has developed an SPM toolbox.

This collective solution whereby researchers teach colleagues has assumed an interesting alternative form. Because people are constantly joining and leaving research groups, laboratory manuals have been prepared which contain basic lessons on how to use certain neuroimaging applications. Whenever new students or researchers come in, there will be easily accessible information for those lacking the knowledge of such packages. For example, at the University of Sao Paulo, Brazil, a research team prepared, in 2001, manuals exploring the BrainVoyager application. These materials are still consulted by newcomers. Another example comes from the D'Or Research Institute, Brazil. When one of the researchers interviewed joined the team, there was a manual available, in both written and electronic format, to speed up the learning process of the FSL and SPM packages. A final example comes from the interview with Dr. Fábio Duran [fab, drn].

[fab, drn]

Dr. Fábio Duran did his Bachelor Degree in Computer Sciences in a small university in the hinterlands of the Brazilian state of Sao Paulo. At the same time, he worked as an IT specialist in a radiology clinic. From time to time, some engineers visited the clinic to perform some maintenance operations in the devices. He was intrigued that when tomographers were opened up, there was always an entry for a hard disk, and when the ultrasound machines were turned around, a network connection cable could be seen. He became interested in those mysterious interfaces. One day he decided to talk to one of the engineers. He was explained that the devices really stored images that could be seen on a computer screen. He was then given a floppy disk with a couple of images. With the help of one of his university lecturers, Fábio was able to visualize the images in a computer. This lecturer invited him to conduct a final Bachelor's Degree report on medical imaging processing. On completion of this small study, which focused on images of the human head, one of his university colleagues told him that a research group of the capital city, Sao Paulo, was receiving applications from prospective Master's Degrees students willing to work on neuroimaging. In 2003, after being selected, he moved to Sao Paulo to begin his Master's Degree. Thirteen years later, he is based in the same institution, now as a laboratory specialist. In his interview, he recalled how a neuroimaging package manual was once prepared in the laboratory.

"[...] there was a student with us who studied everything about the processing of functional resonance [images taken from the brain when certain cognitive or physical tasks are performed] and resting resonance [images of the brain when no task is performed]. But he finished the PhD and left us. But before he left, he prepared a whole tutorial, a whole manual about how to process those images."

"Do you see any problem in the fact that science is becoming so software-intensive and computer-intensive?
Well, I see it as a problem if you don't teach it [...] So it's very important to seriously consider this as, you know, part of any curriculum [...] So I think, yes, the fact is that you have more and more software [...] We need to recognize that we have to change. If you don't teach it, there are two classes of people: the ones who are lucky because either they can learn quickly (and often in the lab they have the resources to teach them), and the ones who don't."*

Dr. Cyril Pernet (Centre for Clinical Brain Sciences, Univ of Edinburgh, UK)

Considering the three gaps described in this section21 (the lack of profound knowledge of programming, processing of biomedical signals, and neuroimaging software packages), as well as the three solutions just described (the provision of formal courses by development teams, the individual effort of researchers who learn to use packages on their own, and the informal teaching initiatives within laboratories), it is possible to advance some reflections on the ways in which current researchers are being formed in today's universities, not only in the neuroimaging area. In spite of the growing use of computers in scientific analyses, many university degrees still provide their students with quite limited knowledge of computing resources. It is curious, to say the least, that on completion of their Master's Degree, or even the PhD, many people still lack even basic programming skills. Nowadays, such skills acquire growing relevance even for researchers acting in humanities and social sciences. As we have seen, a considerable proportion of neuroimaging researchers come from areas such as linguistics or psychology. Devoid of formal training through which they could learn to program, researchers end up relying on paid courses, their own efforts, or informal help from colleagues. In this way, the point made by Dr. Cyril Pernet (Centre for Clinical Brain Sciences, Univ of Edinburgh, UK) is worth being considered: the acquisition of programming skills could be the result of institutional and educational awareness, instead of depending on the good fortune of some students who happen to join laboratories where there will be colleagues offering helping hands and user manuals.

The communicative flows that cross over society can be ingeniously captured by companies, agencies, and institutions (Bicudo 2014). If universities are able to institutionalize the scattered and personal efforts made by their groups and researchers, they will reinforce these communicative initiatives. Thus there will be a broader social basis underpinning a more solid use and design of software packages. In the current state, these communicative possibilities are unassisted and may eventually fade away, not to speak of a possibly precarious use of computer resources.

As shown in the sequence, the absence of clear and formal teaching schemes is but part of the grey zones found in neuroimaging software.

section22 (Bounty of Standards, Dearth of Standards) {

Neuroimaging is a research field of relatively recent constitution (Filler 2009; Kevles 1998; Savoy 2001). One might then expect a lack of well-established standards in terms of concepts and methods. To a large degree, this is true, as neuroimaging researchers are still engaged in a search for the best approaches to collecting and processing data. On the other hand, however, some software packages have managed to impose themselves as widely accepted tools in specific fields of brain studies. Therefore, neuroimaging oscillates between facticity (the presence of rigid rules) and validity (grey areas that open up much leeway for negotiation). This section22 focuses on this paradoxical situation, starting with an analysis of the proliferation of methodological standards.

22.1. Acknowledging the Gold Standard

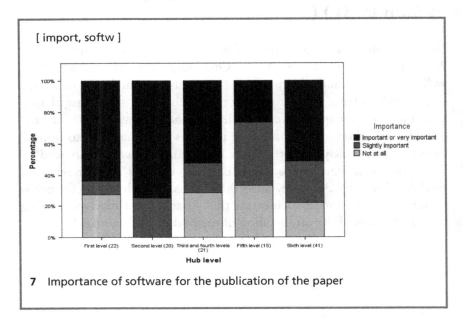

[import, softw]

7   Importance of software for the publication of the paper

Survey 2 helps illustrate the importance that neuroimaging researchers attribute to software [import, softw]. They were asked: "In your opinion, how important was the use of the software for the approval and publication of your 2015 paper?" Most participants consider that the selected software was either important or very important. Interestingly, for the least productive hubs (1st, 2nd, and 3rd levels), the proportion of "important or very important" reached the highest level. This reflects a hopeful stance, as it were, whereby researchers think that by including certain software packages in their analysis methods, they will compensate for their knowledge, financial, or infrastructure limitations, enhancing their capacity to convince peer reviewers of the excellence of their studies. These hopes are less intense at the 5th level (in the Atlantic Coast of the United States) where researchers seem to be more reliant on the reputation of their institutions and the cutting-edge nature of their studies. At the 6th level, the proportion of "important and very important" is heightened, an expression of the massification of software typical of this level. The expectations pertaining to the role played by software in scientific publications are suitably exemplified by the point made by Dr. Fábio Duran (Laboratory for Neuroimaging in Psychiatry, Univ of Sao Paulo, Brazil). Asked whether he would be able to develop new software packages or modules with his programming skills, he claimed: "Yes, but I follow the rules of the laboratory I work in, which ask us to use programs already established in the literature because they are accepted for paper publication more easily."

The value attributed to software, especially in the least productive geographical areas, is an expression of an indisputable phenomenon: software packages have turned into methodological gold standards in neuroimaging. Nowadays, researchers would rarely consider conducting a neuroimaging study as this was done in the 1980s, that is by realizing a visual assessment of the simple, 2D images generated by magnetic resonance scanners. Hence the illustrative response given by Dr. Matthew Brett (School of Biosciences, Univ of Birmingham, UK) when I asked him whether he considers the proliferation of software packages as a good trend in neuroimaging. Puzzled, he declared that my question was actually "a non-question," because "most of the stuff that we do can't be done without a computer."

As explained in section20, many of the current neuroimaging packages are suitable for some specific kinds of data analysis. The scientific literature, where the use of certain applications for certain studies is consistently repeated, helps create analytical standards. Let us consider only the example given by Professor Carlos Garrido (Dept Physics, Univ of Sao Paulo, Brazil): "FreeSurfer. For cortical thickness, everybody uses FreeSurfer as a standard. If there is a second or third or fourth software coming along [...], initially it has to win the competitor, given that in the literature everybody already accepts that FreeSurfer is used to do cortical thickness." Not to forget is the fact that FreeSurfer was initially released in 1999, at an early phase of neuroimaging software. Therefore, some packages, and mainly pioneering ones like FSL, SPM, and AFNI, managed to become gold standards in some neuroimaging niches. This has not prevented new packages from appearing, but it has forced them to either be highly specialized or present innovative features.

In a different context, it is possible to use two concepts that were invoked in chapterThree: facticity and validity. Whereas facticity makes references to situations in which social agents subject themselves to widely recognized facts, behaviours or procedures, validity has to with situations in which language is used to solve tensions, doubts, and disagreements, leading social agents to a state of intercomprehension (Habermas 1996). Some software packages, by gaining a wide recognition and becoming gold standards in certain analysis niches, have fostered the occurrence of facticity in neuroimaging research. Manifestations of validity can also be verified, as highlighted in the sequence.

22.2. Seeking the Gold Standard

In spite of the excitement with which neuroscience research is sometimes targeted by researchers, the media, and policy-makers, the human brain continues to be, to a large extent, a huge mystery. "Although we know considerably more about the brain than we did even several years ago, in many ways it is still rudimentary" (Blank 2007, p. 169). To be sure, neuroscientists have come a long way in terms of understanding some brain conditions and discovering therapies with drugs. However, it is also true that many of these compounds have modest effects and that scientist do not fully understand why they work (Zimmer 2004).

In neuroimaging, and more specifically in the use and development of software packages, knowledge uncertainties can also be pointed to, for two main reasons. First, the data processed by neuroimaging researchers are expressions of signals produced by magnetic resonance scanners. However, there is still some controversy pertaining to the relation between those signals and the actual operations of the brain, in such a way that some researchers are not sure that scanners are really able to detect processes happening in the brain (van Horn et al. 2005). Second, even among the majority of researchers who accept the pertinence of data produced by scanners, there are no clear standards related to how such data should be stored, processed, and interpreted. "Even when there is a clear description of the scientific model employed in a study [...], there may be differences in the algorithmic implementation, hardware platform, compiler, environment configuration, or execution-syntax, which can cause differences in the results even using the same input data" (Dinov et al. 2014, p. 3).

The research account given by Professor Alexandre Franco (Laboratory for Images-Labima, Pontifical Catholic Univ of Rio Grande do Sul, Brazil) is very meaningful. With a group of students, he reviewed the literature to see how magnetic resonance signals can be processed by means of a mathematical approach called graph theory. "[...] when we started looking at publications, I saw that in each publication data were processed in a slightly different way. So 'Okay, but what is the best way? I don't know.' So what did we do? We took a database with more than 200 people and according to each article, we processed those data. Every way led to a different outcome, because of the processing technique [...]." This exercise led to the publication of a paper (Aurich et al. 2015, p. 6) where it is claimed: "[...] there isn't a consensus on which preprocessing strategy is the best to be applied [...]." In his interview, Professor Alexandre Franco stressed the absence of a gold standard, explaining: "If you go to a neuroimaging conference, every person will have a slightly different [data analysis] technique." However, he believes that it is possible for researchers to formulate such standard, in the same way that scientists could arrive at a gold standard to analyse blood samples.

The search for standards is triggering much debate in neuroimaging, and will continue to do so. Methodological proposals have to be formulated, evaluated, and discussed, expressions of validity. For validity manifests itself precisely in moments of uncertainty when agents debate reasons and strive to reach states of intercomprehension (Habermas 1996). Therefore, the development and use of neuroimaging software opens up a leeway for the occurrence of both validity (the search for standards) and facticity (the compliance with standards related to the scientific reputation gained by some software packages). For many researchers and students, this state of things creates technical dependencies. On the one hand, it is necessary to look for guidance on the proper ways to utilize the most prestigious packages. On the other, theoretical orientation is needed to sail through waters that have been navigated but, to a large degree, continue to be unchartered. The uncertainties and disorientations that may occur in neuroimaging research are further explored in the following two sections.

}

section23 (Interactions, Iterations) {

On 30 August 2016, I visited the Laboratory for Images-Labima, Pontifical Catholic University of Rio Grande do Sul, Brazil. My mission was to spend the afternoon there for some interviews. After closing my first conversation, I was supposed to look for the research team leader, Professor Alexandre Franco, who would be the second interviewee. He was sitting at a computer with one of his Master's Degree supervisees, teaching the student to use a certain software functionality. I approached them and asked the name of the software. It was FreeSurfer. "Oh, the beautiful one," I said. He smiled and agreed: "Yes, the beautiful one." He interrupted his teaching task so the interview could begin. On the way to the tiny room where the recorded conversation happened, he explained to me that his student was learning to correct images. Magnetic resonance scanners generate images that are somewhat blurred because of the "noises" caused by the device itself and the subject's head movements. Because of these image imprecisions, the software may get confused, as it were, being unable to identify the frontiers between different brain parts. It is then necessary to use the mouse to manually mark the areas where those frontiers are located. This procedure requires precision and patience, as the user needs to actually draw borderlines as accurately as possible.

Therefore, even in our times of intensive computer usage, there are moments when manual, slow tasks have to be performed. Of course, this is not comparable to old forms of doing neuroimaging when procedures had to be literally manual. For example, in his interview, Professor Li Li Min (Laboratory of Neuroimaging, Univ of Campinas, Brazil) recalled the times when he was a student and had to spend hours painting brain maps manually to highlight a targeted brain region. According to him, software packages have obviously speeded up this kind of task, but students end up failing to acquire an intuition needed to look at brain images and have a feeling for what is going on in terms of brain functioning. This account reminded me of Sennet's (2008) discussion of craftsmanship. Focusing on the work of architects, Sennet compared the times when building projects had to be drawn by hand with current times when computer-aided design (CAD) systems are widely used. For him, the intervention of computers helps architects enhance their projects but impairs the acquisition of an equally important tacit knowledge, an intuition with which architects could, for example, anticipate some practical problems that may occur in a concrete site when the building is eventually constructed. "In the higher stages of skill, there is a constant interplay between tacit knowledge and self-conscious awareness, the tacit knowledge serving as an anchor, the explicit awareness serving as critique and corrective" (Sennett 2008, p. 50). In neuroimaging, the quick diffusion of software packages has also provoked considerable impacts on the ways in which researchers perceive the brain, academic research, and their own position within universities. One of the first manifestations of such changes is precisely the ways in which computers are approached, dealt with, and thought of.

23.1. Habits and Loops

In his autobiography, Linus Torvalds, one of the most prominent people in the history of computer programming and father of the Linux project, remembers the times of his childhood, in the 1980s (Torvalds and Diamond 2001). He used to play with his grandfather's calculator, which took some ten seconds to complete even simple calculations. "With those early devices you knew that what they did was *hard*." (6) For us, contemporary people used to watch computers accurately receive and store the characters of large books in fractions of seconds, this sense of hardship has vanished. As it is no longer necessary to wait seconds for the completion of computing tasks, not only has computers' velocity become natural for us but it is also framed as an obvious duty of our machines.

In the software development jargon, the concept of loop is of paramount relevance. Let us take a reference number, 77.3, and a numeric sequence from 5 to 100. Somebody wants to know the percentage of each number in this sequence in relation to the reference number. The first number (5) would lead to 6.46%, because 5 is 6.46% of 77.3; the second number (6) would lead to 7.76%; the third number (7) would give the outcome of 9.05%; the final number of the sequence (100) would produce 129.36%. With a calculator, it is very easy to find out all the percentages. However, a decisive capacity would be needed: patience. Very few people would not get bored while checking the outcome for all the 96 numbers in this sequence and then taking note of the result. If the C++ programming language was mobilized to perform this task, the following code could be used:

```
float result;
for (x=5; x<=100; x++) {
        float=x * 100 /77.3;
        cout<<result<<endl;
}
```

By putting this code to run in today's average computer, the user would very probably need less than one second to see all the 95 results printed on the screen. Therefore, all the tedious clicks on the calculator buttons are no longer necessary because the computer executes the whole process in a loop, that is, it repeats the very same operation several times. At each repetition (or iteration in the computer science jargon), only a small fraction of the task is changed. In our example, each iteration is performed with a new number: 5, 6, 7, 8, and so on, until the computer reaches the final number (100) and the loop is interrupted. Loops are one of the most central and common strategies of computer programming. We are dealing with the issue of automation because whenever a loop is realized, the computer is made responsible for tedious work that people would take long to complete, being prone to making many errors.

"Do you perceive that those medicine people understand what you do?
[Pause.] Not always. [Laughter] Not always. I mean, sometimes they think it's easy and quick. We would take the software, press a button and it would run it for us. No, it's not always like that. That would be the ideal situation, but it's not always like that [...]. The pre-processing takes three, four hours to get ready. And people think that in ten minutes it's ready [...].

It's a time issue.
Yes. Yes, it's time. 'Oh, do this... Analyse these ten cases for me for tomorrow.' It's not possible to analyse ten for tomorrow. You have to give me one week, honestly, for me to process ten [...]."

Dr. Nathalia Esper (Laboratory for Images-Labima, Pontifical Catholic Univ of Rio Grande do Sul, Brazil)

There is also an issue of naturalization because across the decades, people have become so used to computing efficiency that expectations are sometimes taken to very high levels. Interestingly, those expectations tend to be even higher for people with more superficial knowledge of the coding and technical intricacies of computers. Many of my interviewees carry out collaborative projects in which people with different background are involved. Some of them mentioned the very ambitious and even unrealistic assumptions made by their collaborators, especially those with medical background. When technical knowledge is not very deep, people tend to always expect a formidable performance from computers, which are expected to complete very sophisticated and heavy computations in a very short period. According to Dr. Pedro Moreira (Institute of Life and Health Sciences-ICVS, Univ of Minho, Portugal), "[...] many times [...], there is too high expectation about what you can get. It's a question of not realizing that sometimes processes are very long, because we're working with much data, very complex data, which have very big processing times."

In their turn, software developers and data analysts know the times of computing. Basically, whenever neuroimaging data is analysed, two steps have to be taken. First, the pre-processing phase is where data is prepared and cleaned. Second, the processing phase is where calculations are realized and tests are made to gauge brain activity. José Alves Filho (Laboratory for Images-Labima, Pontifical Catholic Univ of Rio

Grande do Sul, Brazil) explained to me that for pre-processing procedures, it is necessary to wait three hours for each research subject. Before my surprised reaction, he added: "Sometimes it takes the whole day." In the studies conducted by a researcher who is based in the Univ of Sao Paulo, Brazil, the computer is sometimes put to work overnight. He explains that "[...] you don't have to be there all the time; you can trigger the processing and come back on the following day and get the outcome."

It is widely known that across the decades, the capacity of computers has been consistently reinforced. From 1986 to 2007, computer capacity was increased at an annual rate of 58% (Hilbert and López 2011). For those who have somehow worked with computing technologies, the shifts have been quite evident. For example, in his interview Dr. Fábio Duran (Laboratory for Neuroimaging in Psychiatry, Univ of Sao Paulo, Brazil) remembered a research project with 300 subjects conducted around the year 2008. The laboratory's computers worked non-stop for one week to process all the data. According to him, this processing, if repeated on the day of our conversation (year 2016), would be completed in just one day. Therefore, those who design scientific software packages are permanently confronted by a twofold process. On the one hand, computing capacity is taken to higher and higher levels. On the other, scientific datasets are also growing exponentially, requiring quite long processing times.

> *"We've setup a grid system here, for parallel processing, which is something that [the] SPM [package] [...] lacked. At that time, we managed to find a way to distribute the standard processing systems of SPM, so to say, into grids, so each processing phase runs separately, in grids.*
>
> **To speed up the process.**
> *To speed it up, because you imagine, today we have ten projects or twelve projects collecting data, not to speak of those which are at the data processing phase, which are ten or twelve projects more, at different phases. If you look at the volume of data that this represents, it's very big [...]."*
>
> Dr. Ivanei Bramati (Laboratory of Image Processing, D'Or Research Institute, Brazil)

Software speediness and coding efficiency are mantras for neuroimaging software developers. Depending on the coding strategies used by the programmer, the resulting software can perform its tasks more or less rapidly. For example, it was in search of velocity that one of my interviewees, based in King's College London, UK, once decided to translate large pieces of code from MatLab to C++. He explained that "[...] one of the particular projects I was involved with is very computationally heavy to do the processing [...] Well, when I started it, we were taking days to process one image, which was ridiculous. And so I invested a decent bit of time to write that in C++[...]." Another frequent strategy is the design of packages able to converse with other packages, so that users can easily build up pipelines or even perform some parts of the data processing by writing new pieces of code. For example, the IRTK package, designed by Dr. Andreas Schuh (Biomedical Image Analysis Group, Imperial College London, UK) enables the user to write some coding lines in Python so as to customize and enhance some analysis steps. Finally, velocity can be secured at the hardware level, by means of powerful and dynamic processing infrastructure. For particularly productive teams, the installation of systems of high performance turn into a major concern.

If from the software development viewpoint these questions are often considered, they may be overlooked from the viewpoint of software users. When those users have superficial programming knowledge, they are even more oblivious to all those issues. It no longer matters whether the software package recurs to a for-loop or a while-loop; what is important is just to see the output appear on the computer screen.

At this point, researchers reach an obscure terrain where their dependence on computer applications (hence on computer code written by somebody else) can take them to scientific traps. For example, in 2003 a new version of the SPM package, called SPM2, was released, replacing the SPM99 version. According to Dr. Vincent Giampietro (Centre for Neuroimaging Sciences, King's College London, UK), SPM99 contained a coding error that compromised some kinds of analyses, which was corrected in the new version. He says that the SPM developers were humble enough to publicly recognize their failure. "It was something that they publicly announced, you know: 'It's wrong. We've done it wrong. People should stop using it and move to SPM2.' That's great, but what happens to the thousands of papers published with SPM99? I didn't see thousands of retractions." The bug hidden inside the SPM code was particularly problematic for researchers with precarious technical knowledge, as well those lacking time and resources to also correct their studies.

"That's the problem with imaging [...] Except if you do something completely wrong, you will still get results. And the results you get, they may make psychological sense and physiological sense, but they may not be the best results you could get. 'Cause you always get something [...].

So nowadays, as there are more and more software packages, there are maybe some hidden bugs around.
Oh, yeah, like in everything, like in Windows, like in your phone. I mean, there's always hidden bugs [...].

But because of these hidden bugs, there may be many papers being published that contain some... that are not very accurate.
Oh, yeah, yeah. I mean, I remember after SPM99, for a couple of years after that, when you went to conferences, you still had people presenting papers, you know, with SPM99, and when you were asking them 'But do you know you should stop using this, because...?,' either they didn't know or they knew but because so much had been invested already in the analysis...[...] 'Cause people doing neuroimaging as a career, that's one thing. But if you're like a PhD student, you spend two years doing your analysis; the results are good enough, they match your theory, but then somebody says 'Well, there's actually a bug in the software'; well, you don't have time to redo it, so you just move along [...]."

Dr. Vincent Giampietro (Centre for Neuroimaging Sciences, King's College London, UK)

Dr. Vincent Giampietro also remembered a more recent story. In 2016, a paper by Eklund et al. (2016) provoked an earthquake in the area of neuroimaging. According to this paper, the statistical strategies applied by three largely used packages, AFNI, FSL, and SPM, lead to extremely low p-values, thus inflating statistical significance and producing too many false positives. The conclusion was drastic: "This calls into question the validity of countless published fMRI [functional magnetic resonance imaging] studies [...]" (Eklund et al. 2016, p. 7903). In the case of AFNI, the authors identified "a 15-year-old bug" (p. 7904) lurking in the software's code, which was promptly corrected by the AFNI developers. This publication suddenly shocked neuroimaging researchers, who felt scientifically cornered. In 2017, many of them still had a quite defensive stance, which could be realized, for instance, in a certain passage of my interview with Dr. Armin Heinecke (Brain Innovation, Netherlands), head of support of the team responsible for the BrainVoyager package. According to him, after the paper, BrainVoyager also underwent some changes, aimed not to correct errors but to make the application more scientifically robust. "But, of course, what is true is that there are many people who don't have a strong statistic background and sometimes do wrong things, but it's the same in every field. I mean, it's astronomy, biology [...] Not everybody knows everything to the exact extent, power, and depth. So you have to be careful about... It's tough to say: 'The whole field is wrongly reporting effects.'"

Stories of malicious bugs living inside software packages have tempered the excitement of discourses pointing to the marvels of computational analysis. In neuroimaging, the warning voiced by Henson (2005, p. 228) is recalled from time to time: "Statistical packages like SPM are a double-edged sword: they greatly facilitate access to functional imaging data for the novice, but they can also be misused if their principles are not fully understood." If the numbers of cognizant software developers have risen over the last decades, the multiplication of researchers with precarious technical knowledge, potential misusers of software, has also been undeniable. In the remaining parts of this chapterFive, this phenomenon is focused on.

23.2. Distrusting Software

In the area of neuroimaging software, like in other areas of scientific software, it is not always possible to draw a clear line between software developers and users. Frequently, those functions are combined in the same person, as somebody may be the user of a range of packages while designing an original package. However, there is also a growing number of examples for which such distinction becomes possible. Over the decades, there has been a process of massification whereby students are urged to join projects, run studies, and complete degrees very quickly, often lacking the time necessary to acquire more than a superficial computing knowledge. The geographical face of this process can be seen in the following three maps. The indicator shown in these maps is aimed to show how widespread the use of software is in each geographical area. It was calculated with this formula[5]:

$$\sum_{hub} = (A_1 \times I_1) + (A_2 \times I_2) + \cdots + (A_n \times I_n)$$

where:

$\sum_{hub}$: use indicator for the hub

n: number of neuroimaging papers citing an application (software package or toolbox)

A_1: number of applications cited in the first paper identified

I_1: number of institutions (universities, companies or government agencies) involved in the first paper identified

---

[5]More details about the composition of maps are given in the Methodological Appendix.

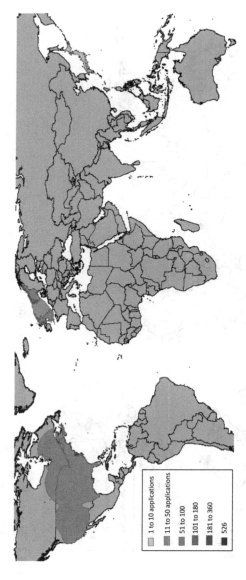

Map 23.1 Geographical hubs and software use indicator: 1995 (*Source Survey 1*)

1 to 10 applications
11 to 50 applications
51 to 100
101 to 180
181 to 360
526

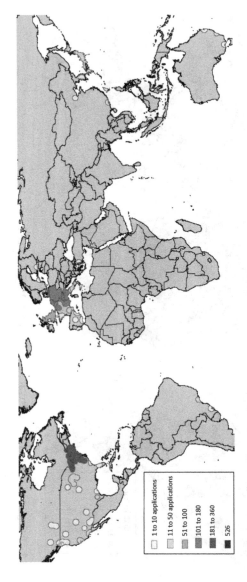

Map 23.2 Geographical hubs and software use indicator: 2005 (*Source Survey 1*)

Map 23.3 Geographical hubs and software use indicator: 2015 (*Source Survey 1*)

In 1995 (Map 23.1), only three hubs were identified, one comprising most of the United States and part of Canada, another one formed by the United Kingdom and Norway, and the last one corresponding to the Japanese territory. All these three hubs displayed a low software use intensity (indicator ranging from 1 to 50). Ten years later (Map 23.2), a considerable multiplication of hubs had taken place, especially in the United States and Western Europe. In addition, some hubs appeared in countries such as China, Brazil, and Turkey. The area with the highest use intensity (indicator ranging from 181 to 360) comprised part of the United States and part of Canada. In 2015 (Map 23.3), hubs were still proliferating, and pioneering hubs appeared in countries such as Mexico, South Africa, and India. The area with the highest use intensity is now the Central European one, encompassing nine countries and displaying the impressive indicator of 526.

This evolution reflects the aforementioned massification of neuro-imaging software. Meaningfully, in the area with the biggest indicator for 2015, most research institutions do not usually conduct the most cutting-edge and high-impact studies. If the six geographical levels presented in Map 23.3 are considered, it is in the two least productive levels (1st and 2nd levels) that the biggest proportions of neuroimaging researchers with background in physics, computer sciences, or engineering were found [rsrch, backg]. Generally, people with this kind of formation are the ones with the most advanced programming skills, being therefore able to use the most advanced features of software packages. Therefore, the hubs where software is the most intensely used are also the ones with the biggest proportion of researchers with relatively more superficial programming knowledge. A typical example was observed in my fieldwork in Netherlands (6th level) where I visited the Donders Institute for Cognitive Neuroimaging, an institution that in spite of its scientific excellence, is largely staffed with people with psychology background and modest programming skills.

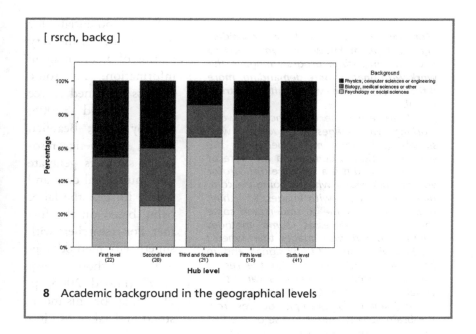

[rsrch, backg]

8   Academic background in the geographical levels

According to Gold et al. (1998, p. 74), "[...] it is [...] difficult for a researcher to fully understand a particular package unless the individual helped to design it." Obviously, no single individual will have deep knowledge of the whole range of software packages available today. However, as soon as a researcher has experience helping design one package, it is easier to grasp the potential analytical challenges brought about by other packages. Thus the person becomes less likely to fall prey to the methodological traps of software. Such technical awareness tends to be reduced wherever a massification of software (that is the usage of packages by crowds of researchers) is verified.

> *"For you, in technical terms, considering the type of knowledge you need to have in computing, programming, is the work with images more demanding, more difficult [than the work with electrical signals]?*
> [...] There are some differences between working with images and working with signals [...] from my experience, the work with signals demands a bit more of abstraction and it is a bit more difficult for you to understand what is going on compared to the work with images. So I think it was easier for me to understand some things, some processes, some methods when I worked with images than when I worked with signals. I think signals end up being very mathematical and the results that come from our methods cannot be visualized [...] the EEG [electroencephalogram] signal, for example, derives from the brain activity, and, because of the EEG signal's features, we cannot understand very well what it means in terms of brain functioning. Of course, there are theories saying that a certain wave in a certain frequency suggests that the brain is working in a certain way, but this is not very direct; there is much abstraction required for you to understand why the signal has those features and extrapolate this to the brain functioning, whereas in structural image, in magnetic resonance, you have a brain image and see what is going on [...]."
>
> Dr. Fabrício Simozo (Computational Group for Medical Signals and Images, Univ of Sao Paulo, Brazil)

Some potential for obscurity lies in the very nature of neuroimaging information. As some authors explained (Joyce 2006; Prasad 2005; Waldby 2000; Beaulieu 2001), magnetic resonance scanners generate both numerical data and bodily images, the latter being based on the former. For researchers with superficial methodological skills, those *"image data"* (Prasad 2005, p. 292) may be misunderstood and taken as "just images," which amounts to covering the complexity of data under a visual veil. The software package is precisely the tool which "does the trick," turning "murky" numbers into "transparent" images. In this sense, it is possible to point to the naturalization of software, a process through which: "The computer, software and algorithm are seen as tools that allow the data to speak for themselves" (Baren-Nawrocka 2013, p. 8). This naturalization has two main expressions. First, neuroimaging researchers may overlook the statistical disputes and conundrums lying behind images, forgetting that brain scans are not photographs but highly manipulated images depicting "statistically significant activation" (Savoy 2001, p. 31) of brain areas. Second, those researchers, attracted by the intuitive and apparently uncontroversial nature of brain images, may eventually confirm the dictum that "seeing is believing" (Burri 2013, p. 372), taking images as their preferred medium because their interpretation is more immediate when compared with numbers or non-intuitive signals.

Aware of this dark side of software, some analysts began to question the naïve reliance on the technologies mobilized in neuroimaging. For Kennedy (2005, p. 19), they "give an unjustified sense of precision" and threaten to become "a kind of post-modern phrenology." Eventually, brain images themselves, which lie at the core of neuroimaging, are looked at with reservation by analysts for whom images fail to fully grasp the nature of the diseases and processes under study (Miller 1996; Miller and Keller 2000).

Obviously, such suspicions attitude end up targeting software packages, especially the user-friendly ones, hence the formulation of questions such as the one raised by Fuchs (2000, p. 491): "[...] are point-and-click user interfaces and sophisticated graphics output enough to ensure that the capabilities of such programs are exploited in the most productive – and scientifically sound – way?" The FieldTrip package,

> *"I'll be frank with you. I have more and more criticisms. This thing of pipeline hides away much data. I know it's a practical thing, you can put things in a production line, you need to get your results, but I'm used to one thing: I do it image by image, manually, I need to see it. I need to see every image [...] I'm at such low level in terms of methods that I say: 'I doubt all of this. I don't even trust the data' [Laughter.] [...] I think that those pipelines that people do, which are recycled and reutilized, they hide away too many things, they make up data... It is difficult.*
>
> ***But for instance, if you develop a big project, with many subjects, do you do it image by image?***
>
> *[...] I'll be honest: I trust only my own pipeline [...] It's like: when you know how sausages are made, you don't eat sausages any more [...] I know what people do with those data analyses. I look at it and say: 'I don't trust this' [...]."*
>
> Brazilian researcher

whose development is coordinated by Professor Robert Oostenveld (MR Techniques in Brain Function group, Donders Institute for Brain, Cognition and Behaviour, Netherlands), has no graphical user interface. For him, the addition of such an interface would represent "a deterioration of the software," because it would prevent users from having a deeper knowledge of the package's features, as well as from developing "an algorithmic view." "And that's why I think that a researcher should actually be encouraged to learn how to program [...] And the graphical user interfaces basically make people, I think, lazier. And I think researchers should not be lazy, but, like, they really should learn skills that are necessary for doing research." In this way, widely used resources, such as graphical interfaces and analysis pipelines, come be targeted by some critical voices. Such criticisms enhance the value attributed to open source packages, because as explained by Professor Carlos Garrido (Department of Physics, Univ of Sao Paulo, Brazil), they allow the user to exactly know what is done with research data, as the source code can be checked with no restrictions.

Some of my interviewees, and particularly those with deep programming knowledge and large data analysis experience, voiced concerns about the fast diffusion of some software packages. At this point, most of today's packages, which are often user-friendly, are said to be threatening sound and critical science. For a researcher based in King's College London, UK, neuroimaging software "[...] makes things much easier, but it also can lead people doing

> *"[...] I've been there to conferences where, you know (less these days), when you have psychiatrists presenting images and somebody is asking a question about what they did in the analysis: 'How did you do this? How did you... What filter did you use?' and they didn't even know what it means, because they're just [sound of knocks on the table, simulating somebody pressing buttons]. They have a recipe. They start, somebody say 'Okay, that's how you're going to analyse data. Open the software.' Open the software. 'Click here.' Click here. 'Open this.' So that's what we're fighting against, basically. We try to teach people here. And of course everybody has their own knowledge, more or less details, but they try to understand, so that they can explain what they do."*
>
> Dr. Vincent Giampietro (Centre for Neuroimaging Sciences, King's College London, UK)

analysis that they don't understand and therefore reaching erroneous conclusions just because they pressed the buttons which seemed obvious but they weren't necessarily trained in the implications of that [...] And it's been too easy to press the buttons and get a [brain] map that looks very believable, but is it necessarily correct?" The danger is considerable because having recourse to brain images is a quick means to make studies appear sound and accurate (Henderson 1998; Latour 1990; Lynch and Edgerton 1988; Dumit 2004). According to Burri (2013), the "visual power" of images is frequently mobilized in conferences.

These phenomena, if looked at from the viewpoint of the actor-network theory (Latour 1987), would lead us to denounce the "black box" nature of neuroimaging software. One would then denounce the "multitude of assumptions" that "[...] 'black boxed' in the incredibly sophisticated computer packages that turned data from voxels in a three-dimensional space into simulated images that had a compelling realism" (Rose 2016, pp. 147–148). To a large degree, this stance is not very different from the "software-as-state-nature" approach described in chapterTwo. For it assumes that behind the orderly, closed façade of scientific productions and theories, lies the actual, original core of science, full of self-centred disputes. From this process, some actors would eventually emerge as winners, managing to somehow impose their truths. The growing numbers of analysts of science subscribing to this approach would then have the responsibility of revealing, behind the elegant art film of science, its mundane, almost grotesque and shameful soap opera.

Controversies, doubts, and uncertainties do not eliminate the communicative relevance of software, which does not need to be harmonious and peaceful in order to make sense. From a communicative point of view, scientific productions (such as scientific software) do not follow crazy trajectories through which states of true messiness would be followed by states of fake order in which "people being convinced stop discussing with one another" (Latour 1987, p. 15). Instead, they can be described as productions around which constant social discussion takes place, and social discussion, instead of being a state of disarrangement or uncertainty, is the only firm certainty people can rely on. Social agents know that whenever disagreements and misunderstandings emerge, which is an everyday event in social life, discussion is the only reliable resource to be mobilized to reach new and more robust states of intercomprehension. In the software domain, manifestations of distrust and the discovery of hidden bugs are components of a social process through which researchers and developers are negotiating the direction to be taken by software use and development. There is no need to look for winners and losers in this endless language-game.

Nevertheless, the same communicative space that fosters dialogue and critical stances opens up some leeway for the generation of trustful and uncritical stances, as shown in the sequence.

23.3. Trusting Software

> *"So nowadays there are many pieces of code published on the internet and people usually take those pieces of code and use them. Is there a risk that by using these pieces of code found on the internet, people are learning bad practices in terms of coding?*
>
> *[...] what I am [...] worried about is some students taking the pieces of code and just reusing them without knowing what they are doing [...] So it's like: 'I want to do x.' And then they pick up a bit of code [...] I mean, they [...] think it does x but it doesn't. And so the issue with all these codes I'm aware of is, people just copy and paste but they don't understand what they're doing [...]."*
>
> Dr. Cyril Pernet (Centre for Clinical Brain Sciences, Univ of Edinburgh, UK)

According to Professor Li Li Min (Laboratory of Neuroimaging, Univ of Campinas, Brazil), in the 1990s most researchers were suspicious about neuroimaging software and preferred not to use them, whereas nowadays those packages have turned into unavoidable tools even though most people lack deep computating knowledge. In this way, he concludes: "You have to believe in software." Indeed, growing numbers of software users are frequently unable to understand the algorithms they increasingly depend on. In technical matters, when knowledge is lacking, it may be replaced with belief. In some instances, a blind faith in software can even be pointed to, as researchers (but mainly students) with little understanding about their research data uncritically frame the results provided by software as binding truths. In his manifesto, Dr. Matthew Brett (2010) names this attitude as the "garbage in, gospel out" approach. In his interview, he spoke of this problematic stance. "[...] I think what happens is, using the computer tends to make people credulous [...] I teach people by getting them to build little modules of what the imaging software does, so that [...] they're in a position where they can think about what that package might be doing [...] and what the problems might be. If that is not the case, it seems to me that you tend to get into that sort of rather helpless and unscientific state where you're basically sort of passing your data to some magical machine which will pass you a magic answer out [...]."

Even researchers with slightly more advanced programming skills may fall prey to trust, when they reuse pieces of computer code without fully understanding the rationale behind them. A researcher based in King's College London, UK, manifested his uneasy with researchers constantly moving between neuroimaging laboratories and leaving behind pieces of computer which might be uncritically reused in the future. This concern also pertains to the lines of code he was writing at the moment of the interview, as he does not know if they will be properly inherited. "[...] essentially there are people, in this department, who have to trust me that I've written my code correctly, but at least I'm here and I'm still working on it, and if somebody spots something that doesn't make sense, they can come and ask why it doesn't make sense. But some day I'm going to leave."

If researchers and students have come to trust software so firmly, this is not simply due to scientific naivety or technical ignorance. To a great extent, this is also provoked by the social legitimacy gained by software and computing resources. Eventually, software is trusted because it is largely framed as a credible tool which cannot be disposed of in our "big data times." In this way, even if research papers and conference presentations are fraught with methodological imprecisions or even errors, they can still be trusted because of their reference to certain software packages, symbols of scientific precision. Therefore, albeit focusing on a particular scientific domain, we are dealing with large communicative processes through which social agents have learnt to acknowledge the methodological and technical soundness of software. This is also why brain images can be increasingly used, by the media, health policy-makers, and caregivers, as convincing messages: these images are credible insofar as people know that their production is made possible by putatively accurate MRI scanners, computers, and software packages. The customary use of such tools, as well the discursive insistence on their soundness, creates a situation in which brain images become a sort of binding social rule. "[…] a person goes by a sign-post only in so far as there exists a regular use of sign-posts, a custom" (Wittgenstein 1963, p. 80). The science behind brain images is highly complex but the images themselves can be intuited by even people with no scientific background. In this way, brain images constitute a mediation between computer code and the everyday languages of society. Their power resides in that they are clear enough to be socially sanctioned while being abstruse enough to remain largely uncontested.

}

section24 (Being a Software User, Being Used by Software) {

According to the story told by Friedman (1992), American physicist Howard Aiken constructed, in 1951, a so-called coding machine. If some buttons were pressed, the device would engrave some computer code on a paper tape, which could be inserted into certain computers so specific functions were triggered. In this way, people, instead of having to memorize pieces of source code, could simply carry code in their pockets. This story shows that the programming logic has to do with human cognitive skills but also with the ability to transfer some human capacities (like mnemonic ones) to machines. Aware that coding machines could be used, many people would simply have recourse to them, instead of making efforts to acquire coding knowledge. Thus at the moment when coding machines began to incorporate some programming skills that were in the past mastered by people only, the constructors of those machines began to possess some knowledge of restricted social diffusion, which they could use in accordance with their needs and intentions. Nowadays, coding machines are no longer used but a concentration of programming skills can still be noticed, as pointed out in chapterFour. In this way, whereas most researchers simply rely on software packages which quickly deal with all the data generated in their studies, a smaller group of researchers becomes responsible for engaging in the social discussion shaping the analysis approaches incorporated by software packages and offered to a growing user base.

Neuroimaging software, as a research instrument, can indeed be framed as a topic thematized in social negotiations and discussions, as explored in chapterTwo. Depending on the outcomes of these debates, software packages will be produced via flexible or semi-flexible schemes, assimilate particular statistical approaches, give priority to certain brain conditions, among other things. However, at a moment when research is speeded up by institutional demands, the group of researchers deeply engaged in these negotiations may shrink more or less dramatically. When this happens, software, from the viewpoint of those excluded from "development elites," slowly ceases to be objects of discussion to become promoters of scientific behaviour. This phenomenon is analysed in this section24.

24.1. Technical and Cognitive Data Ownership

For the developers of any software that gains wide popularity, a challenge imposes itself: the application has to work in different analysis contexts. Every research group has its own procedures for generating, analysing, and storing data. Irrespective of such particularities, the software package must be operational and useful. Therefore, it has to be endowed with analytical malleability while enabling accurate data processing. FSL is one of such packages that gained international popularity. As claimed by Professor Mark Jenkinson (Oxford Centre for Functional Magnetic Resonance Imaging of the Brain-FMRIB, Univ of Oxford, UK), who coordinates the FSL development team: "It's one thing to write a tool which will work on data coming from your own lab with a particular [analysis] protocol and everything else. It's a much harder task to write something which works on anybody's data from any lab with any protocol. Now obviously we don't achieve that with hundred per cent of success [...]."

Precisely because success is not always guaranteed, much discussion is needed to identify problems and adjust the software. This is so because many software users have limited computing knowledge, as explained in previous sections of this chapterFive, but also because some analysis problems prove so tricky that their solution requires intimate and profound knowledge of the software being used. For obvious reasons, the members of the team designing the package are the ones with the closest and most consistent contact with that package, thus being very likely to be able to solve different kinds of analysis hurdles. In Naur's (2001, p. 230) terms, software developers are the ones "having the theory" of the package, being therefore capable of applying "the central laws" of the software to "other similar aspects," that is to the scripts, questions, and data coped with by users.

Thus software users may sometimes need to have direct contact with the team designing the software they utilize. Encounters between users and developers can happen at concrete places. For example, some workshops and courses have been organized by developers, enabling users to have personal guidance. We can once again turn to the example of the FSL package, whose developers give occasional courses on the software. According to Professor Christian Beckmann (Statistical Imaging Neuroscience group, Donders Institute for Brain, Cognition and Behaviour, Netherlands): "[...] on many of the FSL courses that I've been, people actually arrive with... with their own data and come to meet the developers and discuss their particular research project rather than only discussing kind of stuff that was designed as part of the course." A similar phenomenon is verified in the courses organized by Brain Innovation, the company responsible for the BrainVoyager package. In the past, the company offered only standardized courses, but nowadays these courses can also be tailored to students' needs. There is always a hands-on part in which students analyse their own data.

In the many instances where personal encounters are not possible, neuroimaging researchers can use a series of online channels currently available. In his study of open source development communities, Wagstrom (2009, p. 6) explained: "Many [development] projects now feature user-friendly web forums where users can easily post questions and receive answers." Online support is particularly common in research areas to which software is crucial, including bioinformatics (Fuchs 2000) and neuroimaging.

When the technical problems faced by users are too complicated, developers may need to see the dataset provoking the failure. The transfer of neuroimaging data, via online means, has been somewhat problematic and slow, because of the large memory space they usually occupy. Nevertheless, this technical limitation has been overcome over the last years, thanks to improvements made to internet connections and file formats (see the Empirical example in the end of this chapter-Five). This makes it possible for users to send datasets to be analysed by the designers of the software at use. A researcher based in the Univ of Coimbra, Portugal, has sent some data to the BrainVoyager development team, based in Maastricht, Netherlands. "Yes, yes, I have, many times [...] Armin is particularly... he's very helpful [...] He has at some points accessed my computer. When it was not possible to send the data, I gave him remote access to my computer. He accessed it and saw what was going on." This Portuguese interviewee was referring to

Dr. Armin Heinecke (Brain Innovation, Netherlands), whose work consists precisely in giving a helping hand to software users, including advice on dealing with data analysis problems [armn, hnck].

[armn, hnck]

Dr. Armin Heinecke has been at Brain Innovation since 2001. He is German and lives in the German city of Aachen, close to the Dutch border. When he was doing the Bachelor Degree in Germany, he was a student of Professor Rainer Goebel. When Armin completed his PhD and a post-doctoral project, Professor Goebel invited him to join Brain Innovation. Now working as Head of Support, Armin is responsible for helping software users cope with several difficulties, a responsibility which allowed him to develop an intuition for identifying users' needs.

"[...] if you get a complaint or an error message or a question, first you have to understand: 'Okay, what is this person doing, actually? Do I have all the information I need or do I need to ask more?' And: 'At which point do I need to ask for, maybe, sample data?' Because we also do a lot of analysis over here. So sometimes people send us datasets, even small datasets, or in some cases... Over the years we've had actually somebody sending us over the whole dataset, so like gigabytes of data and we ana-lyse it here just to find 'Okay, actually, in this case something weird is hap-pening in subject 10 or so,' which you would not find out by just asking around, so you have to see the data, very often [...]."

According to Professor Rainer Goebel (Brain Innovation, Netherlands), this detailed support is one of the strategies used by the company to attract users towards a proprietary software like BrainVoyager. Such support is also verified for many open source packages such as AFNI, whose development team uses the software's website for this purpose, "[...] including the ability for users to upload datasets if needed to help us understand their issues" (Cox 2012, p. 746). The FSL development team has provided users with the same type of support. Such practice is frequent for teams responsible for semi-flexible software packages but can also be realized by designers of flexible packages such as the community responsible for the Fieldtrip software. In this case, datasets are generally requested by the developers when users ask for the implementation of new features, as explained by the Fieldtrip project's coordinator, Professor Robert Oostenveld (MR Techniques in Brain Function group, Donders Institute for Brain, Cognition and Behaviour, Netherlands).

Thus it is interesting to reflect on the meaning of data ownership in software use and development. Frequently, research groups frame their datasets as a scientific property, as it were, a stance that has limited the extent to which research data has been shared between different groups. From this viewpoint, the data files stored by the group are considered as one of the most valuable resources held by researchers. However, technical data ownership (the storage of datasets in one's computer files) is not substantially more crucial than cognitive data ownership (the possession of the computing and coding knowledge required to process the data). Access to an analysis software may eventually prove futile if there is no access to what Naur (2001) called "the theory" of the software, constituted by the profound knowledge possessed by those who helped create the software. As I claimed elsewhere (Bicudo 2018), "big data" does not necessarily generate "big knowledge."

In chapterTwo, it was seen that in order for software packages to be developed, a great deal of negotiation, interaction, and discussion (in a word communication) is needed. When people do not join this process, even if this happens at only the script level, then this communicative logic is replaced by an instrumental logic that imposes a repetitive and automatic behaviour. If this process reaches very advanced stages, even the most basic capacity enjoyed by researchers (that is, the capacity to ask scientific questions and design research projects) may be replaced with the biases and tendencies that software packages incorporate. This is analysed below.

24.2. Between Behaviour and Action

In chapterThree, the creative activities entailed by software development were focused on. It was shown that such task amounts to a legitimate action, in the sense that something new is brought to the world. "To act, in its most general sense, means to take an initiative, to begin

[…], to set something into motion […]" (Arendt 1998, p. 177). The action represented by the design of new software requires discourse and negotiation, hence the manifold interactions and debates highlighted in chapterTwo. Indeed, as explained by both Habermas (1996) and Arendt (1998), communication and action are entangled concepts. "No other human performance requires speech to the same extent as action" (Arendt 1998, p. 179).

In some economic theories, such as that proposed by Schumpeter (1954), innovation is thought of as a genuine product of initiatives taken by individuals. However, the capacity and willingness to initiate something new depends not only on personal characteristics but has also much to do with the configuration of the social context. In scientific research the same applies: innovative scientists are the ones who come with disruptive ideas but also the ones who experience contexts where innovation is fostered. If the opposite is the case, that is if ideological and institutional requisites force people to be imitative rather than innovative, the most likely outcome will be automatic behaviour (expression of the instrumental logic) rather than creative action (expression of the communicative logic). Over the last decades, academic institutions have been

> *"So over the last years, there's a pro-liferation of these neuroimaging software packages […] They are becoming very popular. Do you see this as a good trend?*
> *[…] Now what we've got is, as you said yourself, we've got all these packages. Every week on Twitter, there's a new package, somebody: 'Hey, I released a toolbox to be doing this, to be doing that.' And people jump at it and use that. And this worries me […] I used to be, you know, a systems administrator here, and I had requests from people saying: 'Can you install this…' I'm talking about ten years ago. 'Can you install this toolbox that I need to use for my project?' And I went checking into the toolbox. There was no reference. This toolbox was just released and there was like one abstract at one conference ten years before, and nothing else. And then I was going back to the user saying: 'Are you sure you want to use this toolbox? Do you know… Are you sure it works?' They are not sure. They don't know […] Now it's been like 'Boom!', this prolif-eration of methods coming from every-where. And some of them will be fine, but some will probably not be fine […] And because of the social media, the best ones*

> *may not be the ones people are using. They will be using sort of the sexiest one, the one which is advertised or still the one which is generating more posts on social media, or user-friendly, rather than something that actually works well [...]."*
>
> Dr. Vincent Giampietro (Centre for Neuroimaging Sciences, King's College London, UK)

largely permeated by this instrumental rationality. To a large degree, they have consented to the intrusion of productivity mandates, as well as the audit rationale described by Power (1999). In neuroscience, this is expressed, for example, by the researchers' tendency to follow methodological fashions and adopt new software packages in a most uncritical way.

The capacity that computers have to modify and sometimes even disrupt the dynamics of companies and universities was already perceived in the 1960s (Ensmenger 2010). Nowadays, with more software packages being used more frequently, this capacity can only be multiplied. One could argue that there is no reason to be concerned with these tendencies because software packages would be only neutral research tools used to collate and store data. However, such neutrality, if it really exists, might become less problematic in the years to come, as some academics start to design research projects based on the data analysis approaches made possible by the available pool of software packages. In my interviews, I could confirm that many neuroimaging researchers, when designing their studies, take into account the methodological options offered by current software packages. This is, for example, an issue that a prospective PhD student would discuss with a supervisor. By the way, in an interview with a PhD student based in King's College London, UK, this topic was addressed. I asked this researcher what comes first: the formulation of the research question or the selection of the software package. "Ideally, you would have a question and you would use whatever package, pipeline, methods are available to you. But it's true that, in practice, sometimes we work backwards [...] It's true, sometimes it happens backwards [...]." It is very hard to assess the frequency with which researchers "work backwards," that is, the frequency with which research questions are shaped by the methods embedded in software packages. Arguably, most people, willing to protect their image of serious researchers, would deny such inversions. However, it is possible to suppose that in times when students are urged to complete their degrees very quickly, and researchers urged to be increasingly productive, the reliance on fast and accurate software may be more appealing than the desire to formulate brilliant or promising questions.

The diffusion and popularization of neuroimaging software can surely trigger a set of communicative actions. However, there is also the danger that research protocols and methods reach high degrees of standardization, with researchers repeating procedures automatically, which is more likely to happen when people lack the knowledge necessary to fully understand how software works. For researchers based in countries with scarce production of software packages and toolboxes, this represents the compliance with analysis methods implemented in distant places. "It requires less effort to conduct research similar to that carried out in developed countries, because methods are already available [...], there are numerous references, and it is easier to publish in journals abroad" (Sartorius 1998, p. 346). From this viewpoint, software use provokes the political side effects described by Santos (2002, p. 81): "Actions are increasingly precise while also being increasingly blind, as they obey foreign projects." In his interview, Professor Christian Beckmann (Statistical Imaging Neuroscience group, Donders Institute for Brain, Cognition and Behaviour, Netherlands) spoke of a clear dissemination of software. "And the danger there is that lots of that research is going to move towards kind of me-too science." These threats are typical of times in which academic study and research are subjected to massification trends. "[...] the more people there are, the more likely they are to behave and the less likely to tolerate non-behavior" (Arendt 1998, p. 43).

To be sure, software use and development still open much leeway for creative action. However, for a growing number of academics, the uncritical and heckles recourse to software threatens such spontaneity, taking researchers close to the "instrumentalization of action" described by Arendt (1998, p. 230). The threat is serious in times when digital technologies acquire a growing power to influence people's behaviour, turning into resources that, in parallel to facilitating many everyday tasks, "make social life amenable to intervention" (Marres 2017). In universities, the cautious and reflected stance that characterizes academic research may turn into a quasi-natural imprudence. "'Quasi-naturalness here refers to a society assimilated to the law-like regularities of objectified nature, because it forces social interactions below the level of free action through an inversion of freedom [...]" (Habermas 2008, p. 195). Once freedom has been swept away from academic research, a curious state is produced where software users may eventually be used by software, having their scientific creativity limited by the options incorporated into computer applications.

}

section25 Empirical Example: DICOM & NIfTI

The technical developments of the 1970s represented a major spur to computational technologies, leading authors to point to this decade as a watershed in the history of information (Castells 2004; Lojkine 1992; Santos 2000). However, in the 1980s users of computer applications, based in both commercial and academic settings, faced a great difficulty that Alshawi (2007, p. 7) described as the formation of "islands of automation": "These applications more often lack the capability to communicate and exchange data between them. When managers started to realise that they could gain further benefits from integrating such packages they were confronted with the lack of interoperability of the different packages." In academic institutions, the range of software packages at use has expanded rapidly, forcing researchers to find ways to move data from one package to others (van Essen 2012). Things are not different for neuroscientists, who have been engaged in the search for common data formats and strategies for data conversion since the 1990s (Beltrame and Koslow 1999).

At the end of this decade, Gold and collaborators (1998) published a study comparing eight neuroimaging software packages developed across the 1990s. The authors had to convert images from their original formats to different formats required by the packages under study. In the early 2000s, this state of things persisted: neuroimaging researchers were using various packages and coping with the issue of file formats. Therefore, they were facing the challenge to which bioinformaticians are used: "For many analyses [...], various techniques and tools are desired across multiple software packages, which invariably results in the process of exporting data from one package and importing it into another" (Jones et al. 2006, p. 535). Not always are such migrations really straightforward.

Because of these historic difficulties, two institutions, the American College of Radiology and the National Electrical Manufacturers Association, joined forces to formulate a standard file format for medical images. Some initial and not completely successful attempts were made. In 1993, finally, those institutions released the Digital Imaging and Communications in Medicine (DICOM) file format, which was intended to be used by researchers producing and storing different sorts of medical images. Slowly, the format was supported by several software packages, including neuroimaging ones.

However, scientists working on brain images still felt the need to address some specific needs of their field, where images produced by very particular devices (magnetic resonance scanners) are worked on. Indeed, those scanners, when used to analyse the human brain, require special adjustments, generating data whose processing demands particular procedures. Once again, two American institutions, this time the National Institute of Mental Health and the National Institute of Neurological Disorders and Stroke, organized a series of meetings, involving both neuroimaging specialists and developers of software packages, to formulate a new file format. In 2004 the Neuroimaging Informatics Technology Initiative (NIfTI) released the NIfTI format, specializing in brain images. The main goal was to promote the "interoperability of neuroimaging research software" (Cox 2012, p. 744), a goal that had not been firmly pursued until that point, as stated on NIfTI's website[6]:

> In the use of functional magnetic resonance imaging (fMRI), [...] many existing tools have been developed piecemeal, by scientists who are interested in answering particular neuroscience questions rather than in producing software products that are optimized for meeting the many and varied needs of the broader research community.

Because of this concern, the NIfTI committee was "filled with representatives from major functional neuroimaging software packages," as explained by Jenkinson and colleagues (2012, p. 788). Jenkinson himself, representing the FSL development team, played an important role in the production of NIfTI. He stresses that NIfTI "has been very successful in allowing data to be exchanged between packages and therefore

---

[6]www.nifti.nimh.nih.gov

increases the scope of what can be done scientifically" (Jenkinson et al. 2012, pp. 788–789).

Another participant of the NIfTI committee was Professor John Ashburner (Wellcome Trust Centre for Neuroscience, Univ College London, UK). In his interview, he explained to me that in the first years of the twenty-first century, different meetings were organized. However, the initiative lost momentum in the next period, and meetings finally ceased to happen. Nowadays, he explains, DICOM continues to be the file format the most frequently used in neuroimaging, in spite of its technical frailties. He doubts that NIfTI will manage to become a gold standard, as originally planned.

Even though the initial effervescence slowly disappeared, in terms of both meetings and funding, NIfTI has certainly had considerable impacts. In 2011 an enhanced version, called NIfTI-2 was released. In the following year, Rainer Goebel, main developer of the BrainVoyager package, was producing a new brain atlas aimed to assess the variability of specialized brain regions. He stressed that the atlas could be downloaded in the NIfTI format (Goebel 2012). Nowadays, in addition to BrainVoyager, NIfTI is supported by all the major neuroimaging packages, such as SPM, FSL, AFNI, FreeSurfer, MRIcro, MIPAV, and others.

From a communicative perspective, the most remarkable deed of both DICOM and NIfTI has been the reinforcement of the systems of communicative actions present in the neuroimaging domain. In order for researchers to establish collaborations, share data, discuss the coding routes taken by software packages, and other tasks, it is crucial that the data analysis work carried out in a certain laboratory can be checked and assessed in other laboratories. This is the distant communication that DICOM and NIfTI enable.

Interestingly, the difficult conversation that prevailed before the release of NIfTI is described, on NIfTI's website, as a "Tower of Babel problem." Professor John Ashburner (Wellcome Trust Centre for Neuroscience, Univ College London, UK) declared: "I think NIfTI was useful in terms of allowing people to mix and match different software, different visualizations [...] So I think it simplified things *a lot* for the kind of neuroscientists-users, so they didn't have to spend a lot of their time converting between different file formats [...]." Indeed, the virtual conversation between different neuroimaging software packages is now considerably easier than it was in the in first years of the twenty-first century. As a consequence, the actual conversation between researchers, whether it happens in concrete places or via the internet, has been facilitated, enabling communication to occur in parallel with the conduct of technical research tasks.

}

References

Aguirre, Geofrey K. 2012. "FIASCO, VoxBo, and MEDx: Behind the code." *NeuroImage* 62:765–767.

Alshawi, Mustafa. 2007. *Rethinking IT in construction and engineering: Organisational readiness.* New York: Taylor & Francis.

Arendt, Hannah. 1998. *The human condition.* 2nd ed. Chicago: University of Chicago Press.

Ashburner, John. 2012. "SPM: A history." *NeuroImage* 62:791–800.

Aurich, Nathassia K., José O. Alves Filho, Ana M. Marques da Silva, and Alexandre Rosa Franco. 2015. "Evaluating the reliability of different preprocessing steps to estimate graph theoretical measures in resting state fMRI data." *Frontiers in Neuroscience* 9, article 48:1–10.

Bashford, Bruce. 1999. *Oscar Wilde: The critic as humanist.* London: Associated University Presses.

Beaulieu, Anne. 2001. "Voxels in the brain: Neuroscience, informatics and changing notions of objectivity." *Social Studies of Science* 31 (5):635–680.

Beltrame, Francesco, and Stephen H. Koslow. 1999. "Neuroinformatics as a megascience issue." *IEEE Transactions on Information Technology in Biomedicine* 3 (3):239–240.

Bicudo, Edison. 2012. "Globalization and ideology: Ethics committees and global clinical trials in South Africa and Brazil." PhD thesis, Department of Political Economy, King's College London. Available at https://kclpure.kcl.ac.uk/portal/en/theses/globalization-and-ideology-ethics-committees-and-global-clinical-trials-in-south-africa-and-brazil(ab94b2ce-b023-4ab3-9dd4-93616db5c455).html.

Bicudo, Edison. 2014. *Pharmaceutical research, democracy and conspiracy: International clinical trials in local medical institutions.* London: Gower and Routledge.

Bicudo, Edison. 2018. "'Big data' or 'big knowledge'? Brazilian genomics and the process of academic marketization." *BioSocieties* 13 (1):1–20.

Blank, Robert H. 2007. "Policy implications of the new neuroscience." *Cambridge Quarterly of Healthcare Ethics* 16:169–180.

Braudel, Fernand. 1979. *Le temps du monde. Vol. III, Civilisation matérielle. Economie et capitalisme.* Paris: Armand Collin.

Brett, Matthew. 2010. Research methods in the twenty-first century. Available at http://matthew.dynevor.org/manifesto.html.

Briggs, J. C. 2002. "Virtual reality is getting real: Prepare to meet your clone." *Futurist* 36 (3):34–41.

Brooks, Frederick P. 1995. *The mythical man-month.* Reading: Addison-Wesley.

Burri, Regula Valérie. 2013. "Visual power in action: Digital images and the shaping of medical practices." *Science as Culture* 22 (3):367–387.

Castells, Manuel. 2004. *The network society: A cross-cultural perspective.* Cheltenham: Edward Elgar.

Cox, Robert W. 2012. "AFNI: What a long strange trip it's been." *NeuroImage* 62:743–747.

Das, Samir, Tristan Glatard, Leigh C. MacIntyre, Cecile Madjar, Christine Rogers, Marc-Etienne Rousseau, Pierre Rioux, Dave MacFarlane, Zia Mohades, Rathi Gnanasekaran, Carolina Makowski, Penelope Kostopoulos, Reza Adalat, Najmeh Khalili-Mahani, Guiomar Niso, Jeremy T. Moreau, and Alan C. Evans. 2016. "The MNI data-sharing and processing ecosystem." *NeuroImage* 124:1188–1195.

Dinov, Ivo D., Petros Petrosyan, Zhizhong Liu, Paul Eggert, Sam Hobel, Paul Vespa, Seok Woo Moon, John D. va Horn, Joseph Franco, and Arthur W. Toga. 2014. "High-throughput neuroimaging-genetics computational infrastructure." *Frontiers in Neuroinformatics* 8:1–11.

Dumit, Joseph. 2004. *Picturing personhood: Brain scans and biomedical identity.* Princeton: Princeton University Press.

Eklund, Anders, Thomas E. Nichols, and Hans Knutsson. 2016. "Cluster failure: Why fMRI inferences for spatial extent have inflated false-positive rates." *Proceedings of the National Academy of Sciences* 113 (28):7899–7905.

Ensmenger, Nathan. 2010. *The computer boys take over: Computers, programmers, and the politics of technical expertise.* Cambridge and London: MIT Press.

Filler, Aaron G. 2009. "The history, development and impact of computed imaging in neurological diagnosis and neurosurgery: CT, MRI, and DTI." *Nature Precedings*:1–76. Available at http://dx.doi.org/10.1038/npre.2009.3267.5.

Fischl, Bruce. 2012. "FreeSurfer." *NeuroImage* 62:774–781.

Friedman, Linda Weiser. 1992. "From Babbage to Babel and beyond: A brief history of programming languages." *Computer Languages* 17 (1):1–17.

Fuchs, Rainer. 2000. "Analyse this... or: Intelligent help for the rest of us." *Bioinformatics* 16 (6):491–493.

Galloway, Patricia. 2012. "Playpens for mind children: Continuities in the practice of programming." *Information & Culture* 47 (1):38–78.

Ghosh, Rishab Aiyer. 2005. "Understanding free software developers: Findings from the FLOSS study." In *Perspectives on free and open source software*, edited by Joseph Feller, Brian Fitzgerald, Scott A. Hissam, and Karim R. Lakhani, 23–46. Cambridge: MIT Press.

Goebel, Rainer. 2012. "BrainVoyager: Past, present, future." *NeuroImage* 62:748–756.

Gold, Sherri, Brad Christian, Stephan Arndt, Gene Zeien, Ted Cizadlo, Debra L. Johnson, Michael Flaum, and Nancy C. Andreasen. 1998. "Functional MRI statistical software packages: A Comparative analysis." *Human Brain Mapping* 6:73–84.

Habermas, Jürgen. 1996. *Between facts and norms: Contributions to a discourse theory of law and democracy.* Studies in contemporary German social thought. Cambridge, MA: MIT Press.

Habermas, Jürgen. 2008. *Between naturalism and religion.* Cambridge: Polity Press.

Henderson, Kathryn. 1998. *On line and on paper: Visual representations, visual culture, and computer graphics in design engineering.* Cambridge: MIT Press.

Henson, Richard. 2005. "What can functional neuroimaging tell the experimental psychologist?" *The Quarterly Journal of Experimental Psychology* 58A (2):193–233.

Hilbert, M., and P. López. 2011. "The world's technological capacity to store, communicate, and compute information." *Science* 332 (6025):60–65.

Jenkinson, Mark, Christian F. Beckmann, Timothy E. J. Berens, Mark W. Woolrich, and Stephen M. Smith. 2012. "FSL." *NeuroImage* 62:782–790.

Jones, Matthew B., Mark P. Schildhauer, O. J. Reichman, and Shawn Bowers. 2006. "The new bioinformatics: Integrating ecological data from the gene to the biosphere." *Annual Review of Ecology, Evolution, and Systematics* 37:519–544.

Jones, S. 2006. "Reality© and virtual reality©—When virtual and real worlds collide." *Cultural Studies* 20 (2–3):211–226. https://doi.org/10.1080/09502380500495692.

Joyce, Kelly. 2006. "From numbers to pictures: The development of magnetic resonance imaging and the visual turn in medicine." *Science as Culture* 15 (1):1–22.

Kennedy, Donald. 2005. "Neuroimaging: Revolutionary research tool or a post-modern phrenology?" *American Journal of Bioethics* 5 (2):19.

Kevles, Bettyann Holtzmann. 1998. *Naked to the bone: Medical imaging in the twentieth century*. New York: Basic Books.

Kitchin, Rob. 2017. "Thinking critically about and researching algorithms." *Information, Communication & Society* 20 (1):14–29.

Lakhani, Karim R., and Robert G. Wolf. 2005. "Why hackers do what they do: Understanding motivation and effort in free/open source software." In *Perspectives on free and open source software*, edited by Joseph Feller, Brian Fitzgerald, Scott A. Hissam, and Karim R. Lakhani, 3–22. Cambridge: MIT Press.

Latour, Bruno. 1987. *Science in action: How to follow scientists and engineers through society*. Cambridge: Harvard University Press.

Latour, Bruno. 1990. "Drawing things together." In *Representation in scientific practice*, edited by Michael Lynch and Steve Woolgar. Cambridge: MIT Press.

Loginova, O. 2009. "Real and virtual competition." *Journal of Industrial Economics* 57 (2):319–342. https://doi.org/10.1111/j.1467-6451.2009.00380.x.

Lojkine, Jean. 1992. *La révolution informationnelle*. Paris: Presses Universitaires de France.

Lynch, Michael, and Samuel Y. Edgerton. 1988. "Aesthetics and digital image processing: Representational craft in contemporary astronomy." In *Picturing power: Visual depiction and social relations*, edited by G. Fyfe and J. Law, 184–220. London: Routledge.

Malecki, E. J. 2017. "Real people, virtual places, and the spaces in between." *Socio-Economic Planning Sciences* 58:3–12. https://doi.org/10.1016/j.seps.2016.10.008.

Marres, Noortje. 2017. *Digital sociology: The reinvention of social research*. Cambridge: Polity Press.

Miller, Gregory A. 1996. "How we think about cognition, emotion, and biology in psychopathology." *Psychophysiology* 33 (6):615–628.

Miller, Gregory A., and Jennifer Keller. 2000. "Psychology and neuroscience: Making peace." *Current Directions in Pshychological Science* 9 (6):212–215.

Mueller, Susanne G., Michael W. Weiner, Leon J. Thal, Ronald C. Petersen, Clifford Jack, William Jagust, John Q. Trojanowski, Arthur W. Toga, and Laurel Becket. 2005. "The Alzheimer's disease neuroimaging initiative." *Neuroimaging Clinics of North America* 15 (4):869–877.

Naur, Peter. 2001. "Programming as theory building." In *Agile software development*, edited by Alistair Cockburn, 227–239. Boston: Addison-Wesley.

Nieuwenhuys, Rudolf, Cees A. J Broere, and Leonardo Cerliani. 2015. "A new myeloarchitectonic map of the human neocortex based on data from the Vogt-Vogt school." *Brain Structure & Function* 220 (5):2551–2573.

Power, Michael. 1999. *The audit society: Rituals of verification.* New York: Oxford University Press.

Prasad, Amit. 2005. "Making images/making bodies: Visibilizing and disciplining through magnetic resonance imaging." *Science, Technology & Human Values* 30 (2):291–316.

Raymond, Eric S. 2001. *The cathedral & the bazaar: Musings on Linux and open source by an accidental revolutionary.* Sebastopol: O'Reilly.

Ribeiro, Andre Santos, Luis Miguel Lacerda, and Hugo Alexandre Ferreira. 2015. "Multimodal Imaging Brain Connectivity Analysis (MIBCA) toolbox." *PEERJ* 3 (e1078):1–28.

Robertson, Roland. 1992. *Globalization: Social theory and global culture.* London: Sage.

Rose, Nikolas. 2016. "Reading the human brain: How the mind became legible." *Body & Society* 22 (2):140–177.

Saad, Ziad S., and Richard C. Reynolds. 2012. "SUMA." *NeuroImage* 62:768–773.

Santos, Milton. 2000. *La nature de l'espace: technique et temps, raison et émotion.* Paris: L'Harmattan.

Santos, Milton. 2002. *A natureza do espaço: técnica e tempo, razão e emoção, Milton Santos collection 1.* Sao Paulo: Edusp.

Sartorius, Norman. 1998. "Scientific work in Third World countries." *Acta Psychiatrica Scandinavica* 98:345–347.

Savoy, Robert L. 2001. "History and future directions of human brain mapping and functional neuroimaging." *Acta Psychologica* 107:9–42.

Schumpeter, Joseph. 1954. *Capitalisme, socialisme et démocratie.* Paris: Payot.

Schwarz, Michael, and Yuri Takhteyev. 2010. "Half a century of public software institutions: Open source as a solution to hold-up problem." *Journal of Public Economic Theory* 12 (4):609–639.

Sennett, Richard. 2008. *The craftsman.* New Haven: Yale University Press.

Simondon, Gilbert. 1969. *Du mode d'existence des objets techniques, Analyses et Raisons 1*. Paries: Aubier.

Skog, Knut. 2003. "From binary strings to visual programming." In *History of Nordic computing*, edited by Janis Bubenko Jr., John Impagliazzo, and Arne Solvberg, 297–310. Boston: Springer.

Stallman, Richard M. 2002a. "Free software: Freedom and cooperation." In *Free software, free society: Selected essays of Richard M. Stallman*, edited by Joshua Gay, 155–186. Boston: GNU Press.

Stallman, Richard M. 2002b. "Why software should not have owners." In *Free software, free society: Selected essays of Richard M. Stallman*, edited by Joshua Gay, 45–50. Boston: GNU Press.

Swade, D. 2003. "Virtual objects: The end of the real?" *Interdisciplinary Science Reviews* 28 (4):273–279. https://doi.org/10.1179/030801803225008686.

Torvalds, Linus, and David Diamond. 2001. *Just for fun: The story of an accidental revolutionary*. New York: HarperCollins.

van Baren-Nawrocka, Jan. 2013. "The bioinformatics of genetic origins: How identities become embedded in the tools and practices of bioinformatics." *Life Sciences, Society and Policy* 9 (7):1–18.

van Essen, David C. 2012. "Cortical cartography and Caret software." *NeuroImage* 62:757–764.

van Horn, John Darrel, John Wolfe, Autumn Agnoli, Jeffrey Woodward, Michael Schmitt, James Dobson, Sarene Schumacher, and Bennet Vance. 2005. "Neuroimaging databases as a resource for scientific discovery." *International Review of Neurobiology* 66:55–87.

von Hippel, Eric, and Georg von Krogh. 2003. "Open source software and the 'private-collective' innovation model: Issues for organization science." *Organization Science* 14 (2):209–223.

Wagstrom, Patrick Adam. 2009. "Vertical interaction in open software engineering communities." PhD, Carnegie Insitute of Technology/School of Computer Science, Carnegie Mellon University.

Waldby, Catherine. 2000. "The Visible Human Project: Data into flesh, flesh into data." In *Wild science: Reading feminism, medicine and the media*, edited by J. Marchessault and K. Sawchuk, 24–38. New York: Routledge.

Weber, Steven. 2004. *The success of open source*. Cambridge: Harvard University Press.

Wittgenstein, Ludwig. 1963. *Philosophische Untersuchungen / Philosophical Investigations*. Oxford: Basil Blackwell.

Zimmer, Carl. 2004. *Soul made flesh: The discovery of the brain—And how it changed the world*. London: William Heinemann.

chapterSix Final Words

The Instrumental Logic of Neuroimaging Software Development

The force and meaning acquired by brain images in our current historic period are undisputable. To some extent, those images have reinforced the process of medicalization by supporting claims according to which more and more psychotherapeutic medicines are needed because, as brain scans show, "mental disorders have a physical home in the physical brain" (Zimmer 2004, p. 286). The scientific aura of brain images is often seen as a justification for them to be used for a growing range of purposes such as forensic ones (Rose 2016). "What a shame it would be if the valuable technology of neuroimaging came to look, in such applications, like a kind of post-modern phrenology" (Kennedy 2005, p. 19).

Brain images may be subjected to even more unexpected and problematic kinds of use if they get intensely explored by market players. So far this exploration has been underway mainly in the field of magnetic resonance scanners, which is dominated by a handful of multinational corporations (Cowley et al. 1994; Mallard 2003; Filler 2009). If market rationales and concentrations are strong in the

© The Author(s) 2019
E. Bicudo, *Neuroimaging, Software, and Communication*,
https://doi.org/10.1007/978-981-13-7060-1_6

generation of brain images (magnetic resonance scanners), the domain of data processing (neuroimaging software) continues to be largely dominated by a non-commercial dynamic. Only two proprietary software packages (LCModel and, especially, BrainVoyager) are widely used by neuroimaging researchers, even though they do not outweigh the relevance of open source packages such as SPM, FSL, and FreeSurfer. Very powerful proprietary packages tend to be present in academic domains of considerable methodological consolidation, such as economics and statistics. In domains of relatively recent emergence, like neuroimaging but also bioinformatics, neurogenetics, and others, open source software tends to prevail, because it enables researchers to have access to computer code, analyze it, modify it, and implement the new functionalities needed when methodologies are still being tested and consolidated.

However, even these relatively new research domains reflect the struggle described by Rifkin (2014): that between the collaborative dynamics of the open source philosophy (which foster democratization) and market rationales that try to gain terrain (promoting monopolization). This tension occurred, for example, in bioinformatics, when companies tried to sell analysis software embedded in DNA sequencing devices, a strategy that failed thanks to the resistance put into practice by academic researchers (Harvey and McMeekin 2010). In neuroimaging, two market trends are looming today. First, powerful manufacturers of magnetic resonance scans might decide to enhance the software that is already embedded in those devices, turning it into a full-fledged analysis application that would then be sold in association with scanners of skyrocketing prices. Second, neuroimaging data analysis might also be commercially targeted by large and famous IT corporations. Some of them have already come up with services of much commercial appeal, including "[…] Amazon, Microsoft, and Google services which, for a fee, users can provision data storage and multi-processor virtual systems upon which to configure and perform neuroimaging or genetics analyses" (Dinov et al. 2014, pp. 8–9). If successful, this new market will probably get segmented, in accordance with the purchase power of different research institutions. If one of these two suppositions (the enhancement of software embedded in magnetic scanners, and the proliferation of commercial data analysis services) prove effective and viable, very strong hierarchies will surely invade the territory of neuroimaging software.

However, irrespective of whether commercial logics come to cominate neuroimaging or not, it is important to consider that hierarchies are already present in this domain. It is not correct to suppose that social and geographical imbalances are disseminated by commercial players only. As seen in chapterFour, software developers based in different institutions and countries have found different conditions for carrying out their coding activities. Even when some initiatives seem to promote collaboration and data sharing, they may turn out to foster further hierarchization. For example, the International Consortium for Brain Mapping (ICBM), a repository for neuroscience data sharing, ends up reinforcing the dominance of its leading institutions, two of which are located in the United States (the University of California, and the University of Texas at San Antonio), one in Canada (the Montreal Neurological Institute), and one in Germany (Heinrich Heine University). Similar initiatives have been launched (Dinov et al. 2014), having similar effects, especially when they manage to acquire large international recognition.

At the level of computer code, hierarchies are manifested through the division between "code" and "script" analyzed in chapterFour and chapterFive. Frequently, code writing is framed as a highly intellectual and creative task. However, it was seen that some code writers have been devoted to the production of very simple code lines, which often reflects their technical and knowledge limitations. In this way, when speaking of computer programming, it is crucial to consider not only "high-level (architectural) software design" (Heliades and Edmonds 1999), that is the sophisticated definition of a software's general structure, and the composition of its most basic pieces of code, as seen in chapterThree. It is equally crucial to consider some recent trends of programming massification whereby some programmers, mainly those who are based in less dynamic countries and universities, become heavily dependent on the coding pillars set up by other programmers working in other places and other universities.

In this context, the difference between action and behaviour becomes very meaningful, as outlined in section24. The creative role that academic researchers are supposed to play is nowadays very much dependent on the mastery of data analysis skills, which, in their turn, are now largely associated with programming skills. Without such mastery, researchers can only have recourse to script writing or, more simply, to the functionalities provided by intuitive, user-friendly software packages. In Stallman's (2002b, p. 164) bold words, your inability to fully understand and modify a software package "makes you a prisoner of your software." Suárez-Díaz (2010, p. 82) has shrewdly argued that attention must be paid "[...] to the role of automation in biology, including the automation of data collection [...] and the automation of statistical inferences, entrenched in software packages." In addition to that, it is important to shed light on situations in which automation affects the scientific attitude itself, making researchers produce automatic analyses that "play safe" at the methodological level but lack scientific daring and critical view. Here, more than artificiality, the major threat is quasi-naturalness. "Quasi-natural societies function *as if* they were subject to natural laws. Systemic regulation occurs through the still intact medium of free action, but it takes place behind the back of acting subjects and downgrades the subjective consciousness of freedom to an illusion" (Habermas 2008, p. 196). Those risks tend to grow very large in our times when universities are subjected to massification, bibliometric rationales, and funding cuts.

From this viewpoint, the multiplication of software packages, analyzed in chapterFour and chapterFive, turns into a problematic tendency. If advanced or at least intermediate programming knowledge were held by most researchers, this proliferation could result in fruitful conversations and exchange of technical experiences. As most researchers have rather superficial programming knowledge gained by means of personal initiative or informal collaborations, the existence of many packages ends up creating small and isolated islands of users, unable to engage in constructive dialogues. In studies focusing on the internet, authors sometimes point out the issue of "filter bubbles" (Bozdag and Hoven 2015), which appear when websites or search engines give always the same kind of information to internet users, preventing these latter from enlarging their scope of knowledge. By the same token, software users can turn into captives inside "software bubbles" if they cannot either circulate between different software packages or converse with users of alternative packages. In these circumstances, communicative potentialities, so paramount for scientific progress, are seriously jeopardized.

These considerations are important in times when digital technologies and the internet are frequently framed as revolutionary weapons with which social disparities could be overcome. In this vein, authors are willing to underline "[...] the social mores and technical protocols of the net. Interactive. Modifiable. Accessible. Communal. Democratic. Upgradeable" (Barbrook 2003, p. 94). Contrary to those interpretations and expectations, the internet and all the digital creations that make it possible are many times contributing to the emergence and maintenance of imbalances and inequalities. This situation imposes blockages to the systems of communicative actions that underpin social life. However, the analyses provided in this book have also displayed that, on many occasions, those systems continue to play their role in a most effective fashion.

The Communicative Logic of Neuroimaging Software Development

As stated above, and analyzed in chapterFour and chapterFive, there is a division between code writing, on the one hand, and script writing, or "lightweight programming" (Skog 2003, p. 304), on the other. In Skog's (2003) interpretation, the appearance and dissemination of script writing was fostered, to a large extent, by the formation of the hardware and software market. To be sure, this factor has been important, as companies gain commercial terrain by providing people with the possibility of writing less complex code whose mastery demands relatively short learning periods. However, this not the only factor involved. Another crucial factor is that, since their initial production, computers constituted completely malleable machines, in the sense that they are "capable of being applied to [...] an extraordinarily diverse range of purposes" (Ensmenger 2010, p. 5). The computer with which academic researchers write code to design a neuroimaging software package can be used to navigate in social media, write reports, or store cooking recipes. In this way, even people with little computing knowledge may be attracted by computers, and therefore promote some expansion of their technical knowledge by means of an everyday contact with computers. Moreover, the possibility of connecting computers in global networks enables collaboration between people with different computing knowledge, in such a way that computers become not only malleable but also conducive to distant negotiations between those people. This can be achieved by means of code writing but also, although to a lesser degree, through the dissemination of script writing. In this way, programmers and users are provided with the possibility to reach agreements pertaining to the ways in which computers are used, code shared, technical support provided, and so on.

In chapterTwo, it was seen that nowadays software development, especially in open source, is never an individual task. It was shown that programmers always have resource to a range of libraries, pieces of code, and functions developed in the past and made available either within groups or on the internet. In this way, it is hard to agree that "[...] participants in open source software projects use their own resources to privately invest in creating novel software code" (von Hippel and von Krogh 2003, p. 213). If programmers were willing to use only their own resources to design software privately, not only would their work be overwhelmingly complex, but it would also take extremely long to be completed. The collective nature of software development is certainly a question of personal choice but also a technical imperative.

In an effort to explain the nature of software design, Moran and Carroll (1997) proposed the concept of "design rationale," which in the summary given by Heliades and Edmonds (1999, p. 391), refers to: "[...] 'Documentation of (a) the reasons for the design of an artefact, (b) the stages or steps of the design process, and (c) the history of the design and its context'." This last word points to the aspect that is the most relevant from a communicative perspective. Indeed, a rationale emerges from the technical intricacies of software development but also from the social contexts in which computer code is produced and algorithms created. "[...] one key to understanding the politics of algorithms might be not so much to look for essences with consequences but to attend to how the figure of the algorithm is employed and comes to matter in specific situations" (Ziewitz 2016, p. 10).

Moreover, one can think of the design rationale that comes to be when computer code or software is used. It is not only in the field of software that the logic introduced by users can even diverge from the logic originally embedded in the technical production (Santos 2000). In the realm of software, this happens when users act towards "[…] embedding the technology in their lives in all kinds of alternative ways and using it for different means, or resisting, subverting and reworking the algorithms' intent […]" (Kitchin 2017, p. 19). For example, developers, which are frequently also users of software, would realize such kind of algorithmic inversion when some lines of code composing a semi-flexible software package, and therefore consolidating geographical or institutional hierarchies, would be studied and adjusted to become part of the codebase of a flexible package whose production is underpinned by a collaborative and open community.

Therefore, a communicative design rationale, so to say, is created through the story of the software, including the assemblage of different pieces of code, the flows of information necessary to produce the software, the relations between programmers engaged in the project, the material infrastructure mobilized during the process, and the ways in which users come into touch with the package. We are then dealing with actual systems of communicative actions. Atal and Shankar (2015) claim that the trajectory of software has two stages: in stage 1 software is designed whereas in stage 2 it is distributed to end-users. It is possible to say that both stages involve substages, all of which are marked by communicative actions necessary to make software both technically and socially viable.

Innovation, Intercomprehension

Nowadays, in a society sometimes described by means of expressions such as information society, technological society, or digital society, innovation has become a sort of mantra repeated by analysts, economists, policy-makers, and others. It is as though the incapacity to innovate (irrespective of what definition of innovation is subscribed to) would be the source of every kind of social poverty and political doom. However, this discourse is generally voiced so automatically and uncritically that some contestable notions are just taken for granted. One of such notions is the concentrated or hierarchical patterns of innovation, according to which it is natural, or even desirable, that most of what is described as technological innovations depends on the initiatives of only a handful of countries and companies. As a consequence, the champions of capitalist innovation end up praising, in either implicit or explicit ways, the persistence of technological, scientific, and social exclusion. In this context, it is crucial to point out the viability of a "distributed technological innovation and development" (Lakhani and Wolf 2005, p. 3), an alternative to the current tendencies towards concentration.

These considerations are key in analyses focusing on software. All the conditions necessary for the emergence of a balanced and democratic system of software development have been created: highly successful platforms for collaborative software development have appeared, technical incompatibilities between different computes have been circumvented, the same programming languages are used in different parts of the world, efficient informational channels have put programmers into consistent contact, and other phenomena. In a balanced innovation system, software packages of great popularity would be the outcome of thriving negotiations and flows of information comprising programmers based in several countries, instead of being the creation of a couple of huge companies heavily concerned with copyrights and market growth. In a dispersed and balanced innovation model, code sharing and technical support would be common practices, which is already possible thanks to platforms such as GitHub, as analyzed in chapter-Three. Considering the technical possibilities of our times, as well as the programming talents dispersed across the globe, the popularity and dominance that companies such as Microsoft, Google, and Apple have managed to maintain by means of marketing and lobbying campaigns can only be seen as a technical and political aberration.

When this discussion is made from the viewpoint of scientific production, what is at stake is the cognitive vitality of most countries, which could conquer the right to be innovative, surmounting the current limitations that oblige them to only carry out the "imitative research" referred to by Sartorius (1998, p. 346). In neuroimaging and other academic domains where open source software prevails, the constitution of a truly balanced innovation system has been, to a considerable extent, blocked by the ongoing force of semi-flexible packages. As analyzed in chapterFour, a curious process took place whereby the passage from rigidity to flexibility involved an intermediate step (semi-flexibilty) which has turned into a sort of technical levee. Once some institutions managed to launch software packages of international reputation, different technical, programming, and institutional factors came into play, consolidating the dominance of those few packages while preventing other, flexible packages from gaining large recognition. The phenomenon is politically tricky because those semi-flexible packages are not under the protection of companies moved by profit motives; they are instead managed by research institutions whose main purpose is to advance knowledge and promote data sharing but whose scientific reputation and financial robustness make their software productions become overwhelming and intimidating figures on the international stage.

However, one can suppose that, in the years to come, the move from semi-flexibility to flexibility will be finally completed. This is so because many of the current flexible packages, by incorporating innovative coding strategies and technical solutions, have some potential for overshadowing their prestigious semi-flexible counterparts. If this tendency continues to gain force, a period can come along when the benefits described by Sartorius (1998) will be materialized: researchers will be able to conduct research considered as high-quality even in less dynamic countries; diseases and conditions of local prevalence will be everywhere studied satisfactorily; and the scientific prestige of a wide range of institutions and universities will be enhanced. The neuroimaging domain already displays such potentialities, especially because it is now so related to the development, modification, and use of software packages, which are products whose flexibility can be consistently increased. If semi-flexibility really becomes a less powerful phenomenon, and if the threats of commercial expansion described above do not prove viable, then the relatively modest communities that produce today's promising flexible packages can grow very large and international in the years to come. The passage from semi-flexibility to flexibility reinforces the parallel passage from "globalization from above" to "globalization from below" (Appadurai 2000), which is also the passage from the "global order" to the "local order" (Santos 2000).

Hopefully, the analyses presented in this book have shown that such passages and changes are by no means utopic. Many of the techniques, relations, and attitudes required by them are already present. In order for the software innovation system to become balanced and democratic, it does not need to be created anew; rather, it needs to be anchored in the systems of communicative actions already existent. The phenomena and instruments on which social discussions focus may be concentrated, but social discussion itself is always distributed. Even though the most successful software packages currently used are dominated by a couple of institutions (semi-flexibility), many other collaborative and promising development projects have been maintained and created (flexibility). The point is that policy-makers, economic analysts, the media, and other players generally frame innovation as an economic or technical issue. When it is finally realized that innovation is also a social issue, because it happens through a collective search for intercomprehension, then alternative forms of innovation (such as alternative communities of computer programmers) will not only be more visible but also gain more social recognition.

As I stressed at the very beginning of this book, analysts have frequently focused on the reasons leading people to produce non-proprietary software. Programmers themselves have put forward their intentions. For example, Stallman (2002a, p. 41) encouraged developers to produce non-proprietary software "so you can help your neighbour" and "so that the whole community benefits." However, there is no need to feel bound to the identification of personal or ideological motives. These latter, in a communicative interpretation, are less important than the systems of communicative actions in which people engage. In the history of software development, there are many examples of developers who decided to share code within other-centred communities after realizing that this is the best way to reach their self-centred goal of producing efficient, bug-free packages (Atal and Shankar 2015). By the way, this is what happened with the Linux operating system (Torvalds and Diamond 2001), perhaps the most famous historic example of open source project. However, even if a certain developer joins a certain community without having communitarian sentiments, the person indirectly reinforces the prestige of the community, which may formulate a discourse that highlights, for example, the number of times that its code has been accessed, using this as an argument to attract new community members. Therefore, the point to be stressed is not so much people's intentions but the communicative channels connecting different agents, including those for whom communication is but a secondary concern. "The world becomes meaningful by being this common object produced through reciprocal relations that create, at the same time, alterity and communication" (Santos 2002, pp. 316–317). In this constant and slow search for intercomprehension, comes into play "[...] the mysterious power of intersubjectivity to unite disparate elements without eliminating the differences between them" (Habermas 2008, p. 21).

One of the central points made throughout this book is that social negotiation and the search for intercomprehension, in the case of programmers, can be achieved not only by means of personal relations and information flows but also via computer code itself. The expressive externalization of the self through code (chapterThree) is matched by the manifold negotiations and connections through code (chapterTwo). Hence the emergence of code as a proto-language (chapterThree). Mackenzie (2006, pp. 24–25) makes an intriguing point:

> [...] codes make no pretension to be a common language. Despite the quasi-universal resonance of the term 'code,' programmers have not invented any Esperanto. Rather, code has dispersed into a cacophony of different coding languages, sometimes hierarchically related, sometimes not: FORTRAN, Perl, Prolog, C, Pascal, Applescript, Javascript, Actionscript, Hypertalk, LISP, Python, Java, C++, C#, etc. The spawning, mutating and cloning of different idioms of code, and indeed of different versions of similar software applications or 'solutions' generates code babble. There are many banal manifestations of this disintegration. The need to constantly upgrade software to fix bugs and ensure compatibility, the struggle to standardize protocols, and the proliferation of clones and variants of the same application all attest to this code babble and to a multiplication of coding dialects.

Even though many different natural languages continue to be spoken in the world, and even though nobody is very willing to learn Esperanto, communication continues to be a key factor in social life. By the way, one of the main triggers for communicative creativeness is the formulation of tactics to circumvent linguistic differences. People can master different spoken languages but everybody shares in the communicative practice of making sense of social life, and this is what creates a common ground on which society lies. By the same token, communication between programmers can happen in spite of their mastering different programming languages. As seen in chapterThree, those who learn one language find it easy to understand the principles of other languages, as there is a common rationale between different languages. In this way, programmers are always able to circumvent technical differences because they share in a computing logic that is mobilized when code lines are written, a certain codebase analyzed, some pieces of code shared, or the pathways taken by a codebase negotiated. Babble, if it emerges at all, is only temporary. For it is possible to "[…] ascribe to the communicatively acting subjects themselves the normal capabilities for managing those disturbances in communication that arise from mere misunderstandings" (Habermas 1996, p. 20).

Thanks to these communicative capacities, developers of neuroimaging software manage to extend their conversations even outside their specific knowledge domain. For example, Dr. Vincent Giampietro (Centre for Neuroimaging Sciences, King's College London, UK) has had some collaboration with oil companies. He once met people who were trying to map the seabed to facilitate oil prospection. "And we said: 'Well, that's what we do everyday. Send us your data' […] It was just bits of the seabed of the ocean there. But it was the same. It was just three-dimensional. It was exactly the same thing. We had just that, you know, three-dimensional datasets to realign. So we were just using our own [neuroimaging] methods at that time to do it. So there's a lot of cross-fertilization." This collaboration was possible because, potentially, different sorts of three-dimensional images can be processed as if they were brain images.

In this sense, neuroimaging data analysts, precisely because, in addition to being scientists, are data analysts, are capable of engaging in collaborations with different types of players, not only oil prospectors. It is then necessary to shed light on "[...] the social mechanisms that curb or facilitate the incorporation of would-be contributions into the domain of science" (Merton 1968, p. 60). A scientific model that makes viable the search for disciplinary and interdisciplinary intercomprehension is also a system likely to produce solid and quick scientific advances. It is then necessary to think about the collaborative and communicative dimensions of initiatives such as open source software.

A Communicative Approach to Open Source

Not only has non-proprietary software gained a growing number of adepts, but it has stimulated the formulation of different interpretive approaches. It is possible to group these views into four categories, as summarized in the following scheme:

	Instrumental rationality	Communicative rationality
Market politics	Costless software	Freedom-promoting software
Sociotechnical politics	Open source software	Flexible software

   Some analysts stress the sheer fact that some software packages can be accessed and used free of charge. This "costless software" approach can be illustrated by the description given by Atal and Shankar (2015, p. 1382): "[...] individual developers contribute to developing the software but do not hold any copyright on their work and hence cannot prevent their contributions from being copied, modified or used by others." From the viewpoint of developers, this strategy proposes a peaceful coexistence between proprietary packages and non-proprietary ones, as the developers of these latter are said to glean some professional and social benefits by renouncing economic gains. From an interpretive perspective, this description stresses the legal and economic technicalities making it possible for developers to circumvent commercial barriers and proprietary restrictions.

Richard Stallman (2002a, b, d) proposed what I am naming the "freedom-promoting" approach, for which software development and use have strong normative contents. The Free Software Foundation was created to incorporate these values. "[...] I talk about issues of ethic, and what kind of a society we want to live in, what makes for a good society, as well as practical, material benefits [...] That's the free software movement" (Stallman 2002b, p. 167). For this reason, Stallman often points out that "free" refers, here, to the notion of freedom. "To understand the concept, you should think of 'free' as in 'free speech,' not as in 'free beer.'" (Stallman 2002a, p. 41) In this approach, the market is still the main reference, as developers are urged to negate and fight against market logics. From an interpretive point of view, this approach is very much related to a communicative search for balanced social relations.

There is a common feature between the "costless" and the "freedom-promoting" approaches. As both are concerned with the software market, in which companies protect products via legal instruments, they both end up highlighting legal issues. This is why one of the most remarkable deeds of the Free Software Foundation was the creation of the General Public License, an instrument safeguarding the public nature of software (Atal and Shankar 2015; von Hippel and von Krogh 2003; Stallman 2002b, c). As a consequence, these approaches tend to overestimate the force of regulations as creators of social bounds. To be sure, regulations have the capacity to create and maintain social expectations (Luhmann 1983). However, as I claimed elsewhere (Bicudo 2006), regulations can only be produced when social negotiations have reached an advanced stage of development and maturation. Moreover, the limit of these approaches is revealed in the fact that developers of non-proprietary software are frequently not concerned with the economic, professional, and ideological issues put forward by analysts. Often, developers work with non-proprietary packages simply because they are well-written and have good technical performance.

In the "open source software" approach, what is stressed is the fact that the software's codebase is largely published, so everybody can have access to it. A common idea, here, is that a large developer community emerges in which technical problems can be quickly identified and solved. Eric Raymond (2001, p. 27), who campaigned for the diffusion of the expression "open source" and helped create the Open Source Initiative in the late 1990s, encouraged programmers to publish their lines of code, claiming: "[...] your users will diagnose problems, suggest fixes, and help improve the code far more quickly than you could unaided." Here, ideological purposes are dropped for the sake of an explanation foregrounding the technical advantages of open source. Interestingly, the open source current has produced stronger social impacts than the Free Software movement. This is so because the technical language it speaks, devoid of ideological tones, is much more appealing to programmers generally oblivious to political matters.

The practical success of the open source approach is also the source of its interpretive limitation. It fails to shed light on the ideological or normative contents that software development has had and continues to have. As noted by von Hippel and von Krogh (2003, p. 215), "[...] goal statements provided by successful open source software projects vary from technical and narrow to ideological and broad, and from precise to vague and emergent [...]." On the one hand: "Ideology isn't what has sold the open source model" (Torvalds and Diamond 2001, p. 227). On the other hand, however, many software developers persist in recognizing the social benefits of publishing their individual coding achievements. If this normative stance can sometimes consist in a mere discourse through which researchers depict themselves as responsible academics (as seen in chapterThree), it is frequently adopted through an actual involvement in collaborative relations (chapterTwo) or fostered by an actual intent to design useful software for people lacking deep computing knowledge (chapterFive). (In this book, I decided not to explore the discussion on the so-called commons, as this would probably expand too much the scope of my analysis. However, this discussion does not exclude software development, a domain full of political and ideological preoccupations with the management of communal goods (Rifkin 2014; Weber 2004). From this viewpoint, open source software is aimed not only at promoting excellent technical results; the point made by Weber (2004, p. 85) is also valid: "Open source intellectual property aims at creating a social structure that expands, not restricts, the commons.")

The approach proposed in this book has a sociological nature. Like the freedom-promoting approach, it is concerned with the hierarchies of software development but, unlike this approach, it assumes that what can counter such hierarchies is not the revolted reaction of socially committed developers, but the constitution of systems of communication grounded on "man's capacity to act and to act together and in concert" (Arendt 1998, p. 23). The approach proposed here, like the open source approach, considers that many key phenomena are manifested and dealt with at the level of technical issues and coding choices but, unlike this approach, it does not downplay the (specifically social) weight of ideologies and worldviews. Flexible software packages certainly have the potential to defy market laws and expand programmers' technical options; more than this, however, flexible packages reinforce social integration by putting programmers based in different institutions and countries into a vivid and consistent process of negotiation. In addition to requiring some resources, such as the internet and development platforms, they frequently involve a certain degree of political awareness, making researchers acknowledge, for example, the normative importance of cooperation, scientific solidarity, and reciprocal technical support, hence the everyday preservation of the spontaneous code sharing and teaching initiatives analyzed in chapterTwo.

Seen from this vantage point, software development is nurtured by three factors. First, it is a product of creative solutions found by developers, who imbue software with a particular rationale (chapterThree). Since the beginning, I had recourse to Naur's (2001, p. 234) theory building approach to software development, according to which: "The building of the program is the same as the building of the theory of it by and in the team of programmers." Second, there is often much negotiation between people involved in the design of a certain software package (chapterTwo). Third, it is crucial to consider that all "the communities around Open Source" (Wagstrom 2009, p. 5) are sustained not only by personal relations but also, and to an increasing degree, by a set of material and computing resources. Most of today's popular semi-flexible packages (such as FSL, FreeSurfer, and AFNI) were launched in the late 1990s, at a moment when programmers lacked the requisites for the design of fully flexible packages, especially a widely diffused internet, and online collaborative development platforms such as SourceForge and GitHub. All these resources would become largely known only in the 2010s, making it easier to create and maintain developer communities, nurture their collaborative spirit, and produce flexible packages.

Therefore, in addition to recognizing the relevance of people's communities, one has to consider the importance of resources, not only digital ones (such as websites or programming platforms) but also material ones (such as the infrastructure necessary for the internet to work). When speaking of flexible software and systems of communicative actions, I am also claiming for a communicative approach that does not lose sight of materiality. Because communication presupposes the social attitude described by Schutz (1974, p. 59) whereby one considers not only "the existence of intelligent (endowed with consciousness) fellow-men" but also "the experienceability (in principle similar to mine) by my fellow-men of the objects in the life-world." The idea of "systems of communicative actions" announces a theoretical project through which the communicative approach can in the future be more assertive in recognizing the relevance of materiality. This is important because: "We are talking about the spatial and temporal phenomenon of language, not about some non-spatial, non-temporal phantasm" (Wittgenstein 1963, p. 47).

What guarantees social integration is not only a regulatory framework but also, and more importantly, the efforts through which developers try to set collective goals, work in distant cooperation, and solve disagreements. For some readers, these coordinating goals may sound too overwhelming, almost impossible to be pursued by simple hackers. However, these goals are never pursued in accordance with official, explicit programs and timeframes. To a large measure, these coordinating tasks are performed in non-planned, intuitive ways. The actions necessary to reach a good balance between the standards of computational technologies, on the one hand, and social needs, on the other, reflect the nature of the "customs" identified by Raymond (2001, p. 75): "[…] most hackers have followed them without being fully aware of doing so." Weber's (2004, p. 73) explanation is even more precise:

The logic of what open source user-programmers do did not emerge from abstract theory. No one deduced a set of necessary tasks or procedures from a formal argument about how to sustain large-scale, decentralized collaboration. It was a game of trial and error – but a game that was played out by people who had a deep and fine-grained, if implicitly, understanding of the texture of the community they were going to mobilize.

On the one hand, software developers' search for intercomprehension triggers discussions of uncertain conclusion, as people are dealing with grey zones of software progress. On the other hand, much of this discussion is pacified by the presence of largely recognized customs pertaining to programming languages, software engineering, computer hardware, and other technical matters.

It is then possible to reflect on the meaning of the connections between software developers. In chapterTwo, the notion of modularity was reviewed. In a sense, developers are connected in quite the same way as modules are connected within a software package. According to a brilliant and most revealing point made by Parnas (1971, p. 2): "*The connections* between modules *are the assumptions which the modules make about each other.*" In other words, a certain module is connected with the rest of the package because, so to say, it intuits that it produces outputs needed by other modules, intuiting, at the same time, that these other modules generate outputs necessary for its own operations. More or less equally, programmers intuit that their coding work depends on previous work by other people while being vital for additional work to be done by future developers. To a large degree, all these personal connections are guaranteed by the impersonal mediation provided by computer code. Whenever I speak of a developer community engaged in communicative tasks, I am not claiming that such community would exist in an explicit manner, as if it was an official movement following a clear agenda. To be sure, some hackers are fond of joining official movements but on most occasions, connections between them are weakly dependent on intentional actions. Frequently, those connections result from relations and processes that cannot be directly perceived and controlled. On the one hand, some technical requisites must be simply complied with if one is to produce any viable and useful software at all. On the other, relations are not always the product of explicit negotiations, being sometimes greatly shaped by the intuitive customs which, across the decades, came to be solidified in the universe of computer programming. Underpinned by wayward social negotiations and ideological preferences, but also by rigid technical standards, programming strategies, and material resources, software development reveals all the paradoxes typical of times when people oscillate between behaviour and action, between hierarchy and liberation, or, to be more precise, between semi-flexibility and flexibility.

References

Appadurai, Arjun. 2000. "Grassroots globalization and the research imagination." *Public Culture* 12 (1):1–19.

Arendt, Hannah. 1998. *The human condition*. Chicago: University of Chicago Press.

Atal, Vidya, and Kameshwari Shankar. 2015. "Developers' incentives and open-source software licensing: GPL vs BSD." *B.E. Journal of Economic Analysis and Policy* 15 (3):1381–1416.

Barbrook, Richard. 2003. "Giving is receiving." *Digital Creativity* 14 (2):91–94.

Bicudo, Edison. 2006. "Normes, térritories et aménagement: les recherches biotechnologies dans l'Union Européenne." Master, Institut de Gégraphie, University Paris 1 (Panthéon Sorbonne).

Bozdag, Engin, and Jeroen van den Hoven. 2015. "Breaking the filter bubble: Democracy and design." *Ethics and Information Tehcnology* 17:249–265. https://doi.org/10.1007/s10676-015-9380-y.

Cowley, L. Tad, Hope L. Isaac, Stuart W. Young, and Thomas A. Raffin. 1994. "Magnetic resonance imaging marketing and investment: Tensions between the forces and the practice of medicine." *Chest* 105 (3):920–928.

Dinov, Ivo D., Petros Petrosyan, Zhizhong Liu, Paul Eggert, Sam Hobel, Paul Vespa, Seok Woo Moon, John D. va Horn, Joseph Franco, and Arthur W. Toga. 2014. "High-throughput neuroimaging-genetics computational infrastructure." *Frontiers in Neuroinformatics* 8:1–11.

Ensmenger, Nathan. 2010. *The computer boys take over: Computers, programmers, and the politics of technical expertise.* Cambridge and London: MIT Press.

Filler, Aaron G. 2009. "The history, development and impact of computed imaging in neurological diagnosis and neurosurgery: CT, MRI, and DTI." *Nature Precedings*:1–76. Available at http://dx.doi.org/10.1038/npre.2009.3267.5.

Habermas, Jürgen. 1996. *Between facts and norms: Contributions to a discourse theory of law and democracy,* Studies in contemporary German social thought. Cambridge, MA: MIT Press.

Habermas, Jürgen. 2008. *Between naturalism and religion.* Cambridge: Polity Press.

Harvey, Mark, and Andrew McMeekin. 2010. "Public or private economies of knowledge: The economics of diffusion and appropriation of bioinformatics tools." *International Journal of the Commons* 4 (1):481–506.

Heliades, G. P., and E. A. Edmonds. 1999. "On facilitating knowledge transfer in software design." *Knowledge-Based Systems* 12:391–395.

Kennedy, Donald. 2005. "Neuroimaging: Revolutionary research tool or a post-modem phrenology?" *American Journal of Bioethics* 5 (2):19.

Kitchin, Rob. 2017. "Thinking critically about and researching algorithms." *Information, Communication & Society* 20 (1):14–29.

Lakhani, Karim R., and Robert G. Wolf. 2005. "Why hackers do what they do: Understanding motivation and effort in free/open source software." In *Perspectives on free and open source software,* edited by Joseph Feller, Brian Fitzgerald, Scott A. Hissam, and Karim R. Lakhani, 3–22. Cambridge: MIT Press.

Luhmann, Niklas. 1983. *Sociologia do direito I, Biblioteca Tempo Universitário 75.* Rio de Janeiro: Tempo Brasileiro.

Mackenzie, Adrian. 2006. *Cutting code: Software and sociality.* New York: Peter Lang.

Mallard, John R. 2003. "The evolution of medical imaging: From Geiger counters to MRI—A personal saga." *Perspectives in Biology and Medicine* 46 (3):349–370.

Merton, Robert K. 1968. "The Matthew effect in science." *Science* 159 (3810):56–63.

Moran, T. P., and J. M. Carroll. 1997. "Overview of design rationale." In *Design rationale: Concepts, techniques and use,* edited by T. P. Moran and J. M. Carroll. Hillsdale: Lawrence Erlbaum Associates.

Naur, Peter. 2001. "Programming as theory building." In *Agile software development,* edited by Alistair Cockburn, 227–239. Boston: Addison-Wesley.

Parnas, David Lorge. 1971. *Information distribution aspects of design methodology*. Pittsburgh: Computer Science Department, Carnegie-Mellon University. Available at http://repository.cmu.edu/cgi/viewcontent.cgi?article= 2828&context=compsci.

Raymond, Eric S. 2001. *The cathedral & the bazaar: Musings on Linux and open source by an accidental revolutionary*. Sebastopol: O'Reilly.

Rifkin, Jeremy. 2014. *The zero marginal cost society: The internet of things, the collaborative, and the eclipse of capitalism*. New York: Palgrave Macmillan.

Rose, Nikolas. 2016. "Reading the human brain: How the mind became legible." *Body & Society* 22 (2):140–177.

Santos, Milton. 2000. *La nature de l'espace: technique et temps, raison et émotion*. Paris: L'Harmattan.

Santos, Milton. 2002. *A natureza do espaço: técnica e tempo, razão e emoção, Milton Santos collection 1*. Sao Paulo: Edusp.

Sartorius, Norman. 1998. "Scientific work in Third World countries." *Acta Psychiatrica Scandinavica* 98:345–347.

Schutz, Alfred. 1974. *The structures of the life-world*. London: Heinemann.

Skog, Knut. 2003. "From binary strings to visual programming." In *History of Nordic computing*, edited by Janis Bubenko Jr., John Impagliazzo, and Arne Solvberg, 297–310. Boston: Springer.

Stallman, Richard M. 2002a. "Free software definition." In *Free software, free society: selected essays of Richard M. Stallman*, edited by Joshua Gay, 41–44. Boston: GNU Press.

Stallman, Richard M. 2002b. "Free software: Freedom and cooperation." In *Free software, free society: Selected essays of Richard M. Stallman*, edited by Joshua Gay, 155–186. Boston: GNU Press.

Stallman, Richard M. 2002c. "Releasing free software if you work at a university." In *Free software, free society: Selected essays of Richard M. Stallman*, edited by Joshua Gay, 61–62. Boston: GNU Press.

Stallman, Richard M. 2002d. "Why software should be free." In *Free software, free society: Selected essays of Richard M. Stallman*, edited by Joshua Gay, 119–132. Boston: GNU Press.

Suárez-Díaz, Edna. 2010. "Making room for new faces: Evolution, genomics and the growth of bioinformatics." *History and Philosophy of the Life Sciences* 32 (1):65–89.

Torvalds, Linus, and David Diamond. 2001. *Just for fun: The story of an accidental revolutionary*. New York: HarperCollins.

von Hippel, Eric, and Georg von Krogh. 2003. "Open source software and the 'private-collective' innovation model: Issues for organization science." *Organization Science* 14 (2):209–223.

Wagstrom, Patrick Adam. 2009. "Vertical interaction in open software engineering communities." PhD, Carnegie Insitute of Technology/School of Computer Science, Carnegie Mellon University.

Weber, Steven. 2004. *The success of open source*. Cambridge: Harvard University Press.

Wittgenstein, Ludwig. 1963. *Philosophische Untersuchungen / Philosophical Investigations*. Oxford: Basil Blackwell.

Ziewitz, Malte. 2016. "Governing algorithms: Myth, mess, and methods." *Science, Technology & Human Values* 41 (1):3.

Zimmer, Carl. 2004. *Soul made flesh: The discovery of the brain—And how it changed the world*. London: William Heinemann.

Methodological Appendix

My research project was underpinned by five methods: a quantitative analysis of neuroimaging papers published in four different years (1985, 1995, 2005, and 2015); a geoprocessing analysis; qualitative interviews; an internet survey with users of neuroimaging software packages (Survey 1); and another survey with developers of software packages (Survey 2). Each of these methods is described in the four following sections.

1. Analysis of Neuroimaging Papers

My initial target was to become familiar with the software packages currently used in neuroimaging. For so doing, I decided to check neuroimaging papers published in 2015. The idea was to look at the methodological sections (or methodological supplements) of these papers, identifying all the software packages cited as data analysis tools.

The platform used for searching for neuroimaging papers was Web of Science.[1] I decided to use Web of Science, not databases such as

---

[1] http://wok.mimas.ac.uk/

© The Editor(s) (if applicable) and The Author(s), under exclusive license to Springer Nature Singapore Pte Ltd. 2019
E. Bicudo, *Neuroimaging, Software, and Communication*,
https://doi.org/10.1007/978-981-13-7060-1

PubMed-MedLine or Ovid, because these latter are specialized in biomedical studies whereas the former embraces a broader range of disciplines. This is important because neuroimaging is not really a well-defined academic discipline. Rather, it is a quite vague domain that includes all studies focusing on the human brain and using images. In this way, even though neuroimaging is mainly carried out by biomedical researchers, it is also performed by other academics such as psychologists, linguists, and social scientists. By using Web of Science, I would then be able to identify a broader range of studies.

Indeed, a quick comparison enabled me to confirm this supposition. In December 2015, I realized a similar search on PubMed and Web of Science, looking for papers published in 2015 only, and obtaining the following results (Table A).

Table A   Search strategy used to identify neuroimaging papers

Platform	Search strategy	Papers identified
PubMed	(neuroimag*[Title/Abstract]) AND (magnetic resonance[Title/Abstract] OR MR[Title/Abstract])	1173
Web of science	(TS=neuroimag* OR TI=neuroimag*) AND (TS=magnetic resonance OR TI=magnetic resonance OR TS=MR OR TI=MR)	1564

Therefore, the use of Web of Science led to the biggest number of papers, a finding that made me choose this platform for my study. As can be seen in the previous table, my search strategy was very broad, aiming to identify all the papers containing "neuroimaging" in either the topic or the title, as well as all the papers related to magnetic resonance technology. After downloading and analyzing papers published in 2015, I decided to extend this analysis, including the years of 2005, 1995, and 1985, using the same search strategy. In this way, I would have a historic overview of the use of neuroimaging software.

My main inclusion criterion was: studies reporting data collected from human beings, produced with magnetic resonance scanners, and analyzed by means of one or more software packages. Other criteria were as follows:

INCLUDED:

- Studies combining data from magnetic resonance scanners with data from positron emission technology and or computed tomography
- Studies citing software packages generally not used in neuroimaging, such as R and Photoshop
- Abstracts
- Case reports
- Book chapters published in journals
- Retrospective studies involving new analyses of old data.

EXCLUDED:

- Studies focusing on animals only
- Studies focusing on both humans and animals
- Studies where neuroimaging software was used to analyse data produced by positron emission technology and or computed tomography but not magnetic resonance scanners
- Papers that launched a software package or toolbox, even if authors provided an actual data analysis as an example
- Studies that quoted an in-house software without specifying its name
- Papers citing only the software embedded in the magnetic resonance scanner
- Reviews.

Some papers could not be analysed as I failed to have access to them because of problems of institutional access. A minority of papers were not written in English. With my language skills, I was able to analyse papers in five languages: English, Portuguese, Spanish, French, and German. When the paper was written in another language, I had to exclude it from the study. For 2015, I excluded one paper published in Mandarin. For 2005, five papers were excluded (two in Polish, two in Korean, and one in Japanese). Therefore, I obtained the following results (Table B).

Table B Outcomes of searches

Year	Realization of search	Papers identified	Access failure	Papers that met the inclusion criteria
2015	January 2016	1564	48	335
2005	March 2016	892	106	226
1995	May 2016	230	68	16
1985	May 2016	5	2	0

These were the programming languages, software packages, and toolboxes cited in the papers included in my analysis:

2015
See Tables C and D.

Table C Software packages identified for the year 2015

40 software packages		
Rigid	Semi-flexible	Flexible
Analyze, BrainParser, BrainWave, Civet, CRKit (Computational Radiology Kit), Felix Software, GraphPad, LCModel, OsiriX, PANDA (Pipeline for Analyzing Brain Diffusion Images), PMOD	AFNI (Analysis of Functional NeuroImages), BioIMage Suite, BrainVoyager, Caret, DtiStudio, FreeSurfer, FSL (FMRIB Software Library), Java, MatLab, MIPAV (Medical Image Processing, Analysis, and Visualization), TrackVis, Unix Shell, XBAM	3DSlicer, ANTS (Advanced Normalization Tools), BRAINS, BRAT (Brainnetome fMRI Toolkit), CCS (Connectome Computation System), DIPY (Diffusion Imaging in Python), DSI Studio, InVesalius, IRTK, itk-Snap, MRIcro, RSFNC (Analysis of Resting State Functional Connectivity Network developed for ABIDE), SPM, SUMA, XMedCon

Table D   Toolboxes identified for the year 2015

68 toolboxes	
Semi-flexible	Flexible
3dAllineate, 3dBlurToFWHM, 3dClust-Sim, 3dDeconvolve, 3dDespike, 3dFWHMX, 3dFourier, 3dMaskave, 3dRSFC, 3dTshift, 3dttest++, 3dvolreg, AlphaSim, Bbregister, BCT (Brain Connectivity Toolbox), BET (Brain Extraction Tool), BVQXTools, Camino (UCL Camino Diffusion MRI Toolkit), CBS High-Res Brain Processing Tools, Cogent 2000, Diffusion Toolbox, EDDY, FAST (FMRIB Automated Segmentation Tool), FEAT (FMRI Expert Analysis Tool), FILM (FMRIB Improved Linear Model), FIRST, FLIRT (FMRI Linear Image Registration Tool), FNIRT (FMRI Non-linear Image Registration Tool), FsFast (FreeSurfer Functional Analysis Stream), FSLview, Group ICA Toolbox, MELODIC (FSL Mutivariate Exploratory Linear Optimized Decomposition into Independent Components), MIBCA (Multimodal Imaging Brain Connectivity Analysis), NetBrainWork, Neuroelf, PyMVPA (MultiVariate Pattern Analysis in Python), Randomise, Siena, SUSAN, TBSS (Tract-Based Spatial Statistical Tool), TOPUP, Unwarp, Wavelet	AFQ (Automatic Fiber Quantification), Anatomy, ART (Artifact Detection Tools), ASL Data Processing Toolbox, BEaST (Brain Extraction Based on nonlocal Segmentation Technique), BPM (Biological Parametric Mapping), CONN, DARTEL, dcm2nii, DPARSF (Data Processing Assistant for Resting State fMRI), Fieldmap, IBASPM (Individual Brain Atlases Using Statistical Parametric Mapping Software), INRIAlign, LST (Lesion Segmentation Tool), MARINA (Masks for Region of Interest Analysis), MarsBAR, NewSegment, PickAtlas, REST (Resting State fMRI Data Analysis Toolkit), rfxplot, Robust Regression, SUIT, Template-o-Matic, VBM, VSRAD (Voxel-Based Specific Regional Analysis System for Alzheimer's Disease)

2005
See Tables E and F.

Table E Software packages identified for the year 2005

44 software packages		
Gid	Semi-flexible	Flexible
Analyze, Brains (Brain Research: Analysis of Images, Networks, and Systems), ImageTool Software, Photshop, SAGE, SigmaPlot, TDIS-FILE-500, XITE (X-Based Image Processing Tools and Environment)	AIR (Automated Image Registration), ANIMAL (Automated Non-Linear Image Matching and Anatomical Labeling), MatLab, NIH Image Software	SPM

Table F Toolboxes identified for the year 2005

26 toolboxes		
Rigid	Semi-flexible	Flexible
ProbeSV	3dAnova, 3dDeconvolve, 3dDespike, 3dMean, 3dVol-2Surf, 3dvolreg, AlphaSim, BET (Brain Extraction Tool), Diffusion Toolbox, FAST (FMRIB Automated Segmentation Tool), FEAT (FMRI Expert Analysis Tool), FILM (FMRIB Improved Linear Model), FLIRT (FMRI Linear Image Registration Tool), FMRIStat, IPS (Interactive Point Selection), mrAlign, mrVista, SurfMeasures	INRIAlign, MarsBAR, PickAtlas, Robust Regression, ROI, SnPM (Statistical NonParametric Mapping), VBM

1995
See Tables G and H.

Table G   Software packages identified for the year 1995

13 software packages		
Rigid	Semi-flexible	Flexible
Analyze, Brains (Brain Research: Analysis of Images, Networks, and Systems), ImageTool Software, Photoshop, SAGE, SigmaPlot, TDIS-FILE-500, XITE (X-Based Image Processing Tools and Environment)	AIR (Automated Image Registration), ANIMAL (Automated Non-Linear Image Matching and Anatomical Labeling), MatLab, NIH Image Software	SPM

Table H   Toolboxes identified for the year 1995

2 **semi-flexible** toolboxes
ClickFit, ProMatLab

In order to classify these computer applications as rigid, semi-flexible, or flexible, I checked their websites (when they were available), the websites of the universities or companies responsible for their design, or looked for information on various internet sources.

In addition to information pertaining to computer applications, I collected information on the authors' institutions of affiliation. When a certain author indicated two or more institutions, all of them were registered in my analysis. It was then possible to draw the relation between the institutions participating in each publication and the computer applications cited. Collecting data on the geographical location of each institution, I could prepare the maps presented in the book.

The limitations of this method are arguably obvious. Platforms such as Web of Science overestimate the weight of publications made in English, as most journals indexed by them are published in this language. Such platforms greatly reflect the dominance of so-called top journals, which publish, for the most part, papers by authors based in economically dynamic countries. This bias is surely present in the biomedical domain (Ho et al. 2013; Mendis et al. 2003), including brain studies (Patel and Sumathipala 2001; Saxena et al. 2003; Sartorius 1998). Therefore, my analysis may have inherited the geographical bias typical of Web of Science. This concern should not be very serious, though, because in biomedical research most authors, including those based in non-English-speaking countries, seek to publish their studies in English, following a long tradition of this research domain.

2. Geoprocessing Analysis

The readers may have noticed that in my maps, national borders are an important reference, but not the main reference. Instead, I considered geographical units that I named "hubs." The identification of these areas derived from a data processing method that I propose to call Bivariate Interrelationship for Cartographic Units Determination (BICUDO). This method is composed by the following five steps:

1. *Collecting and collating georeferenced quantitative data.* In my research project, each research institution figuring in each paper had a precise location (country and city). Thus it was possible to see how many times each city was involved in the use of certain software packages, by means of publications made by researchers based there.

To store those data, I used Windows Excel. For each software package, I prepared a different spreadsheet. For example, this is the whole spreadsheet pertaining to a neuroimaging software called MRIcro, for the year 2005 (Table I).

Table I Database for the MRIcro package

Country	Institution	Nature	State	City	Articles
France	Hospital Pitie-Salpetriere	HP	Ile-de-France	Paris	1
Germany	Hertie Institute for Clinical Brain Research	RH	Baden-Wurttemberg	Tubingen	1
Germany	University of Bonn	RH	North Rhine-Westphalia	Bonn	1
Germany	University of Tubingen	RH	Baden-Wurttemberg	Tubingen	1
Japan	National Food Research Institute	RH	Ibaraki	Tsukuba	2
United Kingdom	University of Nottingham	RH	East Midlands	Nottingham	1
United States	Boston University	RH	Massachusetts	Boston	1
United States	Harvard Medical School	RH	Massachusetts	Boston	1
United States	McLean Hospital	HP	Massachusetts	Belmont	1
United States	Stanford University	RH	California	Stanford	1
United States	University of California	RH	California	Sacramento	1
United States	University of Pittsburgh	RH	Pennsylvania	Pittsburgh	1
United States	Veterans Affairs Boston Healthcare System	HP	Massachusetts	Boston	1

By processing the data stored in each spreadsheet, I produced a final data frame organized according to cities and showing how many times software packages or toolboxes were associated with each city. These are only the first lines of the final data frame (Table J).

Table J First lines of the final database

City	State/region/ province	Country	Software	Toolboxes	Total
Aachen	North Rhine-Westphalia	Germany	4	2	6
Aarhus	Central Jutland	Denmark	1	0	1
Aglandjia	Nicosia	Cyprus	1	0	1
Akron	Ohio	United States	1	0	1
Al Ahsa	Eastern Province	Saudi Arabia	0	2	2
Albuquerque	New Mexico	United States	1	2	3
Alicante	Valencian Community	Spain	2	0	2
Almelo	Overijssel	Netherlands	1	2	3
Amsterdam	North Holland	Netherlands	16	23	39
Ann Arbor	Michigan	United States	8	5	13
Anyang	Gyeonggi	South Korea	2	0	2
Arhus	Aarhus	Denmark	1	0	1
Arnhen	Gelderland	Netherlands	1	4	5
Assen	Drenthe	Netherlands	1	0	1
Athens	Georgia	United States	1	1	2
Atlanta	Georgia	United States	6	11	17

With this type of data frame, one is ready to begin the actual geo-processing analysis. This I did by using TerraView, open source software developed at the National Institute of Spatial Research of Brazil.[2]

2. *Producing an "inverted concentration index."* The BICUDO approach consists in a bivariate interrelationship. The first variable that forms this relationship is a concentration index. However, the BICUDO map requires that this index be inverted. Thus the lowest the variable under consideration, the bigger the concentration index will be. In my analysis, the biggest concentration indexes were then attributed to the cities with the smallest numbers of computer applications used.

To realize such inversion, an inverted index is calculated by means of the following formula:

---

[2]http://www.dpi.inpe.br/terralib5/wiki/doku.php

$$ii = Mx - (x-1)$$

where:

ii: inverted index

x: the variable (in this case, the total number of software packages and toolboxes associated with the city)

Mx: maximum value reached by the variable x in the entire data frame (in this case, 56)

In this way, cities with only 1 computer application associated with them have a concentration index of 56 whereas cities with 56 computer applications have an index of 1. In my analysis, Montreal and New York were the cities with the biggest number of applications (56) and therefore the smallest inverted concentration index (1).

3. *Drawing an inverted concentration map.* There are different types of concentration maps, which show the occurrence of a certain variable in different places. For the production of a BICUDO map, one must use a map on which the concentration is represented by circles. The bigger the concentration of the variable being considered, the bigger the circle. Drawing on the inverted concentration index calculated in the previous step, a concentration map (or inverted concentration map) is produced.

The outcome is a group of circles with different sizes. When the size of any of these circles is changed, all other circles also change, so the size proportion is kept. In the BICUDO approach, it is the size of the biggest circles that need to be adjusted.

4. *Calculating the average distance between geographical areas.* The second variable in the bivariate relationship is the distance between locations. Each city present in the final data frame is represented by a dot on the world map. The TerraView software allows the user to calculate the distance between a certain dot and its closest dot. In this way, each dot has a number associated to it, showing the distance to the closest dot. With a simple operation, the software calculates the average minimal distance considering all the dots (or cities). For the year 2015, I identified 345 cities, the average distance between them being 133 km.

Thus the biggest circles shown in the concentration map will have a radius corresponding to 133 km (or 266 km of diameter). In my analysis, all the cities for which the inverted index is 1 were covered by a circle with this size. Once the concentration map is drawn by the software, the size of the biggest circle can be adjusted, and smaller circles are automatically corrected.

Taking the United States and part of Canada as an example, the following figure shows how the map looks like at the end of this step.

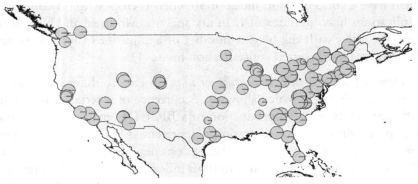

Concentration map in step 4

5. *Drawing the geographical hubs.* Based on an image like the previous one, the geographical hubs are drawn. In the TerraView software, this is possible due to the presence of a plug-in called TerraEdit with which georeferenced polygons can be drawn.[3] For this operation, two simple rules should be complied with: (1) Whenever two or more circles overlap, they form a single polygon (or hub); and (2) When a circle crosses over a country's frontier, the crossing should happen if and only if the circle overlaps with another circle at the other side of the frontier (including when the circle ends up covering some lake or channel).

---

[3]This feature is present in other key geoprocessing applications such as QGIS.

In order for the first rule to be accurately followed, it is frequently necessary to zoom in the image many times because two circles may seem to overlap but when the user zooms in, a tiny gap between them may be revealed. When the final hubs are drawn, it is sometimes necessary to make the hubs' frontiers recede so the separation between two hubs is made clear. The following example is illustrative.

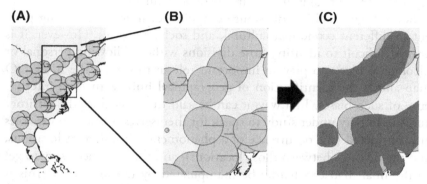

(A)   **(B)**   **(C)**

Drawing the geographical hubs

Figure A shows some of the hubs that cover the American territory and part of Canada. When one zooms in (Figure B), a small gap between two circles becomes clearer. When the hubs are finally drawn (Figure C), their frontiers are taken care of so the separation between the resulting hubs is evident. On Figure B, there is a tiny circle (corresponding to the city of Pittsburgh) that forms a single hub. On Figure C, this circle is enlarged so the hub can be seen more easily.

Note that the final hubs follow the borders defined by the circles but there is no need to be extremely accurate.

With the TerraView software, as well as most geoprocessing software packages currently available, steps 2–4 can be realized with no difficulties. Step 5 becomes lengthy when many hubs have to be drawn.

What is the rationale behind the BICUDO approach? It is aimed to show two phenomena. First, the geographical configuration of some activities (such as software use and design) have a dynamic that is not completely determined by national borders. Some areas within the same country may be excluded from the dynamic of other domestic areas. At the same time, the dynamic of certain areas may benefit from what happens in close cities which are nevertheless located in a different country. This is not to say, however, that the national dynamics and the nation state should be ignored, and this is why the BICUDO map should always include political frontiers, whether they are frontiers separating countries or political units within the same country.

Second, geographical divisions are typical of human life, being present in different economic, cultural, and social activities. However, it is always difficult to identify those divisions without allowing personal or ideological biases to play an important role in this task. The BICUDO map allows the identification of geographical hubs with little interference of such biases. In a way, it can be said that it enables the territory and the activity under study to speak for themselves. Only two variables are considered (the occurrence of a phenomenon in different locations, and the distance between those locations), and the steps necessary to get to the final map can hardly be manipulated by the researcher. This is not to say that we are dealing with a completely objective procedure. It is undeniable that certain choices have to be made, such as the variable whose concentration is assessed, and the kind of location that will be taken into account (cities, states, provinces, regions, and so on).

Finally, it is important to highlight two limitations of the proposed approach. First, the identification of the precise place where a certain phenomenon happens is not always clear-cut. In my analysis, for example, one cannot be completely sure of where software use actually happened. In a paper published by authors based in two different cities, it is plausible that the neuroimaging software was actually used in only one of those institutions. In this way, software use might be overestimated. Second, to a certain degree the BICUDO map downplays the weight of the national scale. On the one hand, it cuts the national territory into different hubs. On the other, it identifies hubs crossing over national borders. In this way, the approach may be considered as inappropriate for the study of activities with heavy national contents, such as oil exploration.

3. Qualitative interviews

The data collection described in this section and the two following sections, as well as the format of their publication, was reviewed and approved by the Research Ethics Committee of King's College London.

For this study, a total of 57 interviews were conducted, in four countries: Brazil, the UK, Netherlands, and Portugal. The choice of these countries was based in the analyses described in the two previous sections. In chapterFour (section15), Map 15.1, shows the six levels of hubs identified for the year 2015. These levels were defined in accordance with the number of neuroimaging software packages and toolboxes used locally. The idea was to select countries at different levels, so different realities, in terms of software development and use, could be accounted for. Brazil belongs to the 1st level (lowest use of software packages); Portugal is in the 2nd level; the UK, in the 5th level; and Netherlands, in the 6th level (with the biggest number of applications used). The following table presents a summary of the interviews (Table K).

Table K   Summary of interviews by location

Period	Country	City	Interviews	Total
February 2015–July 2017	Brazil	Belo Horizonte	1	25
		Ribeirao Preto	7	
		Rio de Janeiro	3	
		Sao Paulo	9[a]	
		Porto Alegre	5	
September 2017	United Kingdom	Birmingham	2[b]	13
		London	10	
		Oxford	1	
October 2017	Netherlands	Maastricht	5	11
		Nijmegen	6	
October 2017	Portugal	Braga	2	8
		Coimbra	3	
		Lisbon	3	
Total				
February 2015–October 2017	4 countries	13 cities	57	

[a]One of these interviewees is based in the city of Natal and was interviewed via Skype
[b]One of these interviewees is based in Edinburgh

For identifying potential interviewees, my main reference was the final data frame described in the item 2 of this Methodological appendix (Geoprocessing analysis). I gave priority to the institutions with high performance in terms of publication of papers. Checking the websites of these institutions, I looked for researchers devoted to neuroimaging. I contacted these researchers by email, inviting them to be interviewed. Whenever I received a positive answer from a Professor with outstanding reputation and large scientific production, I also proposed interviews with students and researchers in the team. This happened, for example, in the research group led by Professor Alexandre Franco (Laboratory for Images-Labima, Pontifical Catholic Univ of Rio Grande do Sul, Brazil), as well as that led by Professor Rainer Goebel (Brain Innovation, Netherlands).

The following table brings information on the career level of my interviewees (Table L).

Table L Summary of interviews by position

Country	Professors and lecturers	Post-doctoral researcher	PhD students	Master's Degree students	BA students	Total
Brazil	9	7	6	1	2	25
United Kingdom	6	6	1	0	0	13
Netherlands	4	2	5	0	0	11
Portugal	3	2	2	1	0	8
Total	**22**	**17**	**14**	**2**	**2**	**57**

These were semi-structured interviews. All of them were recorded with permission from interviewees. They signed an informed consent form, including the free choice of the way they would be identified in my publications: by revealing only the country where they work; by revealing only the name of their institutions; or by revealing both their names and the name of their institutions.

The topics explored in the interviews were the following ones:

1. Interviewee

    Academic career
    Research interests

2. Programming skills

    Programming languages mastered
    Ways in which those languages were learnt
    Preferences in terms of languages

3. Software development

    Existence of (personal or collective) software development projects
    History of the projects
    Developments approaches adopted
    Difficulties faced in software development
    Code sharing practices
    Use of code written by other developers
    Use of online development platforms

4. Neuroimaging software

    Neuroimaging software packages known and used
    Motivations for learning and using those packages
    Favourite packages

5. Team work

    Responsibilities in the software development team
    Relations with other developers

6. Access to magnetic resonance scanners and neuroimaging data

    Types of resonance scanners present in the research institution
    Possible design of new data collecting protocols
    Use of public databases

7. Future projects

Projects for the years to come

In Brazil and Portugal, the interviews were conducted in Portuguese. English was used in the interviews conducted in the UK and Netherlands. The only exception was the interview with Dr. Luís Lacerda (Institute of Child Health, Univ College London), interviewed in Portuguese in London.

According to the interviewee's expertise and time availability, the topics listed above were explored in more or less detail. The shortest interview recording (in Netherlands) lasted only 19 minutes and 28 seconds whereas the longest one (in the UK) lasted 1 hour, 11 minutes and 42 seconds. In average, the interviews lasted 52 minutes and 28 seconds.

I listened to the interview recordings and transcribed the most relevant parts, translating the interviews in which Portuguese was spoken. Each segment was classified according to different codes which were also used for the literature review. In this way, themes to be explored in the final analysis could be identified.

4. Survey 1: Software Use

In order to identify trends and patterns in neuroimaging software use, I carried out an internet survey. People invited to participate were all the authors of the neuroimaging papers included in my analysis (see item 1: Analysis of neuroimaging papers). While analyzing these papers, I collected the email addresses of the corresponding authors. In addition, I used the internet to search for the email addresses of the remaining authors. It was not possible to find such information for a minority of authors.

Based on the knowledge acquired through the literature review and the interviews with Brazilian software developers, I prepared an initial version of a questionnaire. In May 2016, I send the invitation to a small number of researchers, as a pilot study. At that time, my questionnaire was sent as an attachment to my invitation message. Respondents had to download the questionnaire, fill it in, and send it back to me.

As I received a very small number of answers, I decided to prepare an online questionnaire. After revising my questions, I used the JotForm[4] website to produce a questionnaire. With this new approach, my invitation emails contained a clickable link directing potential participants towards the questionnaire.

This is the final version to which participants had access:

By searching neuroimaging papers published in 2015, it was seen that you and colleagues published a paper in which the following software was used: SPM (Statistical Parametric Mapping software)

Please provide some information about your use of this software in this 2015 paper, by replying to the following 17 questions. It is important that you do NOT leave any questions unanswered. For each question, you can select only one option, unless otherwise stated.

I—What is your main background?
[1] Computer sciences
[2] Engineering
[3] Physics
[4] Medical sciences
[5] Biology
[6] Psychology or social sciences
[7] Other

II—When did you first get in touch with the software?
[1] In the 1990s or before
[2] In the beginning of the 2000s
[3] In the mid-2000s
[4] In the end of the 2000s
[5] In the 2010s

III—How did you or your colleagues come to know the software? (SELECT ONE OR MORE)
[1] Reading the scientific literature
[2] Seeing a presentation in a conference or meeting

---

[4]www.jotform.com

[3] Receiving an indication made by colleagues other than my co-authors in the paper
[4] I or my colleagues know the software developer personally
[5] The software is largely known and used at the institution I am based in
[6] Other ways

In case of 6, specify:

IV—How did you learn to use it? (SELECT ONE OR MORE)
[1] I did not actually use it. My colleagues did it for me
[2] I attended a course (in person)
[3] I read one or more books
[4] I was taught by a colleague or supervisor
[5] I used resources found on the internet (manuals, tutorials, videos, courses etc.)
[6] I explored the software on my own, with the knowledge gained by using similar software

V—Previous to the publication of your 2015 paper, how many publications had you made in which the same software was used?
[1] None
[2] From 1 to 5
[3] From 6 to 10
[4] 11 or more

VI—Considering all the software's features, what is your knowledge of it?
[1] Advanced knowledge: I can understand and use almost all the features available with no need of help
[2] Intermediate knowledge: I can use most features, but I do not understand everything and for some operations I need some help
[3] Basic knowledge: I can use and understand only a set of features available, otherwise I would need some help
[4] Very basic knowledge: I can only perform a tiny sequence of commands and analyses without understanding their premises

VII—Have you ever taught somebody to use the software?
[1] No
[2] Yes, just 1 person
[3] Yes, a few people (from 2 to 9)
[4] Yes, several people (10 or more)

VIII—In your opinion the software ...
[1] is the best option of its kind available
[2] is an average software package
[3] is a relatively limited package
[4] I do not have enough knowledge to compare it with other options

IX—Why did you and your colleagues decide to use it? (SELECT ONE OR MORE)
[1] It is free software
[2] It is largely recognized and used in the scientific literature
[3] It is largely recognized and used at my research institution
[4] The source code is available, so it is possible to make modifications and add plug-ins
[5] It possesses features that other software packages do not have
[6] It is user-friendly and its usage does not require advanced computational skills
[7] The resulting images are high-quality
[8] The algorithms, statistical and quantitative processing are high-quality
[9] Indication made by colleagues, supervisors or other peers
[10] It is the only software package of this kind that I know
[11] It is the only software package of this kind that I have access to
[12] I wanted to test its usefulness
[13] There is support via technical services or online fora
[14] Other

In case of 14, specify:

X—In your opinion, how important was the use of the software for the approval and publication of your 2015 paper?
[1] Very important: the reviewers considered its use as a decisive aspect
[2] Important: the reviewers considered its use as a good aspect
[3] Slightly important: the reviewers considered that its use does not compromise the paper
[4] Not important at all: its use was not taken into account by the reviewers

XI—How likely are you to continue to use the software in your future studies?
[1] Very likely: I will definitely use it again
[2] Likely: I will use it again but I will check alternatives

[3] Somewhat likely: I have looked for alternatives but, for the time being, I will use it

[4] Unlikely: I have come across a better alternative but I am either looking for access to it or learning to use it

[5] Very unlikely: I will stop using it because I have found a better alternative, which I can use and to which I have access

XII—Do you have access to the software's source code?
[1] Yes
[2] No
[3] I could have downloaded or required it but this was not necessary

XIII—For your 2015 paper, did you make modifications to the software's source code, install plug-ins or write additional scripts?
[1] Yes
[2] No. I am not able to do it
[3] No. I am able to do it but this was not necessary

XIV—At the time of that study, where was the software installed? (SELECT ONE OR MORE)
[1] In my personal computer
[2] In a colleague's or supervisor's personal computer
[3] At my research institution, in the laboratory or department where I work
[4] At my research institution, in a laboratory or department different from the one where I work
[5] At a research institution different from mine but in the same city
[6] At a research institution located in my country but in a different city or region
[7] In a foreign country

XV—Are you literate in a computer programming language?
[1] No
[2] I used to be but I am no longer literate
[3] Yes, I know 1 language
[4] Yes, I know 2 languages
[5] Yes, I know 3 languages or more

XVI—Pertaining to the issue of free software in neuroimaging studies, which of the following statements reflects your opinion the most closely? (SELECT ONE OR TWO)

[1] It is not an important issue for me, as in my institution I can access all the software packages I need, irrespective of whether they are commercial or free

[2] For scientific or social reasons, the development and usage of free neuroimaging software packages should be encouraged

[3] For technical, scientific or commercial reasons, it is (and will continue to be) difficult (if not impossible) for neuroimaging researchers to work exclusively with free software

4] For technical, scientific or commercial reasons, it is (and will continue to be) difficult (if not impossible) for neuroimaging researchers to work exclusively with commercial software

[5] Every researcher needs to adjust to the free-software era, as this is an unavoidable trend in the future of neuroimaging studies

[6] None of the previous statements makes justice to my view about the issue

Therefore, at the beginning of the questionnaire, I clearly indicated the software package focused on by the questions. The package had already been referred to in my invitation email. This was crucial because in most papers, authors used a set of different neuroimaging software packages. By selecting only one of those, I could collect information about different neuroimaging packages. This also allowed me to process the data by considering the software and its classification as rigid, semi-flexible or flexible.

Taking advantage of a feature offered by JotForm, I made all questions be mandatory, so participants could not submit the questionnaire without answering all the questions. In this way, there was no missing information in my data frame.

The process of sending invitations began in July 2016 and finished the same year, in October. Eventually, 119 questionnaires were received from 25 countries, as shown in the following table (Table M).

Table M   Questionnaires received in Survey 1

Country	Questionnaires
United States	34
Italy	15
Germany	12
Netherlands	9
China	6
United Kingdom	5
Canada	4
France	3
Switzerland	3
Austria	2
Belgium	2
Brazil	2
Denmark	2
Spain	2
India	2
Japan	2
Malasia	2
Mexico	2
Norway	2
Sweden	2
Taiwan	2
South Africa	1
Australia	1
Ireland	1
Israel	1

In order to calculate the response rate, I am not considering the number of researchers invited to participate. For it is important to remember that most papers had more than one author and, according to the instruction sent in my invitation email, it was sufficient that only one author participated in my survey. Therefore, considering the total number of 335 papers included in my study (for the year 2015), the response rate was 35.52%.

The JotForm website generates a PDF file for each completed questionnaire. For pre-processing the quantitative data, I wrote some computer code in the C++ language. In this way, a data frame was produced in the .txt format. Subsequently, this data frame was opened with the SPSS software, where the final data processing was performed.

5. Survey 2: Software Development

Whereas Survey 1 explored software use, Survey 2 focused on software development. I used the list of software packages and toolboxes obtained in the analysis of neuroimaging papers. I then searched for information on these applications on the internet, a task facilitated by the fact that most applications have an official website associated to it. I then collected the emails of the applications' developers. It was sometimes necessary to download the papers that released the application.

I used these email addresses to send invitations to software developers. As I did in Survey 1, the JotForm website was used to store the questionnaire and the answers submitted by developers. This is the questionnaire accessed by participants:

I—Your application (software or toolbox)

1. What is the name of your software or toolbox (as informed in the invitation email you received)?

2. When was the first version of your application released?
In 2016 or after
Between 2006 and 2015
Between 1996 and 2005
Between 1986 and 1995
In 1985 or before

3. When was the most recent version of your application released?
In 2016 or after
Between 2011 and 2015
Between 2001 and 2010
Between 1991 and 2000
Between 1981 and 1990
In 1980 or before

4. In the last five years, how many new versions were launched?
10 or more
Between 5 and 9
Between 2 and 4

1
None

5. Over the last five years, the number of downloads (or purchases) for your application has...
... increased substantially.
... increased.
... remained constant.
... decreased.
... decreased substantially.

II—Your home institution and partners institutions

6. Today, how many institutions participate in the development of your application?
5 or more
4 institutions
3
2
1
None (it is a personal project)

7. What types of institutions are involved in the application's development? (Select one or more)
University
Company
Government agency
Hospital or healthcare practice
Other
None (it is a personal project)

8. In how many countries are those institutions located?
5 or more
4 countries
3
2
1
There is no institution involved (it is a personal project)

III—Funding

9. Where does the funding of your project come from?
Public funding agency
University funds
Company
Personal investment
Other

IV—Data

10. What is (are) the main source(s) of the data used to validate your application? (Select one or more)
Closed database (it was necessary to pay for it)
Open databases
The data were collected at the institutions responsible for the project especially for this development project
The data came from old projects carried out at our institutions
The data were collected by collaborators not involved in the development of the application
Scientific literature
Other

V—Developers

11. How many software developers are there in your project today?
11 or more
6 to 10
3 to 5
2
1

12. In the last five years, how many software developers joined your project?
11 or more
6 to 10
3 to 5
2
1
None

13. In the last five years, how many software developers left your project?
11 or more
6 to 10
3 to 5
2
1
None

14. How many of the original developers are still in the development team?
11 or more
6 to 10
3 to 5
2
1
None

15. Considering the developer(s) who has(have) been in the project for the longest time, when did he (she/they) joined the project?
In 2016 or after
Between 2011 and 2015
Between 2001 and 2010
Between 1991 and 2000
Between 1981 and 1990
In 1980 or before

VI—COLLABORATIONS

16. How many people outside the main development team do give sporadic contributions (in terms of architecture, coding, graphical interface or other crucial things) to your project?
11 or more
6 to 10
3 to 5
2
1
None

VII—DEVELOPMENT

17. Which of the following expressions can be used to describe the ways in which your application is developed? (Select one or more)
Traditional development
Agile software development

Adaptive software development
Agile modelling
Agile unified process
Disciplined agile delivery
Dynamic systems development method
Extreme programming
Feature-driven development
Lean software development
Kanban
Rapid application development
Scrum
Scrumban
A method created by our own development team

18. In average, how many times per month do your developers formally meet to discuss the progress of the project and take joint decisions?
9 or more
5 to 8
3 to 4
2
1
They never meet
There is only one developer

19. How frequently do your developers write code together (sitting at the same computer at the same time)?
Very frequently
Frequently
From time to time
Rarely
They always work on their own
There is only one developer

20. When is the application's architecture decided upon?
Before writing code
During code writing
Before writing code but adjusted during code writing

21. How many developers do take decisions pertaining to the application's architecture?
All developers
Most developers
The minority of developers
Only one developer in the team
There is only one developer

22. Which of the following languages is used to write code? (Select one or more)
Ada
C
C++
C#
C Shell
Delphi
HTML
Java
JavaScript
MATLAB
Pascal
Perl
Python
Ruby
Simula
Unix Shell
Visual Basic
Other

If other, please specifiy:

23. What is(are) the main reason(s) why this(these) language(s) has(have) been chosen for your project? (Select one or two)
Because the developer(s) has(have) deep knowledge of this(these) language(s)
Because this(these) language(s) was(were) used in the original version of the application
Because it is easy to work with modules and plug-ins
Because there are libraries that are good for scientific analyses
Because this(these) language(s) result(s) in good software performance
Because of good graphical results
Because this(these) language(s) is(are) simple to write and read
Because this(these) language(s) force developers to write organized and clear code

Because the developer(s) had successfully used this(these) language(s) in a previous project
Because this(these) language(s) is(are) commonly used in neuroimaging applications
Because this(these) language(s) is(are) widely known and studied by software developers
Because it is relatively easy to detect and handle bugs
Because this(these) language(s) is flexible, dynamic, and cutting-edge
Other

If other, please specify:

24. What is(are) the main action(s) that you use in order to make your application known? (Select one or two)
Scientific publications
Talks at conferences, seminars, workshops, and other meetings
Courses through which people learn to use the application
Advertisements in scientific journals
Flyers or letters sent to potential users
Presential campaigns, visiting potential users
By having research stays at partner institutions
By receiving collaborators from partner institutions
By contacting potential users via telephone, email, or social networks
The application is diffused through word of mouth (spontaneous communication among users)

VIII—SOURCE CODE

25. Is your source code open?
Yes
No

26. In order to share the code, which of the following strategies do you use? (Select one or more)
GitHub
Bitbucket
SourceForge
Another platform like GitHub and Bitbucket
The code can be downloaded on our institution's website
The code is sent via email upon request
Other
The code is not open

27. Are you willing to incorporate modifications made by other people into your code?
Yes
No

28. In the last five years, how many times did people implement modifications to the code and send them back to you for verification and approval?
10 or more
5 to 9
1 to 4
None
I never incorporate modifications from other people

29. How often do your developers utilize, in your project, pieces of code taken from other projects of theirs?
Very frequently
Frequently
From time to time
Rarely
Never

30. How often do your developers utilize, in your project, pieces of code written by other people (whether these people are within or outside the main development team)?
Very frequently
Frequently
From time to time
Rarely
Never

31. How often do your developers utilize, in your project, pieces of code found on the internet?
Very frequently
Frequently
From time to time
Rarely
Never

32. Is your code organized and clear so other developers can understand it easily?
Yes, absolutely
Yes, sufficiently organized
Yes, but it could be better organized
No, it is not
No, absolutely not

33. How important is it for you to make your code organized and clear so other developers can understand it easily?
Absolutely important
Very important
Important
Unimportant
Not important at all

34. What is the amount of comments in your code?
There are comments for almost every block of code
There are comments for all of the most important blocks of code
There are comments for some of the most important blocks of code
There is a small number of comments
There is no comment at all

35. What are the main motivation(s) for you to use comments? (Select one or two)
In order for any developer to read and understand the code easily
In order to guide the work of other developers in our team
In order to guide the work of other developers who may join our team in the future
In order to make the code clear and understandable for ourselves (or myself) in the future
Other
Comments are not used

IX—END USERS

36. How important is for you to make your application be intuitive and user-friendly?
Absolutely important
Very important
Important
Unimportant
Not important at all

37. How frequently do end users provide you with feedback and suggestions?
Very frequently
Frequently
From time to time
Rarely
Never

38. Which of the following means of communication are open to users so they ask questions or make comments about your application? (Select one or more)
Mailing list
Email
Online forum
Social network
Direct communication (in conferences, courses, and so on)
Telephone
Traditional mail
Other
None

39. How frequently is your application modified as a consequence of requests or comments made by end users?
Very frequently
Frequently
From time to time
Rarely
Never

A total of 23 questionnaires were received. In terms of software flexibility, 3 of these questionnaires were related to rigid packages, 7 to semi-flexible packages, and 13 to flexible ones. Therefore, considering the 108 applications identified in the year 2015 (40 software packages and 68 toolboxes), the response rate was 21.29%, inferior to that obtained in Survey 1 (35.52%).

It is important to stress that I used data from Survey 1 and Survey 2 to perform some statistical tests. Considering the methods I used to identify neuroimaging papers, my sample cannot be considered as representative of neuroimaging researchers in general. In order to describe the representativeness of my sample accurately, one has to say: it represents the group of neuroimaging researchers who manage to publish their studies in highly regarded journals. This includes researchers based in scientific leading countries, such as the United States and the UK, as well as those based in emerging economies such as India, China, and Brazil. Therefore, this sample is not representative of neuroimaging researchers working in scientifically marginal countries.

References

Ho, Roger Chun-Man, Kwok-Kei Mak, Ren Tao, Yanxia Lu, Jeffrey R. Day, and Fang Pan. 2013. "Views on the peer review system of biomedical journals: An online survey of academics from high-ranking universities." *BMC Medical Research Methodology* 13 (74):1–15.

Mendis, S., D. Yach, R. Bengoa, D. Narvaez, and X. Zhang. 2003. "Research gap in cardiovascular disease in developing countries." *The Lancet* 361:2246–2247.

Patel, Vikram, and Athula Sumathipala. 2001. "International representation in psychiatric literature: Survey of six leading journals." *British Journal of Psychiatry* 178:406–409.

Sartorius, Norman. 1998. "Scientific work in Third World countries." *Acta Psychiatrica Scandinavica* 98:345–347.

Saxena, Shekhar, Itzhak Levav, Pallab Maulik, and Benedetto Saraceno. 2003. "How international are the editorial boards of leading psychiatry journals?" *The Lancet* 361:609.

References

Adams, Jonathan. 2012. "The rise of research networks." *Nature* 490:335–336.

Aguirre, Geofrey K. 2012. "FIASCO, VoxBo, and MEDx: Behind the code." *NeuroImage* 62:765–767.

Alshawi, Mustafa. 2007. *Rethinking IT in construction and engineering: Organisational readiness*. New York: Taylor & Francis.

Amabile, T. M. 1996. *Creativity in context*. Boulder: Westview.

Appadurai, Arjun. 2000. "Grassroots globalization and the research imagination." *Public Culture* 12 (1):1–19.

Arendt, Hannah. 1978. *The life of the mind: One—Thinking*. London: Secker & Warburg.

Arendt, Hannah. 1998. *The human condition*. Chicago: University of Chicago Press.

Arendt, Hannah. 1998. *The human condition*. 2nd ed. Chicago: University of Chicago Press.

Ashburner, John. 2012. "SPM: A history." *NeuroImage* 62:791–800.

Atal, Vidya, and Kameshwari Shankar. 2015. "Developers' incentives and open-source software licensing: GPL vs BSD." *B.E. Journal of Economic Analysis and Policy* 15 (3):1381–1416.

Aurich, Nathassia K., José O. Alves Filho, Ana M. Marques da Silva, and Alexandre Rosa Franco. 2015. "Evaluating the reliability of different

© The Editor(s) (if applicable) and The Author(s), under exclusive license to Springer Nature Singapore Pte Ltd. 2019
E. Bicudo, *Neuroimaging, Software, and Communication*, https://doi.org/10.1007/978-981-13-7060-1

preprocessing steps to estimate graph theoretical measures in resting state fMRI data." *Frontiers in Neuroscience* 9, article 48:1–10.

Bandettini, Peter. n.d. A short history of Statistical Parametric Mapping in functional neuroimaging.

Barbrook, Richard. 2003. "Giving is receiving." *Digital Creativity* 14 (2):91–94.

Bashford, Bruce. 1999. *Oscar Wilde: The critic as humanist*. London: Associated University Presses.

Bear, Mark F., Barry W. Connors, and Michael A. Paradiso. 2007. *Neuroscience: Exploring the brain*. 3rd ed. Philadelphia: Lippincott Williams & Wilkins.

Beaulieu, Anne. 2001. "Voxels in the brain: Neuroscience, informatics and changing notions of objectivity." *Social Studies of Science* 31 (5):635–680.

Beltrame, Francesco, and Stephen H. Koslow. 1999. "Neuroinformatics as a megascience issue." *IEEE Transactions on Information Technology in Biomedicine* 3 (3):239–240.

Beniger, James R. 1986. *The control revolution: Technological and economic origins of the information society*. Cambridge: Harvard University Press.

Berezsky, Oleh, Grigoriy Melnyk, and Yuriy Batko. 2008. "Modern trends in biomedical image analysis system design." In *Biomedical engineering: Trends in electronics, communications and software*, edited by Anthony N. Laskovski, 461–480. Rijeka: InTech.

Berry, David M. 2011. *The philosophy of software: Code and mediation in the digital age*. New York: Palgrave Macmilllan.

Bicudo, Edison. 2006. "Normes, térritories et aménagement: les recherches biotechnologies dans l'Union Européenne." Master, Institut de Gégraphie, University Paris 1 (Panthéon Sorbonne).

Bicudo, Edison. 2012. "Globalization and ideology: Ethics committees and global clinical trials in South Africa and Brazil." PhD thesis, Department of Political Economy, King's College London. Available at https://kclpure. kcl.ac.uk/portal/en/theses/globalization-and-ideology-ethics-commit-tees-and-global-clinical-trials-in-south-africa-and-brazil(ab94b2ce-b023-4ab 3-9dd4-93616db5c455).html.

Bicudo, Edison. 2014. *Pharmaceutical research, democracy and conspiracy: International clinical trials in local medical institutions*. London: Gower and Routledge.

Bicudo, Edison. 2018. "'Big data' or 'big knowledge'? Brazilian genomics and the process of academic marketization." *BioSocieties* 13 (1):1–20.

Blank, Robert H. 2007. "Policy implications of the new neuroscience." *Cambridge Quarterly of Healthcare Ethics* 16:169–180.

Blau, Peter Michael. 2006. *Exchange and power in social life*. New Brunswick: Transaction.

Blume, Stuart S. 1992. *Insight and industry: On the dynamics of technological change in medicine*. Cambridge: MIT Press.

Bozdag, Engin, and Jeroen van den Hoven. 2015. "Breaking the filter bubble: Democracy and design." *Ethics and Information Tehcnology* 17:249–265. https://doi.org/10.1007/s10676-015-9380-y.

Bradley, William G. 2008. "History of medical imaging." *Proceedings of the American Philosophical Society* 152 (3):349–361.

Braudel, Fernand. 1979. *Le temps du monde. Vol. III, Civilisation matérielle. Economie et capitalisme*. Paris: Armand Collin.

Brett, Matthew. 2010. Research methods in the twenty-first century. Available at http://matthew.dynevor.org/manifesto.html.

Briggs, J. C. 2002. "Virtual reality is getting real: Prepare to meet your clone." *Futurist* 36 (3):34–41.

Brooks, Frederick P. 1995. *The mythical man-month*. Reading: Addison-Wesley.

Bruehl, Annette B. 2015. "Making sense of real-time functional magnetic resonance imaging (rtfMRI) and rtfMRI neurofeedback." *International Journal of Neuropsychopharmacology* 18 (6):1–7.

Burri, Regula Valérie. 2013. "Visual power in action: Digital images and the shaping of medical practices." *Science as Culture* 22 (3):367–387.

Castells, Manuel. 2004. *The network society: A cross-cultural perspective*. Cheltenham: Edward Elgar.

Clark, Margaret, and Judson Mills. 1979. "Interpersonal attraction in exchange and communal relationships." *Journal of Personality and Social Psychology* 37 (1):12–24.

Cowley, L. Tad, Hope L. Isaac, Stuart W. Young, and Thomas A. Raffin. 1994. "Magnetic resonance imaging marketing and investment: Tensions between the forces and the practice of medicine." *Chest* 105 (3):920–928.

Cox, Robert W. 2012. "AFNI: What a long strange trip it's been." *NeuroImage* 62:743–747.

Cusumano, Michael A. 1992. "Shifting economies: From craft production to flexible systems and software factories." *Research Policy* 21 (5):453–480.

Cutanda, Vicente, David Moratal, and Estanislao Arana. 2015. "Automatic brain morphometry and volumetry using SPM on cognitively impaired patients." *IEEE Latin America Transactions* 13 (4):1077–1082.

Das, Samir, Tristan Glatard, Leigh C. MacIntyre, Cecile Madjar, Christine Rogers, Marc-Etienne Rousseau, Pierre Rioux, Dave MacFarlane, Zia Mohades, Rathi Gnanasekaran, Carolina Makowski, Penelope Kostopoulos, Reza Adalat, Najmeh Khalili-Mahani, Guiomar Niso, Jeremy T. Moreau, and Alan C. Evans. 2016. "The MNI data-sharing and processing ecosystem." *NeuroImage* 124:1188–1195.

Deci, E. L., and R. M. Ryan. 1985. *Intrinsic motivation and self-determination in human behavior*. New York: Plenum.

Devanbu, Prem. 2009. "Study the social side of software engineering." *IEEE Software* 26 (1):69.

Dinov, Ivo D., Petros Petrosyan, Zhizhong Liu, Paul Eggert, Sam Hobel, Paul Vespa, Seok Woo Moon, John D. va Horn, Joseph Franco, and Arthur W. Toga. 2014. "High-throughput neuroimaging-genetics computational infrastructure." *Frontiers in Neuroinformatics* 8:1–11.

Dumit, Joseph. 2004. *Picturing personhood: Brain scans and biomedical identity*. Princeton: Princeton University Press.

Dummet, Michael. 1993. *The seas of language*. Oxford: Oxford University Press.

Durkheim, Émile. 1925. *Les formes élémentaires de la vie religieuse*. Paris: Félix Alcan.

Durkheim, Émile. 1932. *De la division du travail social*. Paris: Félix Alcan.

Eklund, Anders, Thomas E. Nichols, and Hans Knutsson. 2016. "Cluster failure: Why fMRI inferences for spatial extent have inflated false-positive rates." *Proceedings of the National Academy of Sciences* 113 (28):7899–7905.

Ensmenger, Nathan. 2010. *The computer boys take over: Computers, programmers, and the politics of technical expertise*. Cambridge and London: MIT Press.

Fazi, Beatrice. 2018. *Contingent computation: abstraction, experience, and indeterminacy in computational aesthetics*. Lanham: Rowman & Littlefield.

Ferreira, Mariana Toledo. 2018. *Centro(s) e periferia(s) na produção do conhecimento em genética humana e médica: um olhar a partir do Brasil*. PhD thesis, Department of Sociology, University of Sao Paulo, Brazil.

Fielding, Roy T. 1999. "Shared leadership in the Apache project." *Communications of the ACM* 42 (4):42–43.

Filler, Aaron G. 2009. "The history, development and impact of computed imaging in neurological diagnosis and neurosurgery: CT, MRI, and DTI." *Nature Precedings*:1–76. Available at http://dx.doi.org/10.1038/npre.2009.3267.5.

Finley, Klint. 2015. The problem with putting all the world's code in GitHub. Wired. Available at https://www.wired.com/2015/06/problem-putting-worlds-code-github/.

Fischl, Bruce. 2012. "FreeSurfer." *NeuroImage* 62:774–781.

Frey, B. 1997. *Not just for the money: An economic theory of personal motivation.* Brookfield: Edward Elgar.

Friedman, Linda Weiser. 1992. "From Babbage to Babel and beyond: A brief history of programming languages." *Computer Languages* 17 (1):1–17.

Fuchs, Rainer. 2000. "Analyse this… or: Intelligent help for the rest of us." *Bioinformatics* 16 (6):491–493.

Galloway, Patricia. 2012. "Playpens for mind children: Continuities in the practice of programming." *Information & Culture* 47 (1):38–78.

Galloway, Alexander R., and Eugene Thacker. 2006. "Language, life, code." *Architectural Design* 76 (5):26–29.

Geertz, Clifford. 1973. *The interpretation of cultures: Selected essays.* New York: Basic Books.

Ghosh, Rishab Aiyer. 2005. "Understanding free software developers: Findings from the FLOSS study." In *Perspectives on free and open source software,* edited by Joseph Feller, Brian Fitzgerald, Scott A. Hissam, and Karim R. Lakhani, 23–46. Cambridge: MIT Press.

Goebel, Rainer. 2012. "BrainVoyager: Past, present, future." *NeuroImage* 62:748–756.

Goering, Richard. 2004. Matlab edges closer to electronic design automation world. *EETimes.* Available at https://www.eetimes.com/document.asp?doc_id=1151422.

Gold, Sherri, Brad Christian, Stephan Arndt, Gene Zeien, Ted Cizadlo, Debra L. Johnson, Michael Flaum, and Nancy C. Andreasen. 1998. "Functional MRI statistical software packages: A Comparative analysis." *Human Brain Mapping* 6:73–84.

Good, Byron. 1994. *Medicine, rationality, and experience.* Cambridge: Cambridge University Press.

Grcar, Joseph R. 2011. "John von Neumann's analysis of Gaussian elimination and the origins of modern numerical analysis." *SIAM Review* 53 (4):607–682.

Habermas, Jürgen. 1984. *The theory of communicative action, vol. 1: Reason and the rationalization of society.* Boston: Beacon Press.

Habermas, Jürgen. 1987. *The theory of communicative action, vol. 2: Lifeworld and system.* Cambridge: Polity Press.

Habermas, Jürgen. 1993. *Justification and application: Remarks on discourse ethics.* Cambridge: Polity Press.

Habermas, Jürgen. 1996. *Between facts and norms: Contributions to a discourse theory of law and democracy.* Studies in contemporary German social thought. Cambridge, MA: MIT Press.

Habermas, Jürgen. 2008. *Between naturalism and religion.* Cambridge: Polity Press.

Harvey, Mark, and Andrew McMeekin. 2010. "Public or private economies of knowledge: The economics of diffusion and appropriation of bioinformatics tools." *International Journal of the Commons* 4 (1):481–506.

Haznedar, M. Mehmet, Francesca Roversi, Stefano Pallanti, Nicolo Baldini-Rossi, David B. Schnur, Elizabeth M. LiCalzi, Cheuk Tang, Patrick R. Hof, Eric Hollander, and Monte S. Buchsbaum. 2005. "Fronto-thalamo-striatal gray and white matter volumes and anisotropy of their connections in bipolar spectrum illnesses." *Biological Psychiatry* 57 (7):733–742.

Heliades, G. P., and E. A. Edmonds. 1999. "On facilitating knowledge transfer in software design." *Knowledge-Based Systems* 12:391–395.

Henderson, Kathryn. 1998. *On line and on paper: Visual representations, visual culture, and computer graphics in design engineering.* Cambridge: MIT Press.

Henson, Richard. 2005. "What can functional neuroimaging tell the experimental psychologist?" *The Quarterly Journal of Experimental Psychology* 58A (2):193–233.

Herbsleb, James D., and Rebecca E. Grinter. 1999. "Archictectures, coordination, and distance: Conway's Law and beyond." *IEEE Software* 16 (5):63–70.

Hilbert, M., and P. López. 2011. "The world's technological capacity to store, communicate, and compute information." *Science* 332 (6025):60–65.

Himanen, Pekka. 2001. *The hacker ethic: And the spirit of the information age.* New York: Randon House.

Ho, Roger Chun-Man, Kwok-Kei Mak, Ren Tao, Yanxia Lu, Jeffrey R. Day, and Fang Pan. 2013. "Views on the peer review system of biomedical journals: An online survey of academics from high-ranking universities." *BMC Medical Research Methodology* 13 (74):1–15.

Hood, Leroy. 1990. No: And anyway, the HGP isn't 'big science'. *The Scientist.* Available at http://www.the-scientist.com/?articles.view/articleNo/11452/title/No--And-Anyway--The-HGP-Isn-t--Big-Science-/. Accessed in February 2015.

Jenkinson, Mark, Christian F. Beckmann, Timothy E. J. Berens, Mark W. Woolrich, and Stephen M. Smith. 2012. "FSL." *NeuroImage* 62:782–790.

Jones, S. 2006. "Reality© and virtual reality©—When virtual and real worlds collide." *Cultural Studies* 20 (2–3):211–226. https://doi.org/10.1080/09502380500495692.

Jones, Matthew B., Mark P. Schildhauer, O. J. Reichman, and Shawn Bowers. 2006. "The new bioinformatics: Integrating ecological data from the gene to the biosphere." *Annual Review of Ecology, Evolution, and Systematics* 37:519–544.

Joyce, Kelly. 2006. "From numbers to pictures: The development of magnetic resonance imaging and the visual turn in medicine." *Science as Culture* 15 (1):1–22.

Joyce, Kelly. 2011. "On the assembly line: Neuroimaging production in clinical practice." In *Sociological reflections on the neurosciences*, edited by Martyn Pickersgill and Ira van Keulen, 75–98. Bingley: Emerald Group.

Kennedy, Donald. 2005. "Neuroimaging: Revolutionary research tool or a post-modem phrenology?" *American Journal of Bioethics* 5 (2):19.

Kevles, Daniel. 1997. *The physicists: The history of a scientific community in modern America*. Cambridge: Harvard University Press.

Kevles, Bettyann Holtzmann. 1998. *Naked to the bone: Medical imaging in the twentieth century*. New York: Basic Books.

Khadilkar, S. V., and S. Wagh. 2007. "Practice patterns of neurology in India: Fewer hands, more work." *Neurology India* 55:27–30.

Kitchin, Rob. 2017. "Thinking critically about and researching algorithms." *Information, Communication & Society* 20 (1):14–29.

Kitchin, Rob, and Martin Dodge. 2011. *Code/space: Software and everyday life*. Cambridge: MIT Press.

Kittler, Friedrich A. 1997. "There is no software." In *Literature, media, information systems: Essays*, edited by Friedrich A. Kittler and John Johnston, 147–155. Amsterdam: OPA.

Kohanski, Daniel. 1998. *Moth in the machine: The power and perils of programming*. New York: St. Martin's Griffin.

Kolk, Anja G. van der, Jeroen Hendrikse, Jaco J. M. Zwanenburg, Fredy Visser, and Peter R. Luitjen. 2013. "Clinical applications of 7T MRI in the brain." *European Journal of Radiology* 82:708–748.

Laal, Marjan. 2013. "Innovation process in medical imaging." *Procedia—Social and Behavioral Sciences* 81:60–64.

Lakhani, Karim R., and Robert G. Wolf. 2005. "Why hackers do what they do: Understanding motivation and effort in free/open source software." In *Perspectives on free and open source software*, edited by Joseph Feller, Brian Fitzgerald, Scott A. Hissam, and Karim R. Lakhani, 3–22. Cambridge: MIT Press.

Lancaster, Jack L., Thomas G. Glass, Bhujanga R. Lankipalli, Hunter Downs, Helen Mayberg, and Peter T. Fox. 1995. "A modality-independent approach to spatial normalization of tomographic images of the human brain." *Human Brain Mapping* 3 (3):209–223.

Lash, Scott. 2007. "Power after hegemony: Cultural studies in mutation?" *Theory, Culture & Society* 24 (3):55–78.

Latour, Bruno. 1987. *Science in action: How to follow scientists and engineers through society*. Cambridge: Harvard University Press.

Latour, Bruno. 1990. "Drawing things together." In *Representation in scientific practice*, edited by Michael Lynch and Steve Woolgar. Cambridge: MIT Press.

Latour, Bruno. 1996. *Aramis, or the love of technology*. Cambridge: Harvard University Press.

Latour, Bruno. 2005. *Reassembling the social: An introduction to Actor-Network Theory*. Oxford: Oxford University Press.

LaToza, Thomas D., Gina Venolia, and Robert DeLine. 2006. "Maintaining mental models: A study of developer work habits." Proceedings of the 28th International Conference on Software Engineering, New York, USA.

Lewis, Jamie, and Andrew Bartlett. 2013. "Inscribing a discipline: Tensions in the field of bioinformatics." *New Genetics and Society* 32 (3):243–263.

Lindenberg, S. 2001. "Intrinsic motivation in a new light." *Kyklos* 54 (2/3):317–342.

Loginova, O. 2009. "Real and virtual competition." *Journal of Industrial Economics* 57 (2):319–342. https://doi.org/10.1111/j.1467-6451.2009.00380.x.

Lojkine, Jean. 1992. *La révolution informationnelle*. Paris: Presses Universitaires de France.

Luhmann, Niklas. 1983. *Sociologia do direito I, Biblioteca Tempo Universitário 75*. Rio de Janeiro: Tempo Brasileiro.

Lynch, Michael, and Samuel Y. Edgerton. 1988. "Aesthetics and digital image processing: Representational craft in contemporary astronomy." In *Picturing power: Visual depiction and social relations*, edited by G. Fyfe and J. Law, 184–220. London: Routledge.

Mackenzie, Adrian. 2003. "The problem of computer code: Leviathan or common power?" Available at http://www.lancaster.ac.uk/staff/mackenza/papers/code-leviathan.pdf.

Mackenzie, Adrian. 2006. *Cutting code: Software and sociality*. New York: Peter Lang.

Malecki, E. J. 2017. "Real people, virtual places, and the spaces in between." *Socio-Economic Planning Sciences* 58:3–12. https://doi.org/10.1016/j.seps.2016.10.008.

Mallard, John R. 2003. "The evolution of medical imaging: From Geiger counters to MRI—A personal saga." *Perspectives in Biology and Medicine* 46 (3):349–370.

Marres, Noortje. 2017. *Digital sociology: The reinvention of social research*. Cambridge: Polity Press.

Maubon, Antoine J., Jean-Michel Ferru, Vincent Berger, Marie Colette Soulage, Marc DeGraef, Pierre Aubas, Patrice Coupeau, Erik Dumont, and Jean-Pierre Rouanet. 1999. "Effect of field strength on MR images: Comparison of the same subject at 0.5, 1.0, and 1.5T." *RadioGraphics* 19:1057–1067.

Mendis, S., D. Yach, R. Bengoa, D. Narvaez, and X. Zhang. 2003. "Research gap in cardiovascular disease in developing countries." *The Lancet* 361:2246–2247.

Merton, Robert K. 1968. "The Matthew effect in science." *Science* 159 (3810):56–63.

Miller, Gregory A. 1996. "How we think about cognition, emotion, and biology in psychopathology." *Psychophysiology* 33 (6):615–628.

Miller, Gregory A., and Jennifer Keller. 2000. "Psychology and neuroscience: Making peace." *Current Directions in Pshychological Science* 9 (6):212–215.

Mockus, Audris. 2009. "Succession: Measuring transfer of code and developer productivity." International Conference on Software Engineering (ICSE 2009), Vancouver, Canada.

Mohamed, Armin, Stefan Eberl, Michael J. Fulham, Michael Kassiou, Aysha Zaman, David Henderson, Scott Beveridge, Chris Constable, and Sing Kai Lo. 2005. "Sequential I-123-iododexetimide scans in temporal lobe epilepsy: Comparison with neuroimaging scans (MR imaging and F-18-FDG PET imaging)." *European Journal of Nuclear Medicine and Molecular Imaging* 32 (2):180–185.

Moler, Cleve. 2004. The origins of MATLAB. MathWorks. Available at https://www.mathworks.com/company/newsletters/articles/the-origins-of-matlab.html.

Moran, T. P., and J. M. Carroll. 1997. "Overview of design rationale." In *Design rationale: Concepts, techniques and use*, edited by T. P. Moran and J. M. Carroll. Hillsdale: Lawrence Erlbaum Associates.

Moser, Ewald. 2010. "Ultra-high-field-magnetic resonance: Why and when?" *World Journal of Radiology* 2 (1):37–40.

Mueller, Susanne G., Michael W. Weiner, Leon J. Thal, Ronald C. Petersen, Clifford Jack, William Jagust, John Q. Trojanowski, Arthur W. Toga, and Laurel Becket. 2005. "The Alzheimer's disease neuroimaging initiative." *Neuroimaging Clinics of North America* 15 (4):869–877.

Naur, Peter. 2001. "Programming as theory building." In *Agile software development*, edited by Alistair Cockburn, 227–239. Boston: Addison-Wesley.

Nieuwenhuys, Rudolf, Cees A. J Broere, and Leonardo Cerliani. 2015. "A new myeloarchitectonic map of the human neocortex based on data from the Vogt-Vogt school." *Brain Structure & Function* 220 (5):2551–2573.

Nofre, David, Mark Priestley, and Gerard Alberts. 2014. "When technology became language: The origins of the linguistic conception of computer programming, 1950–1960." *Technology and Culture* 55 (1):40–75.

O'Reilly, Tim. 2013. "Open data and algorithmic regulation." In *Beyond transparency: Open data and the future of civic innovation*, edited by Brett Goldstein and Lauren Dyson, 289–300. San Francisco: Code for America.

Padma, T. V. 2008. "India plans for interdisciplinary neuroscience research centre." *Nature Medicine* 14 (11):1133.

Parnas, David Lorge. 1971. *Information distribution aspects of design methodology*. Pittsburgh: Computer Science Department, Carnegie-Mellon University. Available at http://repository.cmu.edu/cgi/viewcontent.cgi?article=2828&context=compsci.

Parnas, David Lorge. 1972. "On the criteria to be used in decomposing systems into modules." *Communications of the ACM* 15 (12):1053–1058.

Parnas, David Lorge, and Paul C. Clements. 1986. "A rational design process: How and why to fake it." *IEEE Transactions on Software Engineering* SE-12 (2):251–257.

Pasveer, Bernike. 1989. "Knowledge of shadows: The introduction of X-ray images in medicine." *Sociology of Health and Illness* 11 (4):360–381.

Patel, Vikram, and Athula Sumathipala. 2001. "International representation in psychiatric literature: Survey of six leading journals." *British Journal of Psychiatry* 178:406–409.

Pavlicek, Russell C. 2000. *Embracing insanity: Open source software development*. Indianapolis, IN: Sams.

Plaja, Carme Junqué i, Vera Vendrell, and Jesús Pujol. 1995. "La resonancia magnética funcional: una nueva técnica para el estudio de las bases cerebrales de los procesos cognitivos." *Psicothema* 7 (1):51–60.

Plotinus. 3rd century. *The six Enneads*. Available at http://pinkmonkey.com/dl/library1/six.pdf.

Power, Michael. 1999. *The audit society: Rituals of verification*. New York: Oxford University Press.

Prasad, Amit. 2005. "Making images/making bodies: Visibilizing and disciplining through magnetic resonance imaging." *Science, Technology & Human Values* 30 (2):291–316.

Prasad, Amit. 2014. *Imperial technoscience: Transnational histories of MRI in the United States, Britain, and India*. Cambridge: MIT Press.

Priestley, Mark. 2011. *A science of operations: Machines, logic and the invention of programming*. London: Springer.

Puce, A., R. T. Constable, M. L. Luby, G. McCarthy, A. C. Nobre, D. D. Spencer, J. C. Gore, and T. Allison. 1995. "Functional magnetic resonance imaging of sensory and motor cortex: Comparison with electrophysiological localization." *Journal of Neurosurgery* 83 (2):262–270.

Raymond, Eric S. 2001. *The cathedral & the bazaar: Musings on Linux and open source by an accidental revolutionary*. Sebastopol: O'Reilly.

Ribeiro, Andre Santos, Luis Miguel Lacerda, and Hugo Alexandre Ferreira. 2015. "Multimodal Imaging Brain Connectivity Analysis (MIBCA) toolbox." *PEERJ* 3 (e1078):1–28.

Rifkin, Jeremy. 2014. *The zero marginal cost society: The internet of things, the collaborative, and the eclipse of capitalism*. New York: Palgrave Macmillan.

Robertson, Roland. 1992. *Globalization: Social theory and global culture*. London: Sage.

Rose, Nikolas. 2016. "Reading the human brain: How the mind became legible." *Body & Society* 22 (2):140–177.

Rutt, Brian K., and Donald H. Lee. 1996. "The impact of field strength on image quality in MRI." *Journal of Magnetic Resonance Imaging* 1:57–62.

Ryan, R. M., and E. L. Deci. 2000. "Instrinsic and extrinsic motivations: Classic definitions and new directions." *Contemporary Educational Psychology* 25:54–67.

Saad, Ziad S., and Richard C. Reynolds. 2012. "SUMA." *NeuroImage* 62:768–773.

Santos, Milton. 2000. *La nature de l'espace: technique et temps, raison et émotion*. Paris: L'Harmattan.

Santos, Milton. 2002. *A natureza do espaço: técnica e tempo, razão e emoção, Milton Santos collection 1*. Sao Paulo: Edusp.

Sartorius, Norman. 1998. "Scientific work in Third World countries." *Acta Psychiatrica Scandinavica* 98:345–347.

Savoy, Robert L. 2001. "History and future directions of human brain mapping and functional neuroimaging." *Acta Psychologica* 107:9–42.

Sawle, Guy V. 1995. "Imaging the head: Functional imaging." *Journal of Neurology, Neurosurgery and Psychiatry* 58 (2):132–144.

Sawyer, S., and P. J. Guinan. 1998. "Software development: Processes and performance." *IBM Systems Journal* 37 (4):552–569.

Saxena, Shekhar, Itzhak Levav, Pallab Maulik, and Benedetto Saraceno. 2003. "How international are the editorial boards of leading psychiatry journals?" *The Lancet* 361:609.

Schumpeter, Joseph. 1954. *Capitalisme, socialisme et démocratie*. Paris: Payot.

Schutz, Alfred. 1974. *The structures of the life-world*. London: Heinemann.

Schwarz, Michael, and Yuri Takhteyev. 2010. "Half a century of public software institutions: Open source as a solution to hold-up problem." *Journal of Public Economic Theory* 12 (4):609–639.

Seaver, Nick. 2017. "Algorithms as culture: Some tactics for the ethnography of algorithmic systems." *Big Data & Society* 4 (2):1–12.

Sennett, Richard. 2008. *The craftsman*. New Haven: Yale University Press.

Shapin, Steven, and Simon Schaffer. 1985. *Leviatahn and the air-pump: Hobbes, Boyle, and the experimental life*. Princeton, NJ: Princeton University Press.

Simmel, Georg. 1950 [1903]. "The metropolis and mental life." In *The sociology of Georg Simmel*, edited by Kurt A. Wolff, 409–426. Glencoe: Free Press.

Simmel, Georg. 1997 [1900]. *The philosophy of money*. 2nd ed. London: Routledge.

Simondon, Gilbert. 1969. *Du mode d'existence des objets techniques, Analyses et Raisons 1*. Paries: Aubier.

Skog, Knut. 2003. "From binary strings to visual programming." In *History of Nordic computing*, edited by Janis Bubenko Jr., John Impagliazzo, and Arne Solvberg, 297–310. Boston: Springer.

Stallman, Richard M. 2002a. "Copyright and globalization in the age of computer networks." In *Free software, free society: Selected essays of Richard M. Stallman*, edited by Joshua Gay, 133–154. Boston: GNU Press.

Stallman, Richard M. 2002b. "The danger of software patents." In *Free software, free society: Selected essays of Richard M. Stallman*, edited by Joshua Gay, 95–112. Boston: GNU Press.

Stallman, Richard M. 2002c. "Free software definition." In *Free software, free society: Selected essays of Richard M. Stallman*, edited by Joshua Gay, 41–44. Boston: GNU Press.

Stallman, Richard M. 2002d. "Free software: Freedom and cooperation." In *Free software, free society: Selected essays of Richard M. Stallman*, edited by Joshua Gay, 155–186. Boston: GNU Press.

Stallman, Richard M. 2002e. "Releasing free software if you work at a university." In *Free software, free society: Selected essays of Richard M. Stallman*, edited by Joshua Gay, 61–62. Boston: GNU Press.

Stallman, Richard M. 2002f. "Why software should be free." In *Free software, free society: Selected essays of Richard M. Stallman*, edited by Joshua Gay, 119–132. Boston: GNU Press.

Stallman, Richard M. 2002g. "Why software should not have owners." In *Free software, free society: Selected essays of Richard M. Stallman*, edited by Joshua Gay, 45–50. Boston: GNU Press.

Stein, Lincoln D. 2010. "The case for cloud computing in genome informatics." *Genome Biology* 11 (207):1–7.

Suárez-Díaz, Edna. 2010. "Making room for new faces: Evolution, genomics and the growth of bioinformatics." *History and Philosophy of the Life Sciences* 32 (1):65–89.

Swade, D. 2003. "Virtual objects: The end of the real?" *Interdisciplinary Science Reviews* 28 (4):273–279. https://doi.org/10.1179/030801803225008686.

Tönnies, Ferdinand. 1955. *Community and association*. London: Routledge & Kegan Paul.

Torvalds, Linus, and David Diamond. 2001. *Just for fun: The story of an accidental revolutionary*. New York: HarperCollins.

Tung, Liam. 2018. GitHub rivals gain from Microsoft acquisition but it's no mass exodus, yet. Availalbe at: https://www.zdnet.com/article/github-rivals-gain-from-microsoft-acquisition-but-its-no-mass-exodus-yet/.

van Baren-Nawrocka, Jan. 2013. "The bioinformatics of genetic origins: How identities become embedded in the tools and practices of bioinformatics." *Life Sciences, Society and Policy* 9 (7):1–18.

van Essen, David C. 2012. "Cortical cartography and Caret software." *NeuroImage* 62:757–764.

van Horn, John Darrel, John Wolfe, Autumn Agnoli, Jeffrey Woodward, Michael Schmitt, James Dobson, Sarene Schumacher, and Bennet Vance. 2005. "Neuroimaging databases as a resource for scientific discovery." *International Review of Neurobiology* 66:55–87.

von Hippel, Eric, and Georg von Krogh. 2003. "Open source software and the 'private-collective' innovation model: Issues for organization science." *Organization Science* 14 (2):209–223.

Wagstrom, Patrick Adam. 2009. "Vertical interaction in open software engineering communities." PhD, Carnegie Insitute of Technology/School of Computer Science, Carnegie Mellon University.

Waldby, Catherine. 2000. "The Visible Human Project: Data into flesh, flesh into data." In *Wild science: Reading feminism, medicine and the media*, edited by J. Marchessault and K. Sawchuk, 24–38. New York: Routledge.

Warach, Steven, Jochen Gaa, Bettina Siewert, Piotr Wielopolski, and Robert R. Edelman. 1995. "Acute human stroke studied by whole brain echo planar diffusion-weighted magnetic resonance imaging." *Annals of Neurology* 37 (2):231–241.

Weber, Marx. 1958. "Class, status, party." In *From Marx Weber: Essays in socilogy*, edited by H. H. Gerth and C. Wright Mills, 180–195. New York: Oxford University Press.

Weber, Steven. 2004. *The success of open source.* Cambridge: Harvard University Press.

Weinberg, Gerald M. 1998. *The psychology of computer programming.* New York: Dorset House.

Wittgenstein, Ludwig. 1922. *Tractatus logico-philosophicus.* International Library of Psychology, Philosophy and Scientific Method. London: Kegan Paul.

Wittgenstein, Ludwig. 1963. *Philosophische Untersuchungen / Philosophical Investigations.* Oxford: Basil Blackwell.

Yoxen, Edward. 1987. "Seeing with sound: A study of the development of medical images." In *The social construction of technological systems: New directions in the sociology and history of technology*, edited by Wiebe E. Bijker, Thomas P. Hughes, and Trevor J. Pinch, 281–303. Cambridge: MIT Press.

Ziewitz, Malte. 2016. "Governing algorithms: Myth, mess, and methods." *Science, Technology & Human Values* 41 (1):3.

Zimmer, Carl. 2004. *Soul made flesh: The discovery of the brain—And how it changed the world.* London: William Heinemann.

References to Interviewees

Cases

Alexandre Andrade	193, 224, 254, 302
Alexandre Franco	49, 83, 224, 245, 284, 294, 311, 312, 388
Ana Luisa Raposo	195, 223, 230, 250, 288
André Peres	117, 159
Andreas Schuh	37, 38, 80, 84, 85, 99, 108, 184, 217, 317
Antonio Senra	107, 188, 275
Antonios Makropoulos	78, 293
Armin Heinecke	319, 336
Brunno Campos	152, 195, 239, 253, 275, 302, 304
Carlo Rondinoni	48
Carlos Garrido	122, 183, 275, 284, 309, 328
Christian Beckmann	70, 137, 147, 148, 191, 197, 206, 212, 302, 335, 340
Cyril Pernet	130, 136, 140, 145, 158, 277, 281, 305, 306, 330
Fábio Duran	207, 250, 251, 304, 309, 316
Fabrício Simozo	62, 63, 187, 326
Guillaume Flandin	147, 159, 202
Gustavo Pamplona	278, 279
Hugo Ferreira	177, 178

© The Editor(s) (if applicable) and The Author(s), under exclusive license to Springer Nature Singapore Pte Ltd. 2019
E. Bicudo, *Neuroimaging, Software, and Communication*,
https://doi.org/10.1007/978-981-13-7060-1

Index

Printed in the United States
By Bookmasters